Canadian Mathematical Society
Société Mathématique du Canada

CONFERENCE PROCEEDINGS • Volume 11

REPRESENTATIONS OF FINITE DIMENSIONAL ALGEBRAS

Proceedings of the
Tsukuba International Conference
(Fifth ICRA) held August 13–18, 1990

H. Tachikawa and
V. Dlab, Editors

Published by the American Mathematical Society
for the Canadian Mathematical Society
PROVIDENCE, RHODE ISLAND

CMS CONFERENCE PROCEEDINGS
Series Editors

James B. Carrell and M. Ram Murty

Volume 11

1990 Tsukuba International Conference
on Representations of Algebras and Related Topics
held at Tsukuba University, Japan, August 13–18, 1990

Proceedings of a conference sponsored by the Exchange Fund
of Tsukuba University, Hara Research Fund,
Tsukuba Expo '85 Memorial Fund, and the following companies:
Daiwa Institute of Research Ltd., Hitachi Ltd.,
Mitsubishi Electric Corporation, NEC Corporation,
Oki Electric Ind. Co. Ltd., and Toshiba Corporation.

Accepted for publication by the Canadian Mathematical Society

Published by the American Mathematical Society

1991 *Mathematics Subject Classification* (1985 *Revision*). Primary 16-06;
Secondary 16E30, 16E40, 16G10, 16G20, 16G60, 16G70, 20C10, 20C20.

Library of Congress Cataloging-in-Publication Data

Tsukuba International Conference on Representations of Algebras and Related Topics (1990: Tsukuba University, Japan)

Representations of finite dimensional algebras: proceedings of the Tsukuba International Conference (fifth ICRA) held August 13–18, 1990/H. Tachikawa and V. Dlab, editors.

p. cm.—(CMS conference proceedings; v. 11)

"Tsukuba International Conference on Representations of Algebras and Related Topics held at Tsukuba University, Japan, August 13–18, 1990"
—T.p. verso.

ISBN 0-8218-6016-X (acid-free paper)

1. Associative algebras—Congresses. 2. Representations of algebras—Congresses. I. Tachikawa, Hiroyuki, 1930– . II. Dlab, Vlastimil. III. Title. IV. Series: Conference proceedings (Canadian Mathematical Society); v. 11.

QA151.5.T78 1990 91-24244
512′.24—dc 20 CIP

The paper used in this book is acid-free and falls within the guidelines
established to ensure permanence and durability. ♾
This volume was printed directly from author-prepared copy.
Portions of this publication were typeset using $\mathcal{A}_{\mathcal{M}}\mathcal{S}$-TeX,
the American Mathematical Society's TeX macro system.

10 9 8 7 6 5 4 3 2 1 96 95 94 93 92 91

Contents

Preface

A rapid development of the research on representations of finite-dimensional algebras and group representations, as well as dispersion of the research centers all over the world, require regular meeting in these fields. The Tsukuba International Conference on Representations of Algebras and Related Topics, organized at the University of Tsukuba on August 13–18, 1990, has been the latest in a series of such meetings, reflecting the developments in the theory since the last conference at Carleton University in August 1984 (*Lecture Notes in Mathematics*, Vol. 1177 and 1178, Springer Verlag).

The organizing committee consisted of Professors M. Auslander, R. Bautista, S. Brenner, M. C. R. Butler, V. Dlab, K. R. Fuller, G. Michler, I. Reiten, C. M. Ringel, H. Tachikawa (chairman) and K. Yamagata.

The Tsukuba Conference was preceded by a workshop held on August 9–11, 1990 consisting of eight series of lectures by M. Auslander, G. Azumaya, J. F. Carlson, W. Crawley-Boevy, V. Dlab, Y. A. Drozd, K. Erdmann, and C. M. Ringel indicating some of the present, leading trends in the representation theory and related areas. A separate publication, *Proceedings of the Workshop* (to appear) in *London Mathematical Society Lecture Notes Series*, will contain reports from the workshop.

Recent advances in representation theory were reported in 57 lectures; these are listed on page xix. Not all contributions to the conference appear in these proceedings. On the other hand, some papers which were not presented at the conference are included. All papers that appear in this volume are in final form, and no versions of them will be submitted for publication elsewhere.

We should like to express our thanks to all referees for their assistance.

The organization of the meeting would not have been possible without financial support from Exchange Fund of Tsukuba University, Hara Research Fund, Tsukuba Expo' 85 Memorial Fund, and the following companies: Daiwa Institute of Research Ltd., Hitachi Ltd., Mitsubishi Electric Corporation, NEC Corporation, Oki Electric Ind. Co. Ltd., and Toshiba Corporation. We are most grateful for their assistance.

We wish also to extend our thanks to Dr. H. Fujita, the staffs of Institute of Mathematics, University of Tsukuba; Dr. M. Hoshino, Q. Zeng etc., and students T. Kaita, M. Uematsu, and T. Yoshida who helped so efficiently.

H. Tachikawa and V. Dlab
Tsukuba-Ottawa, December 1990

Brief History of the International Conference on Representation Theory of Algebras

The proceedings of the Fifth International Conference on Representation Theory of Algebras (ICRA V), which was held in August 1990 at Tsukuba University, are being published by the Canadian Mathematical Society. It is natural for the proceedings to have a Canadian publisher, since the developments of the representation theory of finite dimensional algebras have been strongly linked to Canada over the last 20 years.

The first ICRA, held in 1974, brought together mathematicians of different schools, some of whom had not had any contact with one another, and it created an international platform for discussion and collaboration. Three of the four previous ICRA's were held at Carleton University in 1974, 1979, and 1984. The third ICRA was held in 1980 in Puebla, Mexico. All these meetings are well documented in the workshop and conference proceedings which have been published in the *Springer Lecture Notes in Mathematics* series (volumes 488; 831 and 832; 903 and 944; 1177 and 1178). As long as no textbook is available, these volumes, together with the proceedings of the Durham Symposium 1985, (*London Mathematical Society Lecture Notes* series, vol. 116, 1986) and *Topics in Algebra, Part* 1 (Banach Center Publications, vol. 26, 1988), can serve as a main reference for getting to know the subject. Preliminary versions of the Ottawa conferences appeared in the *Carleton University Lecture Notes* series and include additional material.

The next ICRA has been scheduled for 1992; it will again be held at Carleton University, Ottawa.

List of Registered Participants

E. Abe, University of Tsukuba
Tsukuba-Shi, Ibaraki 305, Japan

I. Agoston, Carleton University
Ottawa, K1S 5B6, Canada

T. Asai, Hokkaido University
Kita-Ku, Sapporo 060, Japan

H. Asashiba, Osaka City University
Sugimoto-Ku, Osaka 558, Japan

I. Assem, University of Sherbrooke
Sherbrooke, Quebec, J1K 2R1, Canada

M. Auslander, Brandeis University
Waltham, Massachusetts 02254, U.S.A.

G. Azumaya, Indiana University
Bloomington, Indiana 47405, U.S.A.

Y. Baba, Yamaguchi University
Yamaguchi-Shi, Yamaguchi 753, Japan

R. Bautista, Universidad Nacional Autónoma de México
Mexico 20, D.F. Mexico, Mexico

S. Brenner, University of Liverpool
Liverpool L69 2BX, United Kingdom

M. C. R. Butler, University of Liverpool
Liverpool L69 3BX, United Kingdom

J. F. Carlson, University of Georgia
Athens, Georgia 30602, U.S.A.

F. U. Coelho, University of Liverpool
Liverpool L69 3BX, England

W. W. Crawley-Boevey, University of Oxford
Oxford OX1 3LB, England

T. Dana-Picard, Ben-Gurion University
Beer-Sheva, Israel

A. Dean, Bishop's University
Lennoxville, Quebec, Canada, J1M 1Z7

G. D'Este, Università di Padova
Via Belzoni 7, 35131 Padova, Italy

V. Dlab, Carleton University
Ottawa K1S 5B6, Canada

Ju. A. Drozd, Kiev University
Kiev 252017, U.S.S.R.

S. Endo, Tokyo Metropolitan University
Setagaya-Ku, Tokyo 158, Japan

K. Erdmann, University of Oxford
Oxford OX1 3LB, England

A. Facchini, Università di Udine
Via Zanon, 6-3310 Udine, Italy

R. Farnsteiner, University of Wisconsin
Milwoukee, Wisconsin 53201, U.S.A.

H. Fujita, University of Tsukuba
Tsukuba-Shi, Ibaraki 305, Japan

K. R. Fuller, University of Iowa
Iowa City, Iowa 52240, U.S.A.

K. R. Goodearl, University of Utah
Salt Lake City, Utah 84112, U.S.A.

E. L. Green, Virginia Polytechnic Institute
Blacksburg, Virginia 24061, U.S.A.

A. Hanaki, Chiba University
Chiba-Shi, Chiba 260, Japan

D. Happel, Universität Bielefeld
4800 Bielefeld, Germany

A. Hida, Chiba University
Chiba-Shi, Chiba 260, Japan

Y. Hieda, Osaka City University
Sugimoto-Ku, Osaka 558, Japan

H. Hijikata, Kyoto University
Sakyo-Ku, Kyoto 606, Japan

M. Hikari, Keio University
Minato-Ku, Tokyo 108, Japan

Y. Hirano, Okayama University
Okayama-Shi, Okayama 700, Japan

K. Hirata, Chiba University
Chiba-Shi, Chiba 260, Japan

H. Holden, University of Trondheim
AVH, N-7055 Dragvoll, Norway

M. Hoshino, University of Tsukuba
Tsukuba-Shi, Ibaraki 305, Japan

K. Igusa, Brandeis University
Waltham, Massachusetts 02254, U.S.A.

Y. Iida, Science University of Tokyo
Shinjuku-Ku, Tokyo 162, Japan

T. Ikeda, Hokkaido University of Education
Hakodate-Shi, Hokkaido 040, Japan

T. Ikeyama, Hosei University
Koganei-Shi, Tokyo 184, Japan

T. Inoue, Osaka City University
Sugimoto-Ku, Osaka 558, Japan

H. Ishibashi, Josai University
Sakato-Shi, Saitama 350–20, Japan

H. Ishihara, Hokkaido University
Kita-Ku, Sapporo 060, Japan

T. Ishikawa, Tokyo Metropolitan University
Setagaya-Ku, Tokyo 158, Japan

Y. Iwanaga, Shinshu University
Nagano-Shi, Nagano 380, Japan

T. Izawa, Shizuoka University
Shizuoka-Shi, Shizuoka 422, Japan

J. Kado, Osaka City University
Sugimoto-Ku, Osaka 558, Japan

T. Kaita, University of Tsukuba
Tsukuba-Shi, Ibaraki 305, Japan

T. Kanzaki, Kinki University
Higashiosaka-Shi, Osaka 577, Japan

T. Kato, University of Tsukuba
Tsukuba-Shi, Ibaraki 305, Japan

Y. Kawada, Kyoto University of Technology
Sakyo-Ku, Kyoto 606, Japan

S. Kawata, Osaka City University
Sugimoto-Ku, Osaka 558, Japan

H. Kawai, Kumamoto National College of Technology
Kikuike-Gun, Kumamoto 861–11, Japan

O. Kerner, University of Düsseldorf
4000 Düsseldorf, Germany

K. Kido, Science University of Tokyo
Shinjuku-Ku, Tokyo 162, Japan

E. E. Kirkman, Wake Forest University
Winston-Salem, North Carolina 27106, U.S.A.

K. Kishimoto, Shinshu University
Matsumoto-Shi, Nagano 390, Japan

K. Kitamura, Osaka Kyoiku University
Osaka-Shi, Osaka 543, Japan

Y. Kitamura, Tokyo Gakugei University
Koganei-Shi, Tokyo 184, Japan

M. Kiyota, Tokyo Medical and Dental University
Bunkyo-Ku, Tokyo 113, Japan

M. Kleiner, Syracuse University
Syracuse, New York 13244–1550, U.S.A.

Y. Kobayash, Tokushima University
Tokushima-Shi, Tokushima 770, Japan

O. Konishi, Osaka City University
Sugimoto-Ku, Osaka 558, Japan

S. Koshitani, Chiba University
Chiba-Shi, Chiba 260, Japan

B. Külshammer, Universität Dortmund
4600 Dortmund 50, West Germany

M. Kutami, Yamaguchi University
Yamaguchi-Shi, Yamaguchi 753, Japan

S. Kyuno, Tohoku Gakuin University
Tagajo-Shi, Miyagi 985 Japan

H. Lenzing, Universität-Gesamthochschule Paderborn
4790 Paderborn, Germany

S. Liu, University of Liverpool
Liverpool L69 3BX, England

R. Makino, Fukushima University
Fukushima-Shi, Fukushima 960–12, Japan

M. P. Malliavin, Université de Paris VI
75230 Paris, France

N. Marmaridis, University of Ioannia
45110 Ioannia, Greece

R. Martinez-Villa, Universidad Nacional Autónoma de México
Mexico 20, D.F. Mexico

K. Masaike, Tokyo Gakugei University
Koganei-Shi, Tokyo 184, Japan

A. Masuoka, Nippon Kogyo University
Minamisaitama-Gun, Saitama 345, Japan

H. A. Merklen, Universidade de São Paulo
Postal No. 20.570 São Paulo, Brasil

G. Michler, Universität Essen
4300 Essen, Germany

J. Miyachi, Tokyo Gakugei University
Koganei-Shi, Tokyo 184, Japan

Y. Miyashita, University of Tsukuba
Tsukuba-Shi, Ibaraki 305, Japan

M. Miyauchi, Tokyo Metropolitan College of Aeronautical Engineering
Arakawa-Ku, Tokyo 116, Japan

S. Morimoto, Hagi Koen Gakuin
Hagi-Shi, Yamaguchi 758, Japan

J. Morita, University of Tsukuba
Tsukuba-Shi, Ibaraki 305, Japan

K. Motose, Hirosaki University
Hirosaki-Shi, Aomori 036, Japan

T. Nagahara, Okayama University
Okayama-Shi, Okayama 700, Japan

K. Nakane, Ibaraki University
Mito-Shi, Ibaraki 10, Japan

Y. Ninomiya, Shinshu University
Matsumoto Shi, Nagano 390, Japan

K. Nishida, Nagasaki University
Nagasaki-Shi, Nagasaki 852, Japan

T. Niwasaki, Ehime University
Matsuyama-Shi, Ehime 790, Japan

T. Nozawa, Chiba University
Chiba-Shi, Chiba 260, Japan

Y. Ogawa, Tohoku Institute of Technology
Sendai-Shi, Miyagi 982, Japan

K. Otake, Gunma University
Maebashi-Shi, Gunma 371, Japan

H. Okuno, Juntendo University
Bunkyo-Ku, Tokyo 112, Japan

H. Okuyama, Chiba University
Chiba-Shi, Chiba 260, Japan

T. Okuyama, Osaka City University
Sugimoto-Ku, Osaka 558, Japan

K. Oshiro, Yamaguchi University
Yamaguchi-Shi, Yamaguchi 753, Japan

J. A. de la Peña, Universidad Nacional Autónoma de México
Mexico 20, D.F. Mexico

V. A. Ponomarev, Institute of Biophysics
142292 t. Putschino, U.S.S.R.

I. Reiten, Universitetet I Trondheim
Trondheim 7000, Norway

C. M. Ringel, Universität Bielefeld
4800 Bielefeld, Germany

K. W. Roggenkamp, Universität Stuttgart
7000, Stuttgart 80, Germany

R. Saito, Hokkaido Junior College of Literature and Science
Ebetsu-Shi, Hokkaido 069, Japan

K. Sakano, Yamanashi University
Kofu-Shi, Yamanashi 400, Japan

K. Sanada, Science University of Tokyo
Shinjuku-Ku, Tokyo 162, Japan

H. Sasaki, Yamaguchi University
Yamaguchi-Shi, Yamaguchi 753, Japan

H. Sato, Wakayama University
Wakayama-Shi, Wakayama 640, Japan

M. Sato, Yamanashi University
Kofu-Shi, Yamanashi 400, Japan

H. Sawada, Sophia University
Chiyoda-Ku, Tokyo 102, Japan

M. E. Schaps, Bar-Ilan University
52 900 Ramat-Gan, Israel

K. Sekiguchi, Kokushikan University
Setagaya-Ku, Tokyo 154, Japan

T. Sekiguchi, Chuo University
Bunkyo-Ku, Tokyo 112, Japan

S. A. Sikko, Universitetet I Trondheim
AVH, N-7055 Dragvoll, Norway

A. Skowronski, Uniwersytet Mikolaja Kopernika
87-100 Torun′, Poland

S. O. Smalo, Universitetet I Trondheim
N-7055 Dragvoll, Norway

O. Solberg, Universitetet I Trondheim
Trondheim 7000, Norway

M. Stefanescu, Universitatea 'Al. I. Cuza' Iaşi
6600-Iaşi, Romania

K. Sugano, Hokkaido University
Kita-Ku, Sapporo 060, Japan

T. Sumioka, Osaka City University
Sugimoto-Ku, Osaka 558, Japan

N. Suwa, Tokyo Denki University
Chiyoda-Ku, Tokyo 101, Japan

H. Tachikawa, University of Tsukuba
Tsukuba-Shi, Ibaraki 305, Japan

Y. Tagami, University of Tsukuba
Tsukuba-Shi, Ibaraki 305, Japan

K. Tahara, Aichi University of Education
Kariya-Shi, Aichi 448, Japan

M. Takane, Universidad Nacional Autónoma de México
04510 Mexico, D.F. Mexico

Y. Takegahara, Murroran Institute of Technology
Muroran-Shi, Hokkaido 050, Japan

M. Takeuchi, University of Tsukuba
Tsukuba-Shi, Ibaraki 305, Japan

D. Tambara, Hokkaido University
Kita-Ku, Sapporo 060, Japan

S. Tanabe, Science University of Tokyo
Shinjuku-Ku, Tokyo 162, Japan

M. Tanaka, Kumamoto Institute of Technology
Kumamoto-Shi, Kumamoto 860, Japan

G. Todorov, Northeastern University
Boston, Massachusetts 02115, U.S.A.

A. Tome, Universidad Nacional Autónoma de México
04510 Mexico, D.F. Mexico

H. Tominaga, Okayama University
Okayama-Shi, Okayama 700, Japan

M. Uematsu, University of Tsukuba
Tsukuba-Shi, Ibaraki 305, Japan

T. Ukegawa, Fukuyama University
Fukuyama-Shi, Fukuyama 729-02, Japan

L. Unger, Universität-Gesamthochschule Paderborn
4790 Paderborn, Germany

K. Uno, Osaka University
Toyonaka-Shi, Osaka 560, Japan

Y. Usami, Ochanomizu University
Bunkyo-Ku, Tokyo 112, Japan

T. Wada, Tokyo University of Agriculture and Technology
Fuchu-Shi, Tokyo 183, Japan

T. Wakamatsu, Jobu University
Isezaki-Shi, Gunma 372, Japan

K. Waki, Chiba University
Chiba-Shi, Chiba 260, Japan

A. Watanabe, Kumamoto University
Kumamoto-Shi, Kumamoto 860, Japan

Y. Watanabe, Osaka Women's University
Sakai-Shi, Osaka 590, Japan

A. Wiedemann, Universität Stuttgart
7000 Stuttgart 80, Germany

T. Yamada, Science University of Tokyo
Shinjuku-Ku, Tokyo 162, Japan

K. Yamagata, University of Tsukuba
Tsukuba-Shi, Ibaraki 305, Japan

H. Yamaki, University of Tsukuba
Tsukuba-Shi, Ibaraki 305, Japan

K. Yamashiro, Chiba University
Chiba-Shi, Chiba 260, Japan

K. Yokogawa, Science University of Okayama
Okayama-Shi, Okayama 700, Japan

H. Yoshimura, Yamaguchi University
Yamaguchi-Shi, Yamaguchi 753, Japan

Y. Yoshino, Kyoto University
Kyoto-Shi, Kyoto 606, Japan

Y. Yukimoto, Fukui Institute of Technology
Fukui-Shi, Fukui 910, Japan

D. Zacharia, Syracuse University
Syracuse, New York 13244, U.S.A.

A. G. Zavadskij, Kiev Civil Engineering Institute
Kiev 252037, U.S.S.R.

Q. Zeng, University of Tsukuba
Tsukuba-Shi, Ibaraki 305, Japan

B. Zimmerman-Huisgen, University of Utah
Salt Lake City, Utah 84112, U.S.A.

List of Lectures

I. Agoston (Carleton University)
On the rank of torsion-free modules over tame hereditary algebras

H. Asashiba (Osaka City University)
On local algebras A satisfying $\operatorname{Ext}_A^1(DA, A) = 0$

I. Assem * (University of Sherbrooke)
J. Nehring (Nicholas Copernicus University)
W. Schewe (University of Sherbrooke)
Fundamental domains and duplicated algebras

M. Auslander (Brandeis University)
M. I. Platzeck (Universidad Nacional del Sur)
G. Todorov * (Northeastern University)
Homological theory of idempotent ideals

R. Bautista (Universidad Nacional Autónoma de México)
Almost split sequences and critical partially ordered sets of infinite representation type

S. Brenner * (University of Liverpool)
M. C. R. Butler (University of Liverpool)
Five terms in the middle

W. L. Burt (University of Liverpool)
M. C. R. Butler * (University of Liverpool)
Almost split sequences for bocses

F. U. Coelho (University of São Paulo/University of Liverpool)
Preprojective partitions

W. Crawley-Boevey (Universität Bielefeld/Oxford University)
Algebras with large regular Auslander-Reiten components

T. Dana-Picard (Ben-Gurion University)
Classification of associative algebras with unit

A. P. Dean (Bishop's University)
Infinite dimensional representations of finite dimensional algebras

---

* Denotes the speaker.

G. D′ Este (University of Salerno)
Simple submodules and multiplicities

V. Dlab (Carleton University)
On Bratteli diagrams

Ju. A. Drozd (Kiev University)
On representations of 0*-species*

A. Facchini (University of Ukine)
Mittag-Leffler modules, reduced products and direct products

R. Farnsteiner (University of Wisconsin, Milwaukee)
On the cohomology of Frobenius extensions

K. R. Fuller* (University of Iowa)
W. K. Nicholson (University of Calgary)
J. F. Watters (University of Leicester)
Reflexivity of algebras and bimodules

K. R. Goodearl (University of Utah)
Clique graphs: infinite quivers for noetherian rings

E. L. Green* (Virginia Polytechnic Institute & State University)
E. N. Marcos (Virginia Polytechnic Institute & State University/IME-USP)
The local theory of graded modules

D. Happel (Universität Bielefeld)
On repetitive categories

H. Hijakata (Kyoto University)
On classification of Bass orders

H. Ishibashi (Josai University)
Condition for some subsets of transvections to generate $Sp(V)$

K. Igusa* (Brandeis University)
D. Zacharia (Syracuse University)
On the cyclic homology of monomial relation algebras

O. Kerner (Universität Dusseldorf)
Regular modules over wild hereditary algebras

E. Kirkman* (Wake Forest University)
L. Small (University of California, San Diego)
Q-analogs of harmonic oscillators and related rings

M. Kiyota (Tokyo Medical and Dental University)
T. Wada* (Tokyo University of Agriculture and Technology)
Some remarks on eigenvalues of the Cartan matrix in fintie groups

M. Kleiner (Syracuse University)
Induced comodules and representations of algebras

S. Koshitani (Chiba University)
On p-radical groups in representation theory of finite groups

B. Kulshammer (University of Dortmund)
Conjugacy classes in groups, characters in blocks, and commutators in algebras

S. Kyuno (Tohoku Gakuin University)
M. Stefanescu * (University "Al, I, Cuza")
Correspondences between near-rings and groups, loops or quasigroups

H. Lenzing (Universität Paderborn)
Wild canonical algebras

S. P. Liu (University of Liverpool)
Degrees of irreducible maps and Auslander-Reiten quivers

N. Marmaridis (University of Ioannia)
Polytrivial extensions and tilting modules

R. Martinez-Villa (Universidad Nacional Autónoma de México)
Properties that are left invariant under stable equivalence

H. Merklen (University of São Paulo)
A cokernel functor for triangular matrix algebras

G. Michler (Universität Essen)
On Alperin's weight-conjecture

K. Nishida (Nagasaki University)
New examples of local Bass orders

J. A. de la Peña (Universidad Nacional Autónoma de México)
Varieties of modules for tame and wild algebras

V. A. Ponomarev (Institute of Biological Physics, Pushchino)
Representations of modular lattices attached to equipped graphs. Perfect elements in lattices $L(A_n^{\nu})$

I. Reiten (The University of Trondheim)
The dual of the transpose as a representable functor

C. M. Ringel (Universität Bielefeld)
The category of good modules over a quasi-hereditary algebra has almost split sequences

K. W. Roggenkamp (Universität Stuttgart)
Grothendieck groups of group rings and orders

T. Schaps (Bar-Ilan University)
Separable deformations of modular group algebras

T. Sekiguchi (Chuo University)
N. Suwa* (Tokyo Denki University)
On the unified Kummer-Artin-Schreier-Witt theory

A. Skowronski (University of Torun)
Cycle-finite simply connected algebras

S. O. Smalo (The University of Trondheim)
Functorially finite subcategories

O. Solberg (The University of Trondheim)
Liftings and weak liftings of modules

H. Tachikawa (University of Tsukuba)
Q. Zeng* (University of Tsukuba)
Vanishing of Hochschild cohomologies and generalized biserial local algebras

M. Takane (Universidad Nacional Autónoma de México)
Spectral properties of Coxeter transformations

D. Tambara (Hokkaido University)
Representations of the free matric bialgebra of rank 2

D. Tome (Universidad Nacional Autónoma de México)
On nonnegativity of tits forms and tameness of algebras

L. Unger (Universität Paderborn)
The simplicial complex of partial tilting modules over hereditary algebras

K. Uno (Osaka University)
On the vertices of modules in the Auslander-Reiten quiver of finite groups

T. Wakamatsu (Jobu University)
Grothendieck groups of subcategories

A. Wiedemann (Universität Stuttgart)
Untiled orders of finite global dimension

A. G. Zavadskij (Kiev Civil Engineering Institute)
Some tame linear matrix problems and algorithms for their solving

B. Zimmermann-Huisgen (University of California, Santa Barbara/ University of Utah)
Predicting syzygies over finite-dimensional algebras

Canadian Mathematical Society
Conference Proceedings
Volume 11 (1991)

P-BASES FOR TORSION-FREE REGULAR MODULES

István Ágoston[1] and Andrew P. Dean[1]

Abstract. In [R] a general approach is offered to the theory of infinite dimensional representations over tame hereditary algebras. Among other things many concepts, like that of torsion and torsion-free modules, divisibility, etc., known from the theory of abelian groups, are carried over to the tame hereditary situation. In this approach one of the most important invariants of torsion-free modules is their rank. In our short note we will show that the rank of the torsion-free regular module M, originally defined by Ringel using a certain embedding of M into a divisible module, can be understood as the cardinality of a maximal independent set for a suitably defined dependence relation.

1. Preliminaries. First we recall some definitions and basic results from [R]. Let A be a tame hereditary algebra, finite dimensional over a field k. One can define a torsion theory on the category of (right) A-modules as follows. A module M is called *torsion* if it is spanned by its finite dimensional regular and preinjective submodules, while M is *torsion-free* if every finite dimensional submodule of M is preprojective. The *torsion submodule* of a module M is the largest submodule which is torsion and it will be denoted by $\mathcal{T}(M)$. We call a module M *regular* if it has no finite dimensional preinjective or preprojective direct summands. It can be shown that a module M is regular if and only if $\operatorname{Hom}(M, P) = 0$ for all finite dimensional preprojective modules P and $\operatorname{Hom}(I, M) = 0$ for all finite dimensional preinjective modules I (cf. §4.2 of [R]). The class of torsion regular modules over A forms an exact abelian subcategory, closed under extensions (Theorem 4.4 of [R]).

1980 Mathematics Subject Classification (1985 Revision) 16A46, 16A64
[1] Research partially supported by the NSERC of Canada

A module M is called *divisible* if $\operatorname{Ext}(S, M) = 0$ for every simple regular module S. Actually, if M is divisible then $\operatorname{Ext}(N, M) = 0$ for all regular modules N (Proposition 4.7 of **[R]**). As Ringel has shown, there exists a unique indecomposable torsion-free divisible module, denoted by Q (Theorem 5.3 of **[R]**). The endomorphism ring of Q is a division ring. Every torsion-free module M can be embedded into a direct sum Y of copies of Q in such a way that the quotient Y/M is torsion regular. Moreover, if there is such an embedding into $Y = \bigoplus_I Q$ then the cardinality of the index set I depends only on the module M (Theorem 5.5 of **[R]**). This cardinality is called the *rank* of M and will be denoted by $\operatorname{rk} M$. One can show that the rank is additive on short exact sequences of torsion-free modules (Proposition 2.2 of **[DZ]**). An important fact is that for finite dimensional preprojective modules the rank is just the negative of the defect of the module (Proposition 5.6 of **[R]**).

A submodule N of the module M is called *torsion closed* in M if M/N is torsion-free. The intersection of all torsion-closed submodules of M containing a particular submodule N is called the *torsion-closure* of N in M and will be denoted by $\overline{N}^M$ or simply by $\overline{N}$. Clearly $\overline{N}^M$ is the full preimage of $\mathcal{T}(M/N)$ in M. For any submodule N of M the rank of $\overline{N}$ must satisfy $\operatorname{rk} \overline{N} \leq \operatorname{rk} N$ (Proposition 2.1 of **[ADS]**). The following simple observation will show that taking the torsion closure of a module is of finitary character.

Lemma 1.1. *Let $\mathcal{N} = \{N_i \mid i \in I\}$ be a directed set of torsion-closed submodules of a module M. Then $N = \bigcup_{i\in I} N_i \leq M$ is also torsion-closed in M.*

Proof. Let us take an arbitrary submodule $W \leq M$ such that $N \leq W$ and W/N is finite dimensional. We have to show that W/N is preprojective. Let W' be a finite dimensional preimage of W/N in M, i.e. $W' \leq M$ such that $W' + N = W$. Consider $W' \cap N = W' \cap (\bigcup_{i\in I} N_i) = \bigcup_{i\in I}(N_i \cap W')$. Since W' is finite dimensional and $\mathcal{N}$ is directed, there exists an index $i \in I$ such that $W' \cap N = W' \cap N_i$. Then:

$$W/N = (W' + N)/N \cong W'/W' \cap N = W'/W' \cap N_i \cong (W' + N_i)/N_i \leq M/N_i.$$

Since M/N_i is torsion-free by assumption, W/N is preprojective. Hence N is torsion-closed.

Corollary 1.2. *If $\mathcal{N} = \{N_i \mid i \in I\}$ is a directed set of submodules of a module M then $\overline{\bigcup_{i\in I} N_i} = \bigcup_{i\in I} \overline{N_i}$. In particular for any submodule N of M the* *torsion closure $\overline{N}$ is the union of the torsion closures of all finite dimensional submodules of N.*

2. A dependence relation for torsion-free regular modules. For the rest of the paper, unless otherwise stated, let M be a torsion-free regular module. We shall define a dependence relation on a subset of all submodules of M. Let P be an indecomposable projective module of rank 1 (that is, of defect -1). Define the set $\mathcal{W}_P = \mathcal{W}_P(M)$ as $\mathcal{W}_P = \{W \leq M \mid W \cong P\}$. For $W \in \mathcal{W}_P$ and $\mathcal{P} \subseteq \mathcal{W}_P$ we say that W *depends* on $\mathcal{P}$ if and only if $W \subseteq \overline{\mathcal{P}}^M$, where $\overline{\mathcal{P}}^M$ denotes the torsion closure of the submodule generated by the elements of $\mathcal{P}$. We call a set $\mathcal{P} \subseteq \mathcal{W}_P$ *independent* if no element $W \in \mathcal{P}$ depends on $\mathcal{P} \setminus \{W\}$. The set $\mathcal{P} \subseteq \mathcal{W}_P$ is called a *generating set* for $\mathcal{W}_P$ if every element of $\mathcal{W}_P$ depends on $\mathcal{P}$.

We want to show that this relation satisfies all the standard properties of linear dependence relations. We will need the following observation.

Lemma 2.1. *Let $W \in \mathcal{W}_P$ and $\mathcal{P} \subseteq \mathcal{W}_P$. Then W depends on $\mathcal{P}$ if and only if $W \cap \overline{\mathcal{P}} \neq 0$.*

Proof. The necessity of the condition is obvious from the definition. To show the other direction, assume that $W \cap \overline{\mathcal{P}} \neq 0$. Observe that $\operatorname{rk} W = 1$ implies $\operatorname{rk} \overline{W} = 1$. Since $\overline{W} \cap \overline{\mathcal{P}}$ is torsion closed in $\overline{W}$ and is non-zero by assumption, we must have that $\overline{W} \subseteq \overline{\mathcal{P}}$. Hence W depends on $\mathcal{P}$.

The next proposition shows that our dependence relation is of finitary character.

Proposition 2.2. *Let W be an element of $\mathcal{W}_P$. If W depends on $\mathcal{P} \subseteq \mathcal{W}_P$ then it also depends on a finite subset $\mathcal{F} \subseteq \mathcal{P}$.*

Proof. Assume that $W \subseteq \overline{\mathcal{P}}$. Since $\dim_k W < \infty$, by Corollary 1.2 we get that there is a finite dimensional submodule $N \subseteq \langle \mathcal{P} \rangle$ such that $W \subseteq \overline{N}$. Since there is a finite subset $\mathcal{F} \subseteq \mathcal{P}$ such that $N \subseteq \langle \mathcal{F} \rangle$, we obtain that W depends on $\mathcal{F}$.

Finally, we can show that the exchange property is also satisfied.

Proposition 2.3. *Let $W \in \mathcal{W}_P$ and $\mathcal{P} \subseteq \mathcal{W}_P$. If W depends on $\mathcal{P}$ but does not depend on $\mathcal{P}' = \mathcal{P} \setminus \{X\}$ for some element $X \in \mathcal{P}$ then X depends on $\mathcal{P}'' = (\mathcal{P} \setminus \{X\}) \cup \{W\}$.*

Proof. By assumption we have $\overline{\mathcal{P}'} \subset \overline{\mathcal{P}''} \subseteq \overline{\mathcal{P}}$, hence X does not depend on $\mathcal{P}'$. Thus Lemma 2.1 implies that X embeds into $\overline{\mathcal{P}}/\overline{\mathcal{P}'}$ and the torsion closure of its image is the whole module $\overline{\mathcal{P}}/\overline{\mathcal{P}'}$. Hence we get that $\operatorname{rk}\overline{\mathcal{P}}/\overline{\mathcal{P}'} = 1$. Consequently, we must have $\overline{\mathcal{P}''} = \overline{\mathcal{P}}$. Thus $X \subseteq \overline{\mathcal{P}''}$, that is, X depends on $\mathcal{P}''$.

From the exchange property and the finitary character of our dependence relation we get all the standard theorems on the existence of bases, the invariance of their cardinality etc. A basis for $\mathcal{W}_P$ with respect to this relation will be called a *P-basis* for M.

3. Independent sets, generating sets and the rank. We turn now to the question of characterizing the specific properties of our dependence relation.

Theorem 3.1. *A set $\mathcal{P} = \{P_i \mid i \in I\} \subseteq \mathcal{W}_P$ is independent if and only if*
(i) $P_i \cap \langle P_j \in \mathcal{P} \mid j \neq i\rangle = 0$ for every $i \in I$ (that is, $\bigoplus_{i\in I} P_i \subseteq M$);
(ii) $M/\bigoplus_{i\in I} P_i$ is regular.

Proof. Assume first that the conditions (i) and (ii) are satisfied, thus we have $M/\bigoplus_{i\in I} P_i$ regular. Then consider the following diagram:

$$
\begin{array}{ccccccccc}
0 & \longrightarrow & \bigoplus_{i\in I} P_i & \xrightarrow{\iota} & M & \longrightarrow & M/\bigoplus_{i\in I} P_i & \longrightarrow & 0 \\
 & & \big\downarrow{\scriptstyle \pi_i} & & \swarrow{\scriptstyle \psi} & & & & \\
 & & P_i & & & & & & \\
 & & \big\downarrow{\scriptstyle \iota_i} & & & & & & \\
 & & Q & & & & & &
\end{array}
$$

where π_i is the canonical projection and ι_i is an embedding of P_i into Q. Since $M/\bigoplus_{i\in I} P_i$ is regular, $\operatorname{Ext}(M/\bigoplus_{i\in I} P_i, Q) = 0$, hence there exists a map $\psi : M \longrightarrow Q$ such that $\psi\iota = \iota_i\pi_i$. Thus $\operatorname{Ker}\psi \supseteq \overline{\mathcal{P}\setminus\{P_i\}}$ and $\operatorname{Ker}\psi \cap P_i = 0$. So $\mathcal{P}$ is independent.

Assume now that $\mathcal{P} \subseteq \mathcal{W}_P$ is independent. Thus by Lemma 2.1 we get that $P_i \cap \langle P_j \in \mathcal{P} \mid j \neq i\rangle \subseteq P_i \cap \overline{(\mathcal{P}\setminus\{P_i\})} = 0$. Hence we get (i). To prove condition (ii), assume $\mathcal{P} = \{P_\mu \mid \mu < \kappa\}$ for some cardinal κ. We will prove by transfinite induction that $M_\lambda = M/\bigoplus_{\mu<\lambda} P_\mu$ is regular for each $\lambda \leq \kappa$. The argument we use is essentially from Ringel's proof of Proposition 4.3 of [R], but we include it here for the sake of completeness.

Assume first that $\lambda = \nu + 1$ for some ordinal ν. By assumption $M_\nu = M/\bigoplus_{\mu<\nu} P_\mu$ is regular. Since $\mathcal{P}$ is independent, by Lemma 2.1 we get that P_ν embeds into M_ν as $\hat{P}_\nu$; actually it embeds even into the torsion-free module $\tilde{M}_\nu = M_\nu/\mathcal{T}(M_\nu)$ as $\tilde{P}_\nu$. We will show first that $\tilde{M}_\nu/\tilde{P}_\nu$ is regular. Since M was regular, it is enough to check that $\tilde{M}_\nu/\tilde{P}_\nu$ has no preinjective submodules. Let us assume that $\tilde{W} \subseteq \tilde{M}_\nu$ is a submodule containing $\tilde{P}_\nu$ and $\tilde{W}/\tilde{P}_\nu$ is indecomposable preinjective. But then for the defect of $\tilde{W}$ we get: $\delta(\tilde{W}) = \delta(\tilde{P}_\nu) + \delta(\tilde{W}/\tilde{P}_\nu) = -1 + \delta(\tilde{W}/\tilde{P}_\nu) \geq 0$, and this contradicts the fact that $\tilde{M}_\nu$ is torsion-free. Thus $\tilde{M}_\nu/\tilde{P}_\nu$ is regular. Consider now the following exact sequence:

$$0 \longrightarrow (\mathcal{T}(M_\nu) + \hat{P}_\nu)/\hat{P}_\nu \longrightarrow M_\nu/\hat{P}_\nu \longrightarrow (M_\nu/\hat{P}_\nu)/((\mathcal{T}(M_\nu) + \hat{P}_\nu)/\hat{P}_\nu) \longrightarrow 0.$$

Here the first term is isomorphic to $\mathcal{T}(M_\nu)$, and since it is the torsion submodule of the regular module M_ν, it is regular. On the other hand the last term is isomorphic to $\tilde{M}_\nu/\tilde{P}_\nu$, which was just shown to be regular, too. Thus $M_\nu/\hat{P}_\nu \cong M/\bigoplus_{\mu<\lambda} P_\mu = M_\lambda$, as an extension of two regular modules, is also regular.

Assume now that λ is a limit ordinal and for every $\mu < \lambda$ the module M_μ is regular. Suppose that $M_\lambda = M/\bigoplus_{\mu<\lambda} P_\mu$ has an indecomposable preinjective submodule $W/\bigoplus_{\mu<\lambda} P_\mu$ where $\bigoplus_{\mu<\lambda} P_\mu \subseteq W \subseteq M$. Let W' be a finite dimensional submodule of M satisfying $W = W' + \bigoplus_{\mu<\lambda} P_\mu$. Then $W' \cap \bigoplus_{\mu<\lambda} P_\mu = W' \cap \bigoplus_{\mu<\nu} P_\mu$ for some $\nu < \lambda$. Thus:

$$\begin{aligned} W/\bigoplus_{\mu<\lambda} P_\mu &= (W' + \bigoplus_{\mu<\lambda} P_\mu)/\bigoplus_{\mu<\lambda} P_\mu \cong W'/(W' \cap \bigoplus_{\mu<\lambda} P_\mu) = \\ &= W'/(W' \cap \bigoplus_{\mu<\nu} P_\mu) \cong (W' + \bigoplus_{\mu<\nu} P_\mu)/\bigoplus_{\mu<\nu} P_\mu \subseteq M_\nu, \end{aligned}$$

and this would contradict the regularity of M_ν. Thus M_λ is also regular. This finishes the proof.

Theorem 3.2. *A subset $\mathcal{P} \subseteq \mathcal{W}_P$ is a generating set if and only if $M/\langle\mathcal{P}\rangle$ is torsion, i.e. if and only if $\overline{\mathcal{P}} = M$.*

Proof. We will again use the argument of Ringel from the proof of Proposition 4.3 of [R]. The sufficiency is obvious from the definitions. For the other direction let us assume that $\mathcal{P}$ is a generating set. Assume that $\overline{\mathcal{P}} \neq M$. Then the module $\tilde{M} = M/\overline{\mathcal{P}}$ is a non-zero torsion-free regular module. But then there must exist a non-zero homomorphism $\varphi : P \to \tilde{M}$, otherwise $\tilde{M}$ would become a module over an algebra of finite representation type, and as such it would be a direct sum of finite dimensional indecomposable submodules, contradicting the fact that $\tilde{M}$ is torsion-free and regular. Since $\operatorname{rk} P = 1$, we get that φ must be an embedding. But P is projective, so this also gives an embedding of P into M and the image P_0 is disjoint from $\overline{\mathcal{P}}$. Thus P_0 does not depend on $\mathcal{P}$, contradicting the assumption that $\mathcal{P}$ was a generating set for M.

Finally we want to show that the cardinality of a P-basis for M is $\operatorname{rk} M$.

Theorem 3.3. *Let $\mathcal{P} = \{P_i \mid i \in I\} \subseteq \mathcal{W}_P$ be an independent set. Then $\operatorname{rk} \overline{\mathcal{P}} = |\mathcal{P}|$. In particular if $\mathcal{P}$ is a P-basis for M then $|\mathcal{P}| = \operatorname{rk} M$.*

Proof. Let us take $N = \overline{\mathcal{P}}$. Then the module $N / \bigoplus_{i \in I} P_i$ is the torsion submodule of the module $M / \bigoplus_{i \in I} P_i$, and the latter is regular by Theorem 3.1. Thus $N / \bigoplus_{i \in I} P_i$ must also be regular. If we now embed the module N into a direct sum Y of copies of the module Q in such a way that Y/N is torsion regular, then $Y / \bigoplus_{i \in I} P_i$ is also torsion regular. Hence $\operatorname{rk} N = \operatorname{rk} \bigoplus_{i \in I} P_i$. But clearly $\operatorname{rk} \bigoplus_{i \in I} P_i = |I| = |\mathcal{P}|$, thus we are done.

The previous result shows that the cardinality of a P-basis for M would be the same if we have started with another indecomposable projective module P' which is of rank 1. As a matter of fact we did not use that our set $\mathcal{W}_P$ was "homogeneous": it would have been possible to define our dependence relation for the set $\mathcal{W} = \mathcal{W}(M) = \{N \subseteq M \mid N \text{ is indecomposable projective}, \operatorname{rk} N = 1\}$.

4. Further results. Let us take an independent set $\mathcal{P} = \{P_i \mid i \in I\} \subseteq \mathcal{W}_P$. Then by Theorem 3.1 we get that $\bigoplus_{i \in I} P_i \subseteq M$ and $M / \bigoplus_{i \in I} P_i$ is regular. If we take a projection map $\pi_i : \bigoplus_{i \in I} P_i \to P_i$ and compose it with an embedding $\iota_i : P_i \to Q$ then we can extend this map to a homomorphism $\psi_i : M \to Q$. Let us choose such a homomorphism ψ_i for each $i \in I$. Then we have the following proposition.

Proposition 4.1. *Let $\mathcal{P} = \{P_i \mid i \in I\} \subseteq \mathcal{W}_P$ be an independent set for the torsion-free module M and $\Psi = \{\psi_i \mid i \in I\} \subseteq \operatorname{Hom}(M, Q)$ be a set of homomorphisms as defined above. Then Ψ is independent over $\operatorname{End}(Q)$.*

Proof. Assume that $\alpha_{i_1}\psi_{i_1} + \cdots + \alpha_{i_n}\psi_{i_n} = 0$ for some elements $\alpha_{i_1}, \ldots, \alpha_{i_n} \in \operatorname{End}(Q)$ and some indices $i_1, \ldots, i_n \in I$. We may assume that the indices are all distinct. Since $P_{i_j} \subseteq \operatorname{Ker} \psi_{i_\ell}$ for $j \neq \ell$, we get that $P_{i_j} \subseteq \operatorname{Ker} \alpha_{i_j} \psi_{i_j}$ for $1 \leq j \leq n$. But since $\operatorname{Ker} \psi_{i_j} \cap P_{i_j} = 0$, we must have that $\operatorname{Ker} \alpha_{i_j} \neq 0$ so $\alpha_{i_j} = 0$ for every $1 \leq j \leq n$. Thus Ψ is independent over $\operatorname{End}(Q)$.

As an application of our concepts we will conclude with a statement which generalizes Proposition 6.1.2 of **[R]**.

Proposition 4.2. *Let M be a torsion-free (not necessarily regular) module of finite rank, and let $N \subseteq M$ be a submodule such that M/N is torsion. Then M/N is regular if and only if $\operatorname{rk} N = \operatorname{rk} M$.*

Proof. Assume first that M/N is regular. Then an embedding of M into a direct sum $\bigoplus_{i \in I} Q$ with a torsion regular cokernel yields an embedding of N into the same direct sum with a cokernel term which is also torsion regular. Hence $\operatorname{rk} N = \operatorname{rk} M$.

Assume now that $\operatorname{rk} N = \operatorname{rk} M = n$. Then we have the following diagram:

$$\begin{array}{ccccccccc} 0 & \longrightarrow & N & \xrightarrow{\iota_N} & \bigoplus_1^n Q = Y' & \longrightarrow & Y'/N & \longrightarrow & 0 \\ & & \downarrow{\scriptstyle \varphi'} & & \downarrow{\scriptstyle \varphi} & & \downarrow{\scriptstyle \varphi''} & & \\ 0 & \longrightarrow & M & \xrightarrow{\iota_M} & \bigoplus_1^n Q = Y'' & \longrightarrow & Y''/M & \longrightarrow & 0. \end{array}$$

Here φ' is the embedding of N into M, the modules Y'/N and Y''/M are torsion regular, and the existence of φ follows from $\operatorname{Ext}(Y'/N, Y'') = 0$. Since the modules M/N and Y''/M are torsion, we get that Y''/N is also torsion, hence $\overline{\operatorname{Im} \varphi}^{Y''} = Y''$. This immediately gives that φ is a monomorphism, as otherwise the exact sequence

$$0 \longrightarrow \operatorname{Ker} \varphi \longrightarrow Y' \longrightarrow \operatorname{Im} \varphi \longrightarrow 0$$

would give that $\operatorname{rk} \operatorname{Im} \varphi < \operatorname{rk} Y' = n$, and then we would have that $\operatorname{rk} \overline{\operatorname{Im} \varphi}^{Y''} \leq \operatorname{rk} \operatorname{Im} \varphi < n = \operatorname{rk} Y''$, which contradicts the fact that $\overline{\operatorname{Im} \varphi}^{Y''} = Y''$.

To prove that φ is an epimorphism we will show first that $\operatorname{Cok} \varphi$ is torsion regular. Take a P-basis $\mathcal{P} = \{P_1, P_2, \ldots, P_n\}$ for Y'. Then it embeds into Y'' as $\varphi(\mathcal{P}) = \{\varphi(P_1), \varphi(P_2), \ldots, \varphi(P_n)\} \subseteq \mathcal{W}_P(Y'')$. Since $\overline{\operatorname{Im} \varphi}^{Y''} = Y''$, we must have $\overline{\varphi(\mathcal{P})}^{Y''} = Y''$, so Theorem 3.2 implies that $\varphi(\mathcal{P})$ is a generating set for $\mathcal{W}_P(Y'')$. Since $\operatorname{rk} Y'' = n = |\varphi(\mathcal{P})|$, we get that $\varphi(\mathcal{P})$ is a P-basis for Y''. So $Y''/\langle\varphi(\mathcal{P})\rangle$ is torsion regular by Theorem 3.1. Hence $\operatorname{Cok}\varphi \cong (Y''/\langle\varphi(\mathcal{P})\rangle)/\operatorname{Im}\varphi/\langle\varphi(\mathcal{P})\rangle$ is also torsion regular.

From the fact that $\operatorname{Cok}\varphi$ is torsion regular we get that $\operatorname{Ext}(\operatorname{Cok}\varphi, \operatorname{Im}\varphi) = 0$ hence the embedding $\varphi : Y' \to Y''$ splits. As Y'' is torsion-free, this implies that $\operatorname{Cok}\varphi = 0$, i.e. φ is an epimorphism.

Finally, since $\operatorname{Cok}\varphi = 0$, the Snake Lemma gives us that $\operatorname{Ker}\varphi'' \cong \operatorname{Cok}\varphi' = M/N$. Since φ'' is a homomorphism between torsion regular modules, $\operatorname{Ker}\varphi''$ is torsion regular. Hence M/N is torsion regular, as required.

It is easy to construct examples to show that the assumption on the finiteness of $\operatorname{rk} N = \operatorname{rk} M$ is necessary. Take for instance a module Q_1 isomorphic to Q with two disjoint non-zero finite dimensional submodules: $P_0, P_1 \subseteq Q_1$. (Since $\operatorname{Soc} Q_1$ is not simple, one can obviously find such submodules.) Let us also choose arbitrary non-zero finite dimensional submodules $P_i \subseteq Q_i \cong Q$ for $i = 2, 3, \ldots$. Take $M = \bigoplus_{i=1}^{\infty} Q_i$, $N = \bigoplus_{i=0}^{\infty} P_i$. Then obviously $\operatorname{rk} N = \operatorname{rk} M = \aleph_0$ with $\overline{N} = M$. But M/N cannot be regular as otherwise one could extend the homomorphism $\bigoplus_{i=0}^{\infty} P_i \xrightarrow{\pi_0} P_0 \xrightarrow{\iota} Q$ to a homomorphism $\bigoplus_{i=1}^{\infty} Q_i \xrightarrow{\psi} Q$. Then $\operatorname{Ker}\psi \supseteq \bigoplus_{i=1}^{\infty} P_i$, hence $\operatorname{Ker}\psi \supseteq \overline{\bigoplus_{i=1}^{\infty} P_i}^{M} = M$, contradicting the fact that $\psi \neq 0$, since for example $\operatorname{Ker}\psi \cap P_0 = 0$.

Bibliography

[ADS] I. Ágoston, A.P. Dean and W. Schewe, *Superpurely simple modules*, to appear in Arch. Math.

[R] C.M. Ringel, *Infinite-dimensional representations of finite dimensional hereditary algebras*, Symp. Math. **23** (1979), 321–412

[DZ] A.P. Dean and F. Zorzitto, *Infinite dimensional representations of $\tilde{D}_4$*, Glasgow Math. J. **32** (1990), 25–33

István Ágoston
Dept. of Math. and Stat.
Carleton University
Ottawa, Ontario
Canada, K1S 5B6

Andrew P. Dean
Dept. of Mathematics
Bishop's University
Lennoxville, Québec
Canada, J1M 1Z7

Canadian Mathematical Society
Conference Proceedings
Volume 11 (1991)

THE SELFINJECTIVITY OF A LOCAL ALGEBRA A AND THE CONDITION $\mathrm{Ext}^1_A(DA, A) = 0$

HIDETO ASASHIBA

Dedicated to Professor Hiroyuki Tachikawa on his 60th birthday

ABSTRACT. In this paper we investigate the problem whether the extension condition in the title on a local algebra A implies the selfinjectivity of A. In the case that the cube of the radical of A is zero, we will give a necessary and sufficient condition for local algebras A that are special but do not lose generality, to satisfy this extension condition in terms of square matrices obtained from the structure constants of A. As a corollary we see, in this case, that the problem is solved affirmatively if A is commutative and that there is a counterexample if A is not commutative.

Throughout this paper k is an algebraically closed field and unless otherwise stated A is a finite dimensional algebra with identity over k. J is the Jacobson radical of A. D and $(\ )^*$ denotes the usual selfduality $\mathrm{Hom}_k(-, k)$ and the A-dual $\mathrm{Hom}_A(-, A)$, respectively. All modules are finitely generated.

The Nakayama conjecture says that if each term of the minimal injective resolution of ${}_AA$ is projective, then A is selfinjective. As noticed in Tachikawa [11] and Hoshino [8], this is true iff so are the following two statements:

(1) $\mathrm{Ext}^i_A({}_ADA, {}_AA) = 0$ for all $i \geq 1$ implies that A is selfinjective.

(2) If A is selfinjective and ${}_AM$ is a module with $\mathrm{Ext}^i_A(M, M) = 0$ for all $i \geq 1$, then M is projective.

We are interested in the conjecture (1). This is still open now; solved for instance affirmatively in the case of $J^3 = 0$ [8]. By experience, for local algebras it is expected that the following stronger claim is true:

(1') If A is a local algebra, then $\mathrm{Ext}^1_A({}_ADA, {}_AA) = 0$ implies that A is selfinjective.

1980 *Mathematics subject Classification* (1985 *Revision*). Primary 16-06,16A46,16A52; Secondary 18G15

Key words and phrases. algebra, selfinjective, quasi-Frobenius, extension, Nakayama conjecture

Here we consider the problem (1'). Since the condition $\mathrm{Ext}^1_A({}_ADA, {}_AA) = 0$ is left right symmetric, we simply write this as $\mathrm{Ext}^1_A(DA, A) = 0$ in the sequel.

In section 1, we will show that many left ideals of a local algebra A has to be indecomposable under the condition that $\mathrm{Ext}^1_A(DA, A) = 0$. (Note that a local algebra A is selfinjective iff $\mathrm{soc}\,{}_AA$ is simple iff all non-zero left ideals are indecomposable.)

In section 2, we consider the problem (1') in the case that $J^3 = 0$. We will show that the statement (1') is true for commutative algebras, while a counterexample will be given in the general case. Thus we see that the statement (1') does not hold in general.

In section 3, we pose a problem related to (1') concerning finitistic dimension of A.

Let ${}_AX$ be a module. $|X|$ and $h(X)$ denote the composition length and the height (= Loewy length) of X, respectively. We denote by $\underline{\mathrm{End}}_A(X)$ the factor ring of $\mathrm{End}_A(X)$ modulo the ideal consisting of endomorphisms of X which factor through a projective module. Further $\Omega(X)$ denotes the kernel of the projective cover of X. We set $\Omega := \Omega({}_ADA)$. By $\mathrm{mod}\,A$ we denote the category of finitely generated left A-modules. For a matrix $\boldsymbol{a} = (a_{ij})$, we denote the ith row and the ith column of $\boldsymbol{a}$ by $\underline{a_i}$ and $\overline{a_i}$, respectively. If V is a vector space with a basis $\{x_1, \dots, x_n\}$, we express this as $V = \langle x_1, \dots, x_n \rangle$. For natural numbers m, n, we sometimes identify $n \times m$-matrices over k with linear maps $k^m \to k^n$ defined by the left multiplication.

1. Indecomposable left ideals

Notice that for local algebras A, A is selfinjective iff $\mathrm{soc}\,{}_AA$ is simple iff every non-zero left ideal is indecomposable. In this section, we will show from this point of view that many left ideals are indecomposable under the condition that $\mathrm{Ext}^1_A(DA, A) = 0$. Namely we show the following.

Theorem 1.1. *Let A be a local algebra and $0 \to {}_A\Omega \xrightarrow{\mu} {}_AP \xrightarrow{\varepsilon} {}_ADA \to 0$ a projective cover of ${}_ADA$. Then every left ideal K containing $g\mu(\Omega)$ for some epimorphism $g : P \to A$ is indecomposable if $\mathrm{Ext}^1_A(DA, A) = 0$.*

Remark 1.2. In the above, J satisfies the condition on the K. Hence ${}_AJ$ is indecomposable (whence so is J_A since our condition is left right symmetric) as stated in Hoshino [8]. Thus the above gives a generalization of [8, Corollary 1.7]. Note that this immediately yields that a local algebra A is selfinjective if $J^2 = 0$ and $\mathrm{Ext}^1_A(DA, A) = 0$.

Theorem 1.1 is a direct consequence of the following more general statement, which we will prove below.

Theorem 1.3. *Let A be an artinian ring with identity and ${}_AC$ a colocal module (i.e. $\mathrm{soc}\,C$ be simple). Consider an exact sequence $0 \to Y \xrightarrow{\mu} X \xrightarrow{\varepsilon} C \to 0$ with ε a proper epimorphism which is right minimal (in the sense of Auslander-Reiten [3]). Let $g : X \to Z$ be a proper retraction with Z indecomposable. Then every submodule K of Z containing $g\mu(Y)$ is indecomposable if $\mathrm{Ext}^1_A(C, Z) = 0$.*

To prove Theorem 1.3 we need the following.

Proposition 1.4. *Let A be an artinian ring with identity and ${}_AC$ a colocal module. Consider an exact sequence $0 \longrightarrow Y \xrightarrow{\mu} X \xrightarrow{\varepsilon} C \longrightarrow 0$ with ε a right minimal morphism. If $g : X \longrightarrow Z$ is a proper retraction, then $g\mu(Y) \geq \operatorname{soc} Z$.*

Proof. Let $\pi : Z \longrightarrow M$ be the cokernel of $g\mu$ and $\alpha : C \longrightarrow M$ the map induced from g. By assumption there is a map $h : Z \longrightarrow X$ with $gh = 1_Z$. Since πg is an epimorphism, so is α. If α is a monomorphism, then g must be an isomorphism since ε is right minimal, a contradiction. Thus α is not a monomorphism, whence we have $\alpha(\operatorname{soc} C) = 0$ because C is colocal. This implies $0 = \alpha\varepsilon(\operatorname{soc} X) = \pi g(\operatorname{soc} X)$, and consequently $0 = \pi gh(\operatorname{soc} Z) = \pi(\operatorname{soc} Z)$. □

Proof of Theorem 1.3. Put $f : Y \longrightarrow K$ to be the composite $g\mu$ and α to be the inclusion $K \longrightarrow Z$. Decompose K as $K = K_1 \oplus K_2$. Express $f = (f_1, f_2)^{\mathrm{T}}, \alpha = (\alpha_1, \alpha_2)$ with respect to this decomposition of K. $\mathrm{Ext}_A^1(C, Z) = 0$ guarantees that there is some $g_i : X \longrightarrow Z$ with $\alpha_i f_i = g_i\mu$ for each i. Since $g\mu = (g_1 + g_2)\mu$, we have some $d : C \longrightarrow Z$ such that $g - (g_1 + g_2) = d\varepsilon$. Thus $g - (g_1 + g_2)$ is non-retraction because so is ε. Hence g_1 or g_2 is a retraction, say g_1 is. Since g_1 is not an isomorphism, the above proposition applies to get $g_1\mu(Y) \geq \operatorname{soc} Z$. This shows $\operatorname{soc} Z \leq \alpha_1 f_1(Y) \leq \alpha(K_1)$, thus $K_2 = 0$. □

2. The case where $J^3 = 0$

Here we consider the problem (1') in the case that $J^3 = 0$. Therefore we assume throughout this section that A *is a local algebra with* $J^3 = 0$. Further we may assume that $J^2 \neq 0$ by Remark 1.2. An affirmative answer will be given for commutative algebras while we will give a counterexample in the general case. About five years ago Hoshino obtained the following result (unpublished, oral communication, see Remark 2.5 for the outline of the proof):

Proposition 2.1 (Hoshino). *If $\mathrm{Ext}_A^i(DA, A) = 0$ for $i = 1$ and 2, then A is selfinjective.*

Our investigation in this section started from his proof of this proposition.

Lemma 2.2. *If $\mathrm{Ext}_A^1(DA, A) = 0$, then Ω is indecomposable.*

Proof. As easily verified, for any module ${}_AX$ we have $\underline{\mathrm{End}}_A(X) \cong \underline{\mathrm{End}}_A(\Omega(X))$ if $\mathrm{Ext}_A^1(X, A) = 0$. Apply this for $X = DA$. □

Lemma 2.3. *If $\mathrm{Ext}_A^1(DA, A) = 0$, then $J^2 = \operatorname{soc} {}_AA = \operatorname{soc} A_A$.*

Proof. Clear from Theorem 1.1. □

Proposition 2.4 (Hoshino). *If $\mathrm{Ext}_A^1(DA, A) = 0$, then $|J^2| \leq 2$.*

Proof. For a module ${}_AM$ of height $h(M) = h$, we set $[M] := (|M/JM|, |JM/J^2M|, \dots, |J^{h-1}M|)^{\mathrm{T}}$. Assume that $\mathrm{Ext}^1_A(DA, A) = 0$. Put $[A_A] := (1, n, m)^{\mathrm{T}}$. Then by Lemma 2.3, $m = |J^2| = |\mathrm{soc}\,{}_AA| = |\mathrm{soc}\,A_A|$. Hence we have $[{}_AA] = (1, n, m)^{\mathrm{T}}, [{}_ADA] = (m, n, 1)^{\mathrm{T}} = [DA_A]$. Consider the projective cover $0 \to {}_A\Omega \xrightarrow{\mu} {}_AA^{(m)} \xrightarrow{\varepsilon} {}_ADA \to 0$ of ${}_ADA$. We show that $[\Omega] = \begin{pmatrix} n(m-1) \\ m^2 - 1 \end{pmatrix}$. By Lemma 2.2, $J\Omega = \mathrm{soc}\,\Omega$. Hence $\Omega \cap (J^2)^{(m)} \leq \mathrm{soc}\,\Omega = J\Omega \leq \Omega \cap (J^2)^{(m)}$. As a consequence, $|J\Omega| = |\Omega \cap (J^2)^{(m)}| = |\mathrm{Ker}\,(\varepsilon|(J^2)^{(m)})| = m^2 - 1$. Now since $|\Omega| = n(m-1) + (m^2 - 1)$, we have $|\Omega/J\Omega| = n(m-1)$ as desired. Put $\Omega^* = C_A \oplus S_A$ such that S is semisimple and C has no simple direct summand. Further put $s := |S|$. Next we show that $[\Omega^*] = \begin{pmatrix} m \\ nm(m-1) - s \end{pmatrix}$. First by assumption we have the projective cover $0 \to (DA)^* \xrightarrow{\varepsilon^*} A_A^{(m)} \xrightarrow{\mu^*} \Omega^* \to 0$ of Ω^*. Thus $[\Omega^*] = \begin{pmatrix} m \\ l \end{pmatrix}$ for some l. Since $\mathrm{soc}\,A_A = \mathrm{soc}\,{}_AA$, we easily see that $\mathrm{soc}\,\Omega^* = (\mathrm{top}\,\Omega)^*$. Further $(\mathrm{top}\,\Omega)^* = \mathrm{Hom}_A(\mathrm{top}\,\Omega, \mathrm{soc}\,{}_AA) = \mathrm{Hom}_{A/J}(\mathrm{top}\,\Omega, J^2) \cong (J^2)^{(|\mathrm{top}\,\Omega|)}$. Therefore $|\mathrm{soc}\,\Omega^*| = nm(m-1)$. Hence $|\Omega^* J| = |\mathrm{soc}\,\Omega^*| - |S| = nm(m-1) - s$ as required. Now since $|\mathrm{top}\,C| = |\mathrm{top}\,\Omega^*| - |S| = m - s$, we have an epimorphism $A^{(m-s)} \to C$, which gives rise to an epimorphism $J^{(m-s)} \to CJ = \Omega^* J$. Since $\Omega^* J$ is semisimple, this yields an epimorphism $(J/J^2)^{(m-s)} \to \Omega^* J$. Hence $n(m-s) \geq nm(m-1) - s$. Thus $0 \geq s(1-n)/n \geq m(m-2)$. This gives $m \leq 2$.
□

REMARK 2.5. (1) Proposition 2.1 was proved by computing $[\Omega(\Omega)]$ and $[\Omega(\Omega)^*]$ using the same argument as above again.

(2) By the last inequality $0 \geq s(1-n)/n \geq m(m-2)$, we see the following: If A is not selfinjective and satisfies the condition that $\mathrm{Ext}^1_A(DA, A) = 0$, then $|J^2| = 2$, $|J/J^2| \geq 2$ and Ω^* has no simple direct summand.

We examine algebras A with $|J^2| = 2$ in more detail. First we construct all such algebras by two $n \times n$-matrices.

DEFINITION 2.6. Let $\boldsymbol{a}, \boldsymbol{b}$ be two $n \times n$-matrices over k ($n \geq 2$), which are linearly independent. Then we define an algebra $A = \boldsymbol{A}(\boldsymbol{a}, \boldsymbol{b})$ as follows: $\boldsymbol{A}(\boldsymbol{a}, \boldsymbol{b})$ is a k-vector space with a basis $b := \{1, x_1, x_2, \dots, x_n, s, t\}$ having the multiplication defined by the rule:

- $1 \cdot a = a \cdot 1 = a$ for all $a \in b$.
- $x_i \cdot x_j := a_{ij}s + b_{ij}t$ for all $1 \leq i, j \leq n$.
- $a \cdot a' = a' \cdot a = 0$ for $a = s, t$ and for all $a' \in b \backslash \{1\}$.

It is easy to check that A actually satisfies the associativity.

Note in the above definition that since $\boldsymbol{a}, \boldsymbol{b}$ are linearly independent, $|J^2| = \mathrm{rank} \begin{pmatrix} \underline{a_1} & \underline{a_2} & \cdots & \underline{a_n} \\ \underline{b_1} & \underline{b_2} & \cdots & \underline{b_n} \end{pmatrix} = 2$, (see Lemma 2.9(1) below for the reason).

Hence $A = \boldsymbol{A}(\boldsymbol{a}, \boldsymbol{b})$ is a local algebra with $J^3 = 0$ and $|J^2| = 2$. Conversely let A be an algebra with $J^3 = 0$ and $|J^2| = 2$. Fix some k-basis $\{x_1, \dots, x_n, s, t\}$ $(n \geq 2)$ of J with $J = \langle x_1, \dots, x_n \rangle \oplus J^2$ and $J^2 = \langle s, t \rangle$. Then for every $1 \leq i, j \leq n$ there are unique $a_{ij}, b_{ij} \in k$ such that $x_i \cdot x_j = a_{ij} s + b_{ij} t$. These two $n \times n$-matrices $\boldsymbol{a} := (a_{ij})$, $\boldsymbol{b} := (b_{ij})$ over k are linearly independent because $|J^2| = 2$. And, since $J^3 = 0$, we have $A = \boldsymbol{A}(\boldsymbol{a}, \boldsymbol{b})$.

DEFINITION 2.7. We say that A has a *presentation* $(\boldsymbol{a}, \boldsymbol{b})$ in case $|J^2| = 2$ and $\boldsymbol{a}$, $\boldsymbol{b}$ are defined as above. Clearly $\boldsymbol{A}(\boldsymbol{a}, \boldsymbol{b})$ in the previous definition has a presentation $(\boldsymbol{a}, \boldsymbol{b})$.

REMARK 2.8. Let $A = \boldsymbol{A}(\boldsymbol{a}, \boldsymbol{b})$ and $A = \boldsymbol{A}(\boldsymbol{a}', \boldsymbol{b}')$ for some bases $\{x_1, \dots, x_n, s, t\}$ and $\{x'_1, \dots, x'_n, s', t'\}$ of J as above, respectively. Then $(x'_1, \dots, x'_n) \equiv (x_1, \dots, x_n)P$ modulo J^2 and $(s, t) = (s', t')Q$ for some regular matrices P, Q, and then $(\boldsymbol{a}', \boldsymbol{b}') = (P^{\mathrm{T}} \boldsymbol{a} P, P^{\mathrm{T}} \boldsymbol{b} P) Q^{\mathrm{T}}$.

We describe the structures of ${}_AJ, J_A, {}_ADJ (= D(J_A))$, and $\Omega' := {}_A\Omega \oplus S$ with $(J^2)^{(2)} = S \oplus (\Omega \cap (J^2)^{(2)})$ in terms of $\boldsymbol{a}$ and $\boldsymbol{b}$. Let $A = \boldsymbol{A}(\boldsymbol{a}, \boldsymbol{b})$. Then every left (resp. right) A-module can be presented as a k-vector space U together with k-endomorphisms U_i corresponding to the left (resp. right) multiplications by x_i. If $U = V \oplus W$ and $U_i = \begin{pmatrix} 0 & 0 \\ V_i & 0 \end{pmatrix}$ with respect to this decomposition of U for all i (e.g., if W is the radical of U), then the module U is determined by the n-tuple (V_i), where V_i are k-linear maps $V \to W$. We therefore denote U as

$$\begin{array}{c} V \\ V_1 \downarrow \cdots \downarrow V_n \\ W \end{array} .$$

The following is easily verified.

Lemma 2.9. *Let $A = \boldsymbol{A}(\boldsymbol{a}, \boldsymbol{b})$. Then*

(1) ${}_AJ$ *has the following structure:*

$$\begin{array}{c} \langle x_1, \dots, x_n \rangle \\ L_1 \downarrow \cdots \downarrow L_n \\ \langle s, t \rangle \end{array} ,$$

where $L_i := \begin{pmatrix} \underline{a_i} \\ \underline{b_i} \end{pmatrix}$, *for all i with respect to these bases.*

(2) J_A *has the following structure:*

$$\begin{array}{c} \langle x_1, \dots, x_n \rangle \\ R_1 \downarrow \cdots \downarrow R_n \\ \langle s, t \rangle \end{array} ,$$

where $R_i := \begin{pmatrix} \overline{a_i}^{\mathrm{T}} \\ \overline{b_i}^{\mathrm{T}} \end{pmatrix}$, *for all i with respect to these bases.*

(3) *Let* $\{\hat{1}, \hat{x}_1, \dots, \hat{x}_n, \hat{s}, \hat{t}\}$ *be the dual basis of* $\{1, x_1, \dots, x_n, s, t\}$. *Then*

$$ {}_A DJ = \begin{matrix} \langle \hat{s}, \hat{t} \rangle \\ R_1^{\mathrm{T}} \downarrow \cdots \downarrow R_n^{\mathrm{T}} \\ \langle \hat{x}_1, \dots, \hat{x}_n \rangle \end{matrix} . $$

Lemma 2.10. *Let A have a presentation $(\boldsymbol{a}, \boldsymbol{b})$. Then the following are equivalent:*

(1) $J^2 = \operatorname{soc} A_A$.

(2) *DJ has no simple direct summands.*

(3) $|\operatorname{top} {}_A DA| = 2$.

(4) *The rank of the $n \times 2n$-matrix $(\boldsymbol{a}|\boldsymbol{b})$ is equal to n.*

Proof. The equivalence of (1), (2) and (3) are obvious. The equivalence of (2) and (4) follows from Lemma 2.9(3). □

For algebras $A = \boldsymbol{A}(\boldsymbol{a}, \boldsymbol{b})$ with $\operatorname{rank}(\boldsymbol{a}|\boldsymbol{b}) = n$, we fix a projective cover $0 \to {}_A\Omega \xrightarrow{\mu} {}_A A^{(2)} \xrightarrow{\varepsilon} {}_A DA \to 0$ of ${}_A DA$, where $\varepsilon = (\varepsilon_1, \varepsilon_2)$, and ε_1 and ε_2 are the right multiplications by $\hat{s}$ and $\hat{t}$, respectively.

Lemma 2.11. *Let $A = \boldsymbol{A}(\boldsymbol{a}, \boldsymbol{b})$ with $\operatorname{rank}(\boldsymbol{a}|\boldsymbol{b}) = n$. Let $\{u_1, \dots, u_n\}$ be a basis of the kernel of the linear map $k^{2n} \to k^n$ defined by the matrix $(\boldsymbol{a}|\boldsymbol{b})$. Then the structure of the left A-module $\Omega' := {}_A\Omega \oplus S$ with $(J^2)^{(2)} = S \oplus (\Omega \cap (J^2)^{(2)})$ is as follows:*

$$ \begin{matrix} \langle u_1, \dots, u_n \rangle \\ M_1 \downarrow \cdots \downarrow M_n \\ \langle (s,0), (t,0), (0,s), (0,t) \rangle \end{matrix} , $$

where $M_i = (L_i \oplus L_i)(u_1, \dots, u_n)$ for all i.

Proof. For every $x \in J$, put $\hat{x} := (x, 0) \in J \oplus J$ and $\check{x} := (0, x) \in J \oplus J$. Then $\Omega' = \Omega \cap J^{(2)} = \Omega' \cap (\langle \hat{x}_1, \dots, \hat{x}_n, \check{x}_1, \dots, \check{x}_n \rangle \oplus \langle \hat{s}, \hat{t}, \check{s}, \check{t} \rangle) = R \oplus \langle \hat{s}, \hat{t}, \check{s}, \check{t} \rangle$, where $R := \Omega' \cap \langle \hat{x}_1, \dots, \hat{x}_n, \check{x}_1, \dots, \check{x}_n \rangle$ and we have an isomorphism $\alpha : \operatorname{Ker}(\boldsymbol{a}|\boldsymbol{b}) \to R$ defined as $\alpha(c) = \sum_{i=1}^{n} \hat{x}_i c_i + \check{x}_i c_{n+i}$ for every $c = (c_i)_{i=1}^{2n} \in \operatorname{Ker}(\boldsymbol{a}|\boldsymbol{b})$. Since $\operatorname{rank}(\boldsymbol{a}|\boldsymbol{b}) = n$, $\dim \operatorname{Ker}(\boldsymbol{a}|\boldsymbol{b}) = n$. Choose a basis $\{u_1, \dots, u_n\}$ of $\operatorname{Ker}(\boldsymbol{a}|\boldsymbol{b})$. Then identifying R with $\operatorname{Ker}(\boldsymbol{a}|\boldsymbol{b})$ by α, Ω' is expressed as

$$ \begin{matrix} \langle u_1, \dots, u_n \rangle \\ M_1 \downarrow \cdots \downarrow M_n \\ \langle \hat{s}, \hat{t}, \check{s}, \check{t} \rangle \end{matrix} , $$

where M_i is given by the left multiplication by x_i for all i. Since $\Omega' \leq J^{(2)}$, we have a commutative diagram:

$$ \begin{array}{ccc} \langle u_1, \dots, u_n \rangle & \xrightarrow{\alpha} & \langle \hat{x}_1, \dots, \hat{x}_n, \check{x}_1, \dots, \check{x}_n \rangle \\ {\scriptstyle M_i}\big\downarrow & & \big\downarrow{\scriptstyle L_i \oplus L_i} \\ \langle \hat{s}, \hat{t}, \check{s}, \check{t} \rangle & = & \langle \hat{s}, \hat{t}, \check{s}, \check{t} \rangle \end{array} , $$

for all i. This proves the assertion. □

To characterize properties of an algebra A having a presentation $(\boldsymbol{a}, \boldsymbol{b})$ in terms of $\boldsymbol{a}$ and $\boldsymbol{b}$, we define linear maps given by the matrices $\boldsymbol{a}, \boldsymbol{b}$ below.

DEFINITION 2.12. Let $\boldsymbol{a}, \boldsymbol{b}$ be $n \times n$-matrices over k. $\mathbf{B}(\boldsymbol{a}, \boldsymbol{b}) : k^{4n} \longrightarrow k^{4n}$ is a linear map defined by the matrix

$$\left(\begin{array}{cc|cc} \boldsymbol{a} & & \boldsymbol{a}^{\mathrm{T}} & \\ & \boldsymbol{a} & \boldsymbol{b}^{\mathrm{T}} & \\ \hline \boldsymbol{b} & & & \boldsymbol{a}^{\mathrm{T}} \\ & \boldsymbol{b} & & \boldsymbol{b}^{\mathrm{T}} \end{array}\right).$$

Assume that $\operatorname{rank}(\boldsymbol{a}|\boldsymbol{b}) = n$, and choose some k-basis $\{u_1, \ldots, u_n\}$ of the kernel of the linear map defined by the matrix $(\boldsymbol{a}|\boldsymbol{b})$. Then we define $\mathbf{M}(\boldsymbol{a}, \boldsymbol{b}) : k^{n^2+4} \longrightarrow k^{2n^2}$ as a linear map defined by the matrix

$$\left(\begin{array}{cccc|c|ccc} \boldsymbol{a} & & & & (\boldsymbol{b}|0)u_1 & & & \\ & \boldsymbol{a} & & & (\boldsymbol{b}|0)u_2 & & & \\ & & \ddots & & \vdots & & & \\ & & & \boldsymbol{a} & (\boldsymbol{b}|0)u_n & & & \\ \hline \boldsymbol{b} & & & & & (\boldsymbol{b}|0)u_1 & (\boldsymbol{a}|0)u_1 & (0|\boldsymbol{a})u_1 \\ & \boldsymbol{b} & & & & (\boldsymbol{b}|0)u_2 & (\boldsymbol{a}|0)u_2 & (0|\boldsymbol{a})u_2 \\ & & \ddots & & & \vdots & \vdots & \vdots \\ & & & \boldsymbol{b} & & (\boldsymbol{b}|0)u_n & (\boldsymbol{a}|0)u_n & (0|\boldsymbol{a})u_n \end{array}\right).$$

Finally $\mathbf{L}(\boldsymbol{a}, \boldsymbol{b}) : k^3 \longrightarrow k^{n^2}$ is a linear map defined by the matrix

$$\begin{pmatrix} (\boldsymbol{b}|0)u_1 & (\boldsymbol{a}|0)u_1 & (0|\boldsymbol{a})u_1 \\ (\boldsymbol{b}|0)u_2 & (\boldsymbol{a}|0)u_2 & (0|\boldsymbol{a})u_2 \\ \vdots & \vdots & \vdots \\ (\boldsymbol{b}|0)u_n & (\boldsymbol{a}|0)u_n & (0|\boldsymbol{a})u_n \end{pmatrix},$$

which is a part of the matrix of $\mathbf{M}(\boldsymbol{a}, \boldsymbol{b})$.

REMARK 2.13. (1) In the above definition, $\mathbf{M}(\boldsymbol{a}, \boldsymbol{b})$ and $\mathbf{L}(\boldsymbol{a}, \boldsymbol{b})$ do not depend on the choice of $\{u_i\}$ up to isomorphisms. In fact, suppose that $\{v_i\}$ is obtained by any elementary transformation of $\{u_i\}$, and let $\mathbf{M}'$, $\mathbf{L}'$ be given by $\{v_i\}$. Then the matrices of $\mathbf{M}'$ and $\mathbf{L}'$ can be obtained by certain elementary transformations of matrices of $\mathbf{M}(\boldsymbol{a}, \boldsymbol{b})$ and $\mathbf{L}(\boldsymbol{a}, \boldsymbol{b})$, respectively.

(2) Let A have presentations $(\boldsymbol{a}, \boldsymbol{b})$ and $(\boldsymbol{a}', \boldsymbol{b}')$. Then by using Remark 2.8 we see that $\mathbf{B}(\boldsymbol{a}, \boldsymbol{b}) \cong \mathbf{B}(\boldsymbol{a}', \boldsymbol{b}')$, $\mathbf{M}(\boldsymbol{a}, \boldsymbol{b}) \cong \mathbf{M}(\boldsymbol{a}', \boldsymbol{b}')$ and $\mathbf{L}(\boldsymbol{a}, \boldsymbol{b}) \cong \mathbf{L}(\boldsymbol{a}', \boldsymbol{b}')$. Therefore we denote $\mathbf{B} = \mathbf{B}(A) = \mathbf{B}(\boldsymbol{a}, \boldsymbol{b})$, $\mathbf{M} = \mathbf{M}(A) = \mathbf{M}(\boldsymbol{a}, \boldsymbol{b})$ and $\mathbf{L} = \mathbf{L}(A) = \mathbf{L}(\boldsymbol{a}, \boldsymbol{b})$.

Lemma 2.14. *Let A have a presentation $(\boldsymbol{a},\boldsymbol{b})$ with* $\operatorname{rank}(\boldsymbol{a}|\boldsymbol{b}) = n$. *Then* $J\Omega = \Omega \cap (J^2)^{(2)}$ *iff* $\operatorname{Ker}\mathbf{L}(\boldsymbol{a},\boldsymbol{b}) = 0$.

Proof. Since $\dim \Omega \cap (J^2)^{(2)} = \dim \operatorname{Ker} \varepsilon|_{(J^2)^{(2)}} = 3$, we have $J\Omega = \Omega \cap (J^2)^{(2)}$ iff $\operatorname{rank}(M_1,\dots,M_n) = 3$ by Lemma 2.11. And, $\operatorname{rank}(M_1,\dots,M_n) =$

$$\operatorname{rank}(M_1,\dots,M_n)^{\mathrm{T}} = \operatorname{rank}\begin{pmatrix} (\boldsymbol{a}|0)u_1 & (\boldsymbol{b}|0)u_1 & (0|\boldsymbol{a})u_1 & (0|\boldsymbol{b})u_1 \\ \vdots & \vdots & \vdots & \vdots \\ (\boldsymbol{a}|0)u_n & (\boldsymbol{b}|0)u_n & (0|\boldsymbol{a})u_n & (0|\boldsymbol{b})u_n \end{pmatrix} =$$

$\operatorname{rank}\mathbf{L}$ because $(\boldsymbol{a}|\boldsymbol{b})u_i = 0$ for all i. □

DEFINITION 2.15. We say that a local algebra A is *critical* in case $|J^2| = 2$, $J^2 = \operatorname{soc} A_A$ and $J\Omega = \Omega \cap (J^2)^{(2)}$. Note that an algebra A having a presentation $(\boldsymbol{a},\boldsymbol{b})$ is critical iff $\operatorname{rank}(\boldsymbol{a}|\boldsymbol{b}) = n$ and $\operatorname{Ker}\mathbf{L}(\boldsymbol{a},\boldsymbol{b}) = 0$ by Lemmas 2.10, 2.14.

REMARK 2.16. (1) We already know that if $\operatorname{Ext}^1_A(DA, A) = 0$, then A is critical or selfinjective. Hence in order to verify the statement (1') in the introduction, it is enough to show that $\operatorname{Ext}^1_A(DA, A) \neq 0$ for critical algebras A.

(2) If A is critical, then $[\Omega] = \binom{n}{3}$ as in the proof of Proposition 2.4.

Recall the following ([1]): Let Λ be a finite dimensional k-algebra with the Jacobson radical $\boldsymbol{r}$, and assume that $\boldsymbol{r}^2 = 0$. Put $B := \begin{pmatrix} \Lambda/\boldsymbol{r} & 0 \\ \boldsymbol{r} & \Lambda/\boldsymbol{r} \end{pmatrix}$. Then there is a k-functor $F : \operatorname{mod}\Lambda \to \operatorname{mod} B$ defined as follows: For $M \in \operatorname{mod}\Lambda$, $F(M) := (M/\boldsymbol{r}M, \boldsymbol{r}M, f)$, where f is given by the commutative diagram

$$\begin{array}{ccc} \boldsymbol{r}\otimes_\Lambda M & \longrightarrow & \boldsymbol{r}M \\ \cong\downarrow & & \| \\ \boldsymbol{r}\otimes_{\Lambda/\boldsymbol{r}} M/\boldsymbol{r}M & \xrightarrow{f} & \boldsymbol{r}M \end{array}$$

with the canonical maps. For $g \in \operatorname{Hom}_\Lambda(M, N)$, $F(g) : F(M) \to F(N)$ is the pair (g_1, g_2) given by the commutative diagram

$$\begin{array}{ccccccccc} 0 & \to & \boldsymbol{r}M & \to & M & \to & M/\boldsymbol{r}M & \to & 0 \\ & & g_2\downarrow & & g\downarrow & & g_1\downarrow & & \\ 0 & \to & \boldsymbol{r}N & \to & N & \to & N/\boldsymbol{r}N & \to & 0 \end{array}$$

with exact rows. Hence by definition, $\operatorname{Hom}_B(FM, FN) = \operatorname{Hom}_\Lambda(M,N)/S(M, N)$, where $S(M,N) := \{g \in \operatorname{Hom}_\Lambda(M,N) | g(M) \le \boldsymbol{r}N\} = \operatorname{Hom}_\Lambda(M, \boldsymbol{r}N)$. In particular, $\operatorname{Hom}_\Lambda(M,N) \cong \operatorname{Hom}_B(FM, FN) \oplus \operatorname{Hom}_\Lambda(M, \boldsymbol{r}N)$ as k-vector spaces. We apply this to $\Lambda := A/J^2$ for critical algebras A. Note that in this case, $\boldsymbol{r} = J/J^2$ and $\Lambda/\boldsymbol{r} = k$.

We describe the structures of $F({}_AJ)$, $F({}_ADJ)$ and $\bar{\Omega} := F({}_A\Omega) \oplus Be_2$, where $e_2 := \begin{pmatrix} 0 & 0 \\ 0 & 1 \end{pmatrix} \in B$. Let $A = \boldsymbol{A}(\boldsymbol{a},\boldsymbol{b})$ and B as above. Then every left B-module U is expressed by two k-vector spaces V and W together with linear maps $U_i : V \longrightarrow W$ corresponding to the left multiplications by x_i for all i. We denote this as follows:

$$\begin{array}{c} V \\ U_1 \downarrow \cdots \downarrow U_n \\ W \end{array} .$$

Lemma 2.17. *Let $A = \mathbf{A}(\boldsymbol{a}, \boldsymbol{b})$ be a critical algebra. Then keeping the notations in Lemmas 2.9, 2.11, we have the following.*

(1) $$F(J) = \begin{array}{c} \langle x_1, \dots, x_n \rangle \\ L_1 \downarrow \cdots \downarrow L_n \\ \langle s, t \rangle \end{array} ,$$

(2) $$F(DJ) = \begin{array}{c} \langle \hat{s}, \hat{t} \rangle \\ R_1^{\mathrm{T}} \downarrow \cdots \downarrow R_n^{\mathrm{T}} \\ \langle \hat{x}_1, \dots, \hat{x}_n \rangle \end{array} ,$$

(3) $$\bar{\Omega} = \begin{array}{c} \langle u_1, \dots, u_n \rangle \\ M_1 \downarrow \cdots \downarrow M_n \\ \langle (s,0), (t,0), (0,s), (0,t) \rangle \end{array} .$$

Proof. This follows by Lemmas 2.9, 2.11. □

We will give a criterion for an algebra A having a presentation $(\boldsymbol{a}, \boldsymbol{b})$ to satisfy the condition $\mathrm{Ext}_A^1(DA, A) = 0$, in terms of $\boldsymbol{a}$ and $\boldsymbol{b}$ below. To do this we need the following lemma.

Lemma 2.18. *Let A be a critical algebra. Then*

(1) $\dim(DA)^* - 4 = \dim \operatorname{Ker} \mathbf{B}(A) = (2n+2) - \dim \operatorname{Im} \mu^*$.

(2) $\dim \Omega^* - (2n+2) = \dim \operatorname{Ker} \mathbf{M}(A)$.

Proof. (1) $(DA)^* = \mathrm{Hom}_A({}_A DA, {}_A A) = \mathrm{Hom}_A((DA/\mathrm{soc}\, DA), J) = \mathrm{Hom}_A(DJ, J) \cong \mathrm{Hom}_k(\mathrm{top}\, DJ, J^2) \oplus \mathrm{Hom}_B(F(DJ), F(J))$. Hence $\dim(DA)^* = 4 + \dim \mathrm{Hom}_B(F(DJ), F(J))$. Note by Lemma 2.17 that $\mathrm{Hom}_B(F(DJ), F(J))$ is given by the set of pairs $(\mathbf{X}, \mathbf{Y})$ of an $n \times 2$-matrix $\mathbf{X} = (X_{ij})$ and a $2 \times n$-matrix $\mathbf{Y} = (Y_{ij})$ satisfying

$$(*) \qquad L_i \mathbf{X} = \mathbf{Y} R_i^{\mathrm{T}} \text{ for all } i.$$

These conditions on $\mathbf{X}, \mathbf{Y}$ can be regarded as linear equations on the entries of $\mathbf{X}, \mathbf{Y}$. The matrix of these equations has the form of the matrix of $\mathbf{B}$ by suitable equivalent transformations. In fact, $(*)$ is equivalent to the following:

$$\begin{pmatrix} \underline{a_i \overline{X_1}} & \underline{a_i \overline{X_2}} \\ \underline{b_i \overline{X_1}} & \underline{b_i \overline{X_2}} \end{pmatrix} = \begin{pmatrix} \underline{Y_1 \overline{a_i}} & \underline{Y_1 \overline{b_i}} \\ \underline{Y_2 \overline{a_i}} & \underline{Y_2 \overline{b_i}} \end{pmatrix}, \quad \text{for all } i.$$

This holds iff

$$\begin{pmatrix} \underline{a_i} & & \overline{a_i}^{\mathrm{T}} & \\ & \underline{a_i} & \overline{b_i}^{\mathrm{T}} & \\ \underline{b_i} & & & \overline{a_i}^{\mathrm{T}} \\ & \underline{b_i} & & \overline{b_i}^{\mathrm{T}} \end{pmatrix} \begin{pmatrix} \overline{X_1} \\ \overline{X_2} \\ -\underline{Y_1}^{\mathrm{T}} \\ -\underline{Y_2}^{\mathrm{T}} \end{pmatrix} = 0, \text{ for all } i,$$

iff

$$\mathbf{B}(A) \begin{pmatrix} \overline{X_1} \\ \overline{X_2} \\ -\underline{Y_1}^{\mathrm{T}} \\ -\underline{Y_2}^{\mathrm{T}} \end{pmatrix} = 0.$$

Hence $\mathrm{Hom}_B(F(DJ), F(J)) \cong \mathrm{Ker}\, \mathbf{B}(A)$ as k-vector spaces. The rest follows from $\dim(DA)^* + \dim Im\mu^* = \dim A^{(2)} = 2n + 6$.

(2) $\Omega^* = \mathrm{Hom}_A(\Omega, A) = \mathrm{Hom}_A(\Omega, J) \cong \mathrm{Hom}_k(\mathrm{top}\, \Omega, J^2) \oplus \mathrm{Hom}_B(F(\Omega), F(J))$. Hence $\dim \Omega^* = 2n + \dim \mathrm{Hom}_B(F(\Omega), F(J))$. Further $\dim \mathrm{Hom}_B (F(\Omega), F(J)) = \dim \mathrm{Hom}_B(\bar{\Omega}, F(J)) - 2$. By Lemma 2.17, a similar calculation as above shows that $\mathrm{Hom}_B(\bar{\Omega}, F(J))$ is isomorphic to the kernel of the linear map $\tilde{\mathbf{M}} : k^{n^2+8} \to k^{2n^2}$ defined by the matrix

$$\left(\begin{array}{ccc|c|c} \boldsymbol{a} & & & & \\ & \ddots & & \tilde{\mathbf{L}} & \\ & & \boldsymbol{a} & & \\ \hline \boldsymbol{b} & & & & \\ & \ddots & & & \tilde{\mathbf{L}} \\ & & \boldsymbol{b} & & \end{array}\right).$$

where $\tilde{\mathbf{L}}$ is equal to

$$\begin{pmatrix} (\boldsymbol{a}|0)u_1 & (0|\boldsymbol{a})u_1 & (\boldsymbol{b}|0)u_1 & (0|\boldsymbol{b})u_1 \\ (\boldsymbol{a}|0)u_2 & (0|\boldsymbol{a})u_2 & (\boldsymbol{b}|0)u_2 & (0|\boldsymbol{b})u_2 \\ \vdots & \vdots & \vdots & \vdots \\ (\boldsymbol{a}|0)u_n & (0|\boldsymbol{a})u_n & (\boldsymbol{b}|0)u_n & (0|\boldsymbol{b})u_n \end{pmatrix}.$$

In fact, $\mathrm{Hom}(\bar{\Omega}, F(J))$ is given by the set of pairs $(\mathbf{X}, \mathbf{Y})$ of an $n \times n$-matrix $\mathbf{X} = (X_{ij})$ and a 2×4-matrix $\mathbf{Y} = (Y_{ij})$ satisfying $L_i \mathbf{X} = \mathbf{Y} M_i$ for all i. Putting $K_i := L_i \oplus L_i$, this is expressed as follows:

$$\begin{pmatrix} \underline{a_i}\overline{X_1} & \cdots & \underline{a_i}\overline{X_n} \\ \underline{b_i}\overline{X_1} & \cdots & \underline{b_i}\overline{X_n} \end{pmatrix} = \begin{pmatrix} \underline{Y_1 K_i u_1} & \cdots & \underline{Y_1 K_i u_n} \\ \underline{Y_2 K_i u_1} & \cdots & \underline{Y_2 K_i u_n} \end{pmatrix}, \text{ for all } i.$$

This holds iff

$$\left(\begin{array}{ccc|ccc} \underline{a_i} & & & (K_iu_1)^{\mathrm{T}} & & \\ & \ddots & & \vdots & & \\ & & \underline{a_i} & (K_iu_n)^{\mathrm{T}} & & \\ \hline \underline{b_i} & & & & (K_iu_1)^{\mathrm{T}} & \\ & \ddots & & & \vdots & \\ & & \underline{b_i} & & (K_iu_n)^{\mathrm{T}} & \end{array}\right) \begin{pmatrix} \overline{X_1} \\ \vdots \\ \overline{X_n} \\ -\underline{Y_1}^{\mathrm{T}} \\ -\underline{Y_2}^{\mathrm{T}} \end{pmatrix} = 0, \text{ for all } i,$$

iff

$$\tilde{\mathbf{M}} \begin{pmatrix} \overline{X_1} \\ \vdots \\ \overline{X_n} \\ -\underline{Y_1}^{\mathrm{T}} \\ -\underline{Y_2}^{\mathrm{T}} \end{pmatrix} = 0.$$

This shows that $\dim \mathrm{Hom}_B(\bar{\Omega}, F(J)) = \dim \mathrm{Ker}\, \tilde{\mathbf{M}}$. Using $(\boldsymbol{a}|\boldsymbol{b})u_i = 0$ for all i, we easily see that $\dim \mathrm{Ker}\, \tilde{\mathbf{M}} = \dim \mathrm{Ker}\, \mathbf{M}(A) + 4$. This completes the proof. □

Proposition 2.19. *Let A have a presentation $(\boldsymbol{a}, \boldsymbol{b})$. Then $\mathrm{Ext}^1_A(DA, A) = 0$ iff $\mathrm{Ker}\, \mathbf{B}(\boldsymbol{a}, \boldsymbol{b}) = 0$ and $\mathrm{Ker}\, \mathbf{M}(\boldsymbol{a}, \boldsymbol{b}) = 0$.*

Proof. ($\Leftarrow$). Assume $\mathrm{Ker}\, \mathbf{B}(\boldsymbol{a}, \boldsymbol{b}) = 0$. Then $\mathrm{rank}\,(\boldsymbol{a}|\boldsymbol{b}) = n$, i.e., $\mathbf{M}(\boldsymbol{a}, \boldsymbol{b})$ is defined. Further assume that $\mathrm{Ker}\, \mathbf{M}(\boldsymbol{a}, \boldsymbol{b}) = 0$. Then A is critical. Hence the assertion follows by Lemma 2.18.

($\Rightarrow$). Assume that $\mathrm{Ext}^1_A(DA, A) = 0$. Then A is critical, the sequence $0 \to (DA)^* \xrightarrow{\varepsilon^*} A_A^{(2)} \xrightarrow{\mu^*} \Omega^* \to 0$ is exact, and $[\Omega^*] = \begin{pmatrix} 2 \\ 2n \end{pmatrix}$. Hence Lemma 2.18 guarantees that $\mathrm{Ker}\, \mathbf{B}(\boldsymbol{a}, \boldsymbol{b}) = 0$ and $\mathrm{Ker}\, \mathbf{M}(\boldsymbol{a}, \boldsymbol{b}) = 0$. □

REMARK 2.20. Let $\boldsymbol{a}, \boldsymbol{b}$ be $n \times n$-matrices. If $\mathrm{Ker}\, \mathbf{B}(\boldsymbol{a}, \boldsymbol{b}) = 0$, then as easily seen $\boldsymbol{a}, \boldsymbol{b}$ can not be linearly dependent, i.e., $\boldsymbol{A}(\boldsymbol{a}, \boldsymbol{b})$ can be constructed by Definition 2.6.

Theorem 2.21. *Let A be a commutative local algebra with $J^3 = 0$. Then $\mathrm{Ext}^1_A(DA, A) = 0$ implies that A is selfinjective.*

Proof. Let A be a commutative critical algebra and $(\boldsymbol{a}, \boldsymbol{b})$ a presentation of A. We have only to show that $\mathrm{Ext}^1_A(DA, A) \neq 0$. Since A is commutative, $\boldsymbol{a} = \boldsymbol{a}^{\mathrm{T}}, \boldsymbol{b} = \boldsymbol{b}^{\mathrm{T}}$. Hence as easily seen, the matrix of $\mathbf{B}(A)$ is similar to the following matrix:

$$\left(\begin{array}{cc|cc} \boldsymbol{a} & \boldsymbol{b} & & \\ \hline & & \boldsymbol{a} & \\ & & \boldsymbol{b} & \boldsymbol{a} \\ & & & \boldsymbol{b} \end{array}\right)$$

Thus $\dim \mathrm{Ker}\, \mathbf{B}(A) \geq n$. Hence by Proposition 2.19, $\mathrm{Ext}^1_A(DA, A) \neq 0$. □

REMARK 2.22. If for any $1 \leq i \leq n$, there exists some $c \in k$ such that for any $1 \leq j \leq n$, $x_i \cdot x_j = cx_j \cdot x_i$, then by elementary transformations, we can transform the matrix of $\mathbf{B}(A)$ into the same form as in the commutative case whenever $c \neq 0$ for all i. Hence also in this case, $\operatorname{Ext}_A^1(DA, A) \neq 0$.

Next we investigate the case where the matrix $\boldsymbol{a}$ or $\boldsymbol{b}$ is regular. By exchanging s and t if necessary, we may assume that $\boldsymbol{a}$ is regular.

Proposition 2.23. *Let A have a presentation $(\boldsymbol{a}, \boldsymbol{b})$ with $\boldsymbol{a}$ a regular matrix. Then*

(1) $\operatorname{Ker} \mathbf{B}(\boldsymbol{a}, \boldsymbol{b}) = 0$ *iff the commutator* $[\boldsymbol{b}\boldsymbol{a}^{-1}, \boldsymbol{b}^{\mathrm{T}}\boldsymbol{a}^{-\mathrm{T}}]$ *is a regular matrix.*

(2) $\operatorname{Ker} \mathbf{M}(\boldsymbol{a}, \boldsymbol{b}) = 0$ *iff the degree of the minimal polynomial of* $\boldsymbol{a}^{-1}\boldsymbol{b}$ *is greater than 3.*

(3) *Hence* $\operatorname{Ext}_A^1(DA, A) = 0$ *iff* $[\boldsymbol{b}\boldsymbol{a}^{-1}, \boldsymbol{b}^{\mathrm{T}}\boldsymbol{a}^{-\mathrm{T}}]$ *is regular, and the degree of the minimal polynomial of* $\boldsymbol{a}^{-1}\boldsymbol{b}$ *is greater than 3.*

Proof. We denote by E_m the identity $m \times m$-matrix for a natural number m. (1) Direct calculation shows that $\mathbf{B}$ is isomorphic to the linear map defined by the matrix $E_{3n} \oplus [\boldsymbol{b}\boldsymbol{a}^{-1}, \boldsymbol{b}^{\mathrm{T}}\boldsymbol{a}^{-\mathrm{T}}]$, which gives the assertion.

(2) Define $n \times 1$-matrices $\hat{u}_i, \check{u}_i$ by $(\hat{u}_i|\check{u}_i)^{\mathrm{T}} := u_i$ for all i. Since the matrices $(\boldsymbol{a}|\boldsymbol{b}\hat{u}_i)$ are singular, there exist $n \times 1$-matrices v_i over k such that $(\boldsymbol{a}|\boldsymbol{b})(v_i|\hat{u}_i)^{\mathrm{T}} = 0$ for all i. Further since $\{u_i, \dots, u_n\}$ is a basis of the kernel of the linear map defined by the matrix $(\boldsymbol{a}|\boldsymbol{b})$, there exist some $n \times n$-matrix $\boldsymbol{c} := (c_{ij})$ such that $(v_i|\hat{u}_i)^{\mathrm{T}} = (u_1, \dots, u_n)(c_{1i}, \dots, c_{ni})^{\mathrm{T}}$ for all i. Hence $(v_1, \dots, v_n) = (\hat{u}_1, \dots, \hat{u}_n)\boldsymbol{c}$ and $(\hat{u}_1, \dots, \hat{u}_n) = (\check{u}_1, \dots, \check{u}_n)\boldsymbol{c}$. Using $\boldsymbol{b}\hat{u}_i = -\boldsymbol{a}v_i$ for all i, we see that $\mathbf{M}$ is isomorphic to the linear map defined by the matrix

$$E_{n^2} \oplus \begin{pmatrix} \boldsymbol{b}v_1 & \boldsymbol{b}\hat{u}_1 & \boldsymbol{a}\hat{u}_1 & \boldsymbol{a}\check{u}_1 \\ \vdots & \vdots & \vdots & \vdots \\ \boldsymbol{b}v_n & \boldsymbol{b}\hat{u}_n & \boldsymbol{a}\hat{u}_n & \boldsymbol{a}\check{u}_n \end{pmatrix},$$

Hence $\operatorname{Ker} \mathbf{M} = 0$ iff the set $\{\boldsymbol{b}(v_1, \dots, v_n), \boldsymbol{b}(\hat{u}_1, \dots, \hat{u}_n), \boldsymbol{a}(\hat{u}_1, \dots, \hat{u}_n), \boldsymbol{a}(\check{u}_1, \dots, \check{u}_n)\}$ is linearly independent. Since $\boldsymbol{a}$ is regular we can choose $(\check{u}_1, \dots, \check{u}_n) = E_n$. Then $(\hat{u}_1, \dots, \hat{u}_n) = \boldsymbol{c} = -\boldsymbol{a}^{-1}\boldsymbol{b}$ and $(v_1, \dots, v_n) = \boldsymbol{c}^2$. Accordingly, $\operatorname{Ker} \mathbf{M} = 0$ iff the set $\{\boldsymbol{a}\boldsymbol{c}^3, \boldsymbol{a}\boldsymbol{c}^2, \boldsymbol{a}\boldsymbol{c}, \boldsymbol{a}\}$ is linearly independent iff the set $\{\boldsymbol{c}^3, \boldsymbol{c}^2, \boldsymbol{c}, E_n\}$ is linearly independent. As a consequence, $\operatorname{Ker} \mathbf{M} = 0$ iff the degree of the minimal polynomial of $-\boldsymbol{c} = \boldsymbol{a}^{-1}\boldsymbol{b}$ is greater than 3. □

REMARK 2.24. (1) In the above if $n := |J/J^2| = 3$, then the degree of the minimal polynomial of $\boldsymbol{a}^{-1}\boldsymbol{b}$ is ≤ 3. Hence $\operatorname{Ext}_A^1(DA, A) \neq 0$.

(2) If $n = 2$, then $\operatorname{Ker} \mathbf{L}(A) \neq 0$. Hence also in this case, the statement (1') in the introduction is true. Namely, if A is a local algebra with $J^3 = 0$ and $|J/J^2| \leq 2$, then $\operatorname{Ext}_A^1(DA, A) = 0$ implies that A is selfinjective.

(3) If rank $\boldsymbol{c} \leq 1$ for $\boldsymbol{c} = \boldsymbol{a}$ or $\boldsymbol{b}$, then $\operatorname{Ker}\mathbf{L}(A) \neq 0$. Hence also in this case, we have $\operatorname{Ext}^1_A(DA, A) \neq 0$. [To check (2),(3) of the above, consider a "normal form" of $(\boldsymbol{a}|\boldsymbol{b})$ in the following sense: If we replace the ordered basis $(x_1, \dots, x_n)$ by another basis $(x_1, \dots, x_n)P$, where P is a regular matrix over k, then $(\boldsymbol{a}|\boldsymbol{b})$ changes to $(P^{\mathrm{T}}\boldsymbol{a}P|P^{\mathrm{T}}\boldsymbol{b}P)$ as in Remark 2.8. If rank $\boldsymbol{a} = 1$, then $\boldsymbol{a}$ can be transformed into a matrix $\boldsymbol{a}'$ such that all the columns of $\boldsymbol{a}'$ except for the first are zero and (1,1)-entry is 1. Using this form it is easy to check that $\operatorname{Ker}\mathbf{L}(A) \neq 0$.]

(4) For any $n \times n$-matrices $\boldsymbol{a}, \boldsymbol{b}$, if $[\boldsymbol{b}\boldsymbol{a}^{-1}, \boldsymbol{b}^{\mathrm{T}}\boldsymbol{a}^{-\mathrm{T}}]$ is regular, then necessarily $\boldsymbol{a}, \boldsymbol{b}$ are linearly independent, which means that we can construct $\boldsymbol{A}(\boldsymbol{a}, \boldsymbol{b})$ as in Remark 2.20.

EXAMPLE 2.25. We give a counterexample to the statement (1') in the introduction. Let $\boldsymbol{a}$ be the identity $n \times n$-matrix E_n and $\boldsymbol{b}$ an $n \times n$-matrix over k. Then by Proposition 2.23 and by the above Remark (4), $A := \boldsymbol{A}(\boldsymbol{a}, \boldsymbol{b})$ can be constructed and $\operatorname{Ext}^1_A(DA, A) = 0$ iff $[\boldsymbol{b}, \boldsymbol{b}^{\mathrm{T}}]$ is regular and the degree of the minimal polynomial of $\boldsymbol{b}$ is > 3. Actually there exists such matrix $\boldsymbol{b}$. For instance, let $n = 4$ and $\boldsymbol{b} := \begin{pmatrix} 0 & 0 & 0 & 0 \\ 1 & 0 & 0 & 0 \\ 0 & 0 & 1 & 0 \\ 0 & 0 & 1 & 1 \end{pmatrix}$. Then this $\boldsymbol{b}$ satisfies the required conditions. Hence $\boldsymbol{A}(\boldsymbol{a}, \boldsymbol{b}) \cong k\langle x, y, z, w\rangle / \{(x, y, z, w)^3 + (xy, xz, xw, yz, yw, zx, zy, zw, wx, wy, x^2 - y^2, z^2 - w^2, yx - wz, z^2 - x^2 - yx)\}$ satisfies $\operatorname{Ext}^1_A(DA, A) = 0$ but not selfinjective.

3. A connection with finitistic dimension

We realized that the problem (1') in the introduction is for commutative algebras, and the following remains open.

(1") If A is a local commutative algebra, then $\operatorname{Ext}^1_A(DA, A) = 0$ implies that A is selfinjective.

In connection with this we have another problem concerning finitistic dimension of an algebra. Recall that the *left finitistic dimension* (denoted by l.fin.dim A) of an algebra A is defined as $\sup\{\mathrm{pd}M | M \in \operatorname{mod} A \text{ with } \mathrm{pd}M < \infty\}$, where $\mathrm{pd}M$ denotes the projective dimension of M. To relate this dimension with our problem we define the following:

DEFINITION 3.1. Let (I_i) be the minimal injective resolution of A_A. Then the infimum of the set $\{n \geq 0 | \bigoplus_{i=0}^{n} I_i \text{ is a cogenerator}\}$, denoted by r.cog.dim A, is called the *right cogenerator dimension* of A. By definition r.cog.dim$A = \infty$ if this set is empty. Generalized Nakayama conjecture says that all algebras have finite right cogenerator dimensions.

We use arguments in [7] to obtain the following three statements.

Proposition 3.2. *Let A be an algebra. Then* r.cog.dim$A \leq$ l.fin.dim$A$.

Proof. We may assume that $c :=$ r.cog.dim$A > 0$. Then there is a simple right A-module S such that $\mathrm{Ext}^c_A(S, A) \neq 0$ and $\mathrm{Ext}^i_A(S, A) = 0$ for all i with $0 \leq i \leq c-1$. Let (P_i) be a minimal projective resolution of S. Then we have an exact sequence

$$(*) \qquad 0 \to P_0^* \to P_1^* \to \cdots \to P_c^* \to M \to 0$$

of left A-modules, where M is the cokernel of $P_{c-1}^* \to P_c^*$. Hence l.fin.dim $A \geq c$ because $(*)$ is a minimal projective resolution of M. □

We set $U := A \oplus DA$ and $B := \mathrm{End}_A(U_A) \cong \mathrm{End}_A({}_AU)^{\mathrm{op}}$.

Proposition 3.3. *For any $d \geq 0$, if one of the dimensions* r.fin.dim B, l.fin.dim B, r.cog. dim B *and* l.cog.dim B *is less than or equal to d, then the condition that $\mathrm{Ext}^i_A(DA, A) = 0$ for all i with $1 \leq i \leq d-1$ implies that A is selfinjective.*

Proof. Assume that $\mathrm{Ext}^i_A(DA, A) = 0$ for all i with $1 \leq i \leq d-1$. Let (I_i) be a minimal injective resolution of U_A. Then by applying $(U, -) := \mathrm{Hom}_A({}_BU_A, -)$ we have an exact sequence

$$0 \to B_B \to (U, I_0) \to (U, I_1) \to \cdots \to (U, I_{d-1}) \to (U, I_d).$$

As easily seen (U, I_i) are projective and injective B-modules for all i. Hence if r.cog.dim $B \leq d$, then B is selfinjective, whence so is A. The rest follows by Proposition 3.2 and by the fact that $\mathrm{End}_A({}_AU) = B^{\mathrm{op}}$. □

Applying the above proposition for $d = 0, 1$, we get the following, for which it is essential that the dominant dimension of B is greater than or equal to 2 ([11]).

Corollary 3.4. *The following are equivalent for an algebra A:*

(1) *A is selfinjective.*

(2) *B is selfinjective.*

(3) l.fin.dim $B \leq 1$.

(4) l.fin.dim $B = 0$.

(5) r.cog.dim $B \leq 1$.

(6) r.cog.dim $B = 0$.

(3') r.fin.dim $B \leq 1$.

(4') r.fin.dim $B = 0$.

(5') l.cog.dim $B \leq 1$.

(6') l.cog.dim $B = 0$.

From Proposition 3.3 for $d = 2$, we have the following problem:

Problem. For every local commutative algebra A, is one of the dimensions r.fin.dim B, l.fin.dim B, r.cog.dim B and l.cog.dim B less than or equal to 2?

Acknowledgements

I would like to thank Professor H. Tachikawa who introduced me into this problem and Doctor M. Hoshino who showed me his proof of Proposition 2.1 for their encouragements and for valuable discussions with them.

References

1. Auslander, M. and Reiten, I.: *Stable equivalence of Artin algebras*, Proc. of the Conference on Orders, Group Rings and Related Topics, Springer Lecture Notes **353** (1973), 8–71.
2. Auslander, M. and Reiten, I.: *On a generalized version of the Nakayama conjecture*, Proc. Amer. Math. Soc. **52** (1975), 69–74.
3. Auslander, M. and Reiten, I.: *Representation theory of artin algebras IV*, Comm. Alg.**5** (1977), 443–518.
4. Bass, H.: *Finitistic dimension and a homological generalization of semiprimary rings*, Trans. Amer. Math. Soc. **95** (1960), 466–488.
5. Fuller, K. R. and Zimmermann-Huisgen, B.: *On the Generalized Nakayama conjecture and Cartan determinant problem*, Trans. Amer. Math. Soc. **294** (1986), 679–691.
6. Hoshino, M.: *Modules without self-extensions and Nakayama's conjecture*, Arch. Math. **43** (1984), 493–500.
7. Hoshino, M.: *On Nakayama's Conjecture*, Proc. of Representations of Rings and Duality held at Research Institute for Mathematical Sciences, Kyoto University, Kokyuroku Vol. **628** (1987), 1–8 (in Japanese).
8. Hoshino, M.: *On algebras with radical cube zero*, Arch. Math., Vol.**52** (1989), 226–232.
9. Mueller, B. J.: *The classification of algebras by dominant dimension*, Canad. J. Math. **20** (1968), 398–409.
10. Nakayama, T.: *On algebras with complete homology*, Abh. Math. Sem. Univ. Hamburg **22** (1958), 300–307.
11. Tachikawa, H.: Quasi-Frobenius rings and generalizations, Springer Lecture Notes **351** (1973).

Department of Mathematics
Osaka City University
Sugimoto, Sumiyoshi-ku
Osaka, 558 Japan
E-mail: E51833@JPNKUDPC.bitnet

Canadian Mathematical Society
Conference Proceedings
Volume 11 (1991)

FUNDAMENTAL DOMAINS AND DUPLICATED ALGEBRAS

by

Ibrahim Assem (Sherbrooke), Jerzy Nehring (Toruń) and Wiebke Schewe (Sherbrooke)

ABSTRACT. Let A be a finite dimensional, basic and connected algebra over an algebraically closed field. Denote by T(A) the trivial extension of A by its minimal injective cogenerator and by $\hat{A}$ the repetitive algebra of A. Recall that the push-down functor F_λ: ind $\hat{A} \to$ ind T(A) associated to the covering functor F: $\hat{A} \to$ T(A) is dense whenever T(A) is of polynomial growth. By a fundamental domain for T(A) in $\hat{A}$ we mean a finite full convex and connected subcategory B of $\hat{A}$ such that the restriction of F_λ to ind B is dense. We characterise minimal fundamental domains for T(A) whenever T(A) is representation-infinite of polynomial growth.

INTRODUCTION

Let k denote a fixed algebraically closed field and A be a finite dimensional k-algebra (associative, with an identity), which we shall moreover assume to be basic and connected. As is usual in representation theory, every (locally) finite dimensional

---

A.M.S. 1980 (1985) subject classification: 16A46, 16A64

k-algebra will equally be considered as a k-category [11]. For an algebra A, we shall denote by mod A the category of finite dimensional right A-modules and by ind A the full subcategory of mod A consisting of all indecomposable objects. Recall that the trivial extension T(A) of A by its minimal injective cogenerator bimodule ${}_A DA_A = \mathrm{Hom}_k({}_A A_A, k)$ is the algebra whose vector space structure is that of the k-space $A \oplus DA$ and whose multiplication is defined by

$$(a, f)\,(b, g) = (ab, ag + fb)$$

for $a, b \in A$ and $f, g \in {}_A DA_A$. Then T(A) is a self-injective and, in fact, even a symmetric algebra.

Trivial extension algebras have been extensively investigated in representation theory. Part of their importance comes from their close relation with the derived category of a finite dimensional algebra [15]. Presently, the trivial extension algebras of polynomial growth (in the sense of [26]) have been completely classified (see [7, 20, 21, 22]). Indeed, one can associate to an algebra A its repetitive algebra [18]

$$\hat{A} = \begin{bmatrix} \ddots & & & & 0 \\ \ddots & A_{i-1} & & & \\ & Q_{i-1} & A_i & & \\ & & Q_i & A_{i+1} & \\ 0 & & & \ddots & \ddots \end{bmatrix}$$

where matrices have only finitely many non-zero coefficients, $A_i = A$ and $Q_i = {}_A DA_A$ for all $i \in \mathbb{Z}$, all the remaining coefficients are zero, and multiplication is induced from the bimodule structure of DA and the zero morphism $DA \otimes_A DA \to 0$. Thus $\hat{A}$ is an associative locally finite dimensional self-injective algebra without identity. The identity maps $A_i \to A_{i-1}$ and $Q_i \to Q_{i-1}$ induce an automorphism ν of $\hat{A}$ called the Nakayama automorphism. It is shown in [27] that a complete set of representatives for the isomorphism classes of representation-infinite standard self-injective algebras of polynomial growth is given by the algebras of the form $\hat{A}/G$, where A is a representation-infinite tilted algebra of euclidean type having a complete slice in its preinjective component [16, 24], or a tubular algebra [24], and G is an infinite cyclic group of k-linear automorphisms of $\hat{A}$, which is admissible, that is, its action on the objects is free and has only finitely many orbits. In particular, $\hat{A}$ is a Galois covering [14] of T(A) with the infinite cyclic group generated by ν. Moreover, A can be identified with a complete ν-slice of $\hat{A}$, that is, a full convex subcategory of $\hat{A}$ consisting of a complete set of representatives of the ν-orbits of objects in $\hat{A}$.

It is a natural question to try to relate the representation theory of T(A) to the representation theory of the following QF-3 algebra, called the duplicated algebra of A (see [1, 2])

$$\bar{A} = \begin{bmatrix} A_0 & 0 \\ Q & A_1 \end{bmatrix}$$

where $A_0 = A_1 = A$ and $Q = {}_A DA_A$, and the multiplication is induced from the bimodule structure of DA. It was shown in [1] that, if A is tilted of Dynkin type, then $\bar{A}$ is a fundamental domain for T(A) in $\hat{A}$. Here a fundamental domain for an algebra A in a G-exhaustive [13] Galois covering $\tilde{A}$ with group G is a finite full convex and connected subcategory B of $\tilde{A}$ such that the restriction to ind B of the push-down functor F_λ: ind $\tilde{A}$ $\to$ ind A associated to the given covering [11, 14] is dense, that is, for every indecomposable A-module M, there exists an indecomposable B-module X such that $F_\lambda X \xrightarrow{\sim} M$. The notion of fundamental domain was introduced by Larrión and Salmerón in [19], where they proved their existence for representation-finite algebras having a directed module category. Recently, their existence was shown for an arbitrary representation-finite algebra [25]. For a representation-infinite trivial extension algebra T(A) of polynomial growth, it follows from the results in [6] that fundamental domains for T(A) in $\hat{A}$ exist. Our objective in the present article is to show that the minimal fundamental domains are given in this case by the duplicated algebras of a nice class of algebras.

Before stating the main result, we shall need the following definitions. A period of a self-injective algebra $\hat{A}/G$ of polynomial growth with G = <g> is a full convex and connected subcategory $\mathcal{C}$ of ind $\hat{A}$ such that $\mathcal{C} \cap {}^g\mathcal{C} = \emptyset$ and the restriction to $\mathcal{C}$ of the push-down functor ind $\hat{A} \to$ ind $(\hat{A}/G)$ is dense. As is the case for the representation-finite algebras [1, 12, 18, 25], we can expect the fundamental domains for representation-infinite trivial extension algebras of polynomial growth to be related to the periods of $\hat{A}$. Recall that, if A is a representation-infinite tilted algebra of euclidean type having a complete slice in its preinjective component, or a tubular algebra, then A is a truncated branch extension (branch extension in the sense of [24] (4.4)) of a tame concealed algebra, see [24] (4.9). The notions of truncated branch extension and coextension have been generalised in [5] (see also [7]) to the notion of branch enlargement. We shall say that an algebra A is a domain algebra if either A is a representation-finite tilted algebra of euclidean type or else is a truncated branch enlargement of domestic or tubular type

$$A = {}_{j=1}^{q}[F_j, L_j]\, C\, [E_i, K_i]_{i=1}^{p}$$

of a tame concealed algebra C having the property that E_i and F_j belong to different tubes of mod C for each pair (i, j). Thus, every tilted algebra of euclidean type, and every tubular algebra is a domain algebra. We may now state our main theorem.

THEOREM. Let A be a tilted algebra of euclidean type or a tubular algebra. The following conditions are equivalent:

(i) B is a fundamental domain for T(A) in $\hat{A}$ and, for any $0 \neq e^2 = e \in B$, the quotient algebra $B / \langle e \rangle$ is not.

(ii) B is a fundamental domain for T(A) in $\hat{A}$ and any proper full convex subcategory of B is not.

(iii) B is the support algebra of a period $\mathcal{C}$ of $\hat{A}$.

(iv) $B \xrightarrow{\sim} \bar{E}$, where E is a domain algebra such that $T(E) \xrightarrow{\sim} T(A)$.

(v) ind B contains a full convex subcategory $\mathcal{C}$ isomorphic to a period of $\hat{A}$ such that:

(1) All indecomposable projective B-modules which are not injective are predecessors of $\mathcal{C}$.

(2) All indecomposable projective-injective B-modules lie in $\mathcal{C}$.

(3) All indecomposable injective B-modules which are not projective are successors of $\mathcal{C}$.

It follows from the results in [20, 21] that we obtain in this way fundamental domains for all representation-infinite trivial extension algebras of polynomial growth, and from the results in [27] that this implies the construction of fundamental domains for all the standard representation-infinite self-injective algebras of polynomial growth. As we shall see, domain algebras have global dimension ≤ 2, and thus the standard self-injective algebras of polynomial growth can be described using only algebras of global dimension ≤ 2. Finally, observe that it follows from [22] (4) and [27] that a domain algebra E satisfies $T(E) \xrightarrow{\sim} T(A)$ if and only if E is a complete ν-slice of $\hat{A}$, or equivalently if and only if E is obtained from A by a sequence of reflections. Indeed, if E is a domain algebra, there exists, by [7] (2.6), a truncated branch extension B obtained from E by a finite sequence of reflections (and thus such that $T(B) \xrightarrow{\sim} T(E)$). The statement follows by applying [27] (2.2) and (3.2) (see also [22] (4)) to the algebras A and B.

The article is organised as follows: after a preliminary section (1), we shall describe the periods in section (2). Section (3) is devoted to the description of the

5

duplicated algebras of domain algebras, and section (4) to the proof of the main theorem. Finally, section (5) consists of concluding remarks and examples.

1. PRELIMINARIES

1.1 Notation

For a quiver Q, we shall denote by Q_0 its set of points and by Q_1 its set of arrows. For a (locally) finite dimensional k-algebra A, we shall denote by Q_A its ordinary quiver. For $i \in (Q_A)_0$, we denote by e_i the corresponding primitive idempotent of A and by S(i) the corresponding simple A-module. By P(i) (respectively, I(i)) is denoted the projective cover (respectively, the injective envelope) of S(i). Recall from [18] that $i \in (Q_A)_0$ is called a strong sink if there exists no chain of non-zero non-isomorphisms between indecomposables which is of the form $M_0 \to M_1 \to \dots M_t \to I(i)$, with M_0 injective. Dually, we define a strong source. For an A-module M, its support Supp M is the full subcategory of the k-category A consisting of all objects $i \in A_0$ such that $\mathrm{Hom}_A(P(i), M) \neq 0$. A module M is called sincere if Supp M = A. For a full subcategory $\mathcal{C}$ of ind A, its support Supp ($\mathcal{C}$) is defined to be the full subcategory of A consisting of all objects $i \in A_0$ such that $\mathrm{Hom}_A(P(i), M) \neq 0$ for some $M_A \in \mathcal{C}_0$, and its content Cont ($\mathcal{C}$) is the full subcategory of A consisting of all objects $i \in A_0$ such that $P(i) \in \mathcal{C}_0$. By τ_A (or τ, if there is no danger of confusion) we denote the Auslander-Reiten translation DTr in mod A, and by Γ_A the Auslander-Reiten quiver of A (see [8, 24]). In discussing properties of Γ_A, we shall not distinguish between the (isomorphism class of an) indecomposable module, and the point of Γ_A which represents it.

The one-point extension (respectively, coextension) of an algebra A by an A-module M, see [24], will be denoted by A[M] (respectively, [M]A). Let A be an algebra such that Q_A has no oriented cycles and $i \in (Q_A)_0$ be a sink. The reflection $S_i^+ A$ of A at i is the quotient of the one-point extension A[I(i)] by the two-sided ideal generated by e_i (see [18]). Dually, starting with a source j, we define the reflection $S_j^- A$. It follows from [23, 28] that the trivial extensions and the repetitive algebras of A and $S_i^+ A$ (respectively, A and $S_j^- A$) are isomorphic. The quiver of $S_i^+ A$ is denoted by $\sigma_i^+ Q_A$ and is called a ν-reflection of Q_A. In $\sigma_i^+ Q_A$, the sink i of Q_A is replaced by a source denoted by i'. A (strong) ν-reflection sequence of sinks $i_1, \dots i_t$ is a sequence of points of Q_A such that i_s is a (strong) sink of $\sigma_{i_{s-1}}^+ \cdots \sigma_{i_1}^+ Q_A$ for $1 \le s \le t$.

1.2 Branch enlargements

For branches and branch enlargements, we shall use the results and notations introduced in [5,7]. Recall that an algebra A is a domestic truncated branch extension (respectively, coextension) of a tame concealed algebra if and only if A is a tilted algebra of euclidean type with a complete slice in its preinjective component and no projectives in this component (respectively, a complete slice in its preprojective component and no injectives in this component), see [24] (4.9). Following [27], we shall then call A a euclidean (respectively, co-euclidean) algebra. Also, the tubular (respectively, cotubular) algebras coincide with the tubular truncated branch extensions (respectively, coextensions) of a tame concealed algebra, and it is shown that every tubular algebra is cotubular and conversely [24] (5.2.3).

1.3 Description of ind Â, for A euclidean (see [27])

It follows from [7] (4.3) that there exists a ν-reflection sequence of sinks which forms an enumeration of the objects of A of the form $1, \ldots a_0, a_0 + 1, \ldots a_1, a_1 + 1, \ldots a_2, a_2 + 1, \ldots a_3, \ldots a_7$ such that the following statements hold:

(i) $S^+_{a_7} \ldots S^+_1 A \xrightarrow{\sim} \nu A$ while $A' = S^+_{a_3} \ldots S^+_1 A$ is a euclidean category of the same tubular type as A, and $\{\nu^q A, \nu^q A'\}_{q \in \mathbf{Z}}$ forms a complete family of euclidean full subcategories of Â of the same tubular type as A.

(ii) $D' = S^+_{a_0} \ldots S^+_1 A$ and $D = S^+_{a_4} \ldots S^+_1 A$ are co-euclidean categories of the same tubular type as A, and $\{\nu^q D, \nu^q D'\}_{q \in \mathbf{Z}}$ forms a complete family of co-euclidean full subcategories of Â of the same tubular type as A.

(iii) Let C and C' denote respectively the tame concealed full convex subcategories of A and A'. Then A (respectively, $\nu^{-1}D$) is a branch extension (respectively, coextension) of C, while A' (respectively, D') is a branch extension (respectively, coextension) of C'. Moreover, $\{\nu^q C, \nu^q C'\}_{q \in \mathbf{Z}}$ forms a complete family of tame concealed full convex subcategories of Â.

(iv) $1, \ldots a_0$ is a strong ν-reflection sequence of sinks in A, while $a_3 + 1, \ldots a_4$ is a strong ν-reflection sequence of sinks in A'.

(v) The objects $a_0 + 1, \ldots a_1$ belong to branches of D' corresponding to one tube of mod D'; the objects $a_1 + 1, \ldots a_2$ and $a_2 + 1, \ldots a_3$ belong respectively to branches corresponding to two other tubes of mod D'. Similarly the objects $a_4 + 1, \ldots a_5$, respectively, $a_5 + 1, \ldots a_6$ and $a_6 + 1, \ldots a_7$, belong to branches of D corresponding to different tubes of mod D. (If there are no branches of D', respectively D, then let D' = A', respectively $D = \nu A$).

(vi) ind $\hat{A} = \bigvee_{p \in \mathbb{Z}} (\mathcal{X}_p \vee \mathcal{T}_p)$, where $\mathcal{X}_{2q} = \nu^q(\mathcal{X}_0)$, $\mathcal{X}_{2q+1} = \nu^q(\mathcal{X}_1)$, $\mathcal{T}_{2q} = \nu^q(\mathcal{T}_0)$, $\mathcal{T}_{2q+1} = \nu^q(\mathcal{T}_1)$ and $\mathcal{X}_0$, $\mathcal{X}_1$ are transjective full subcategories of ind $\hat{A}$ containing each at least one projective, and whose stable Auslander-Reiten quiver is of the form $\mathbb{Z}\vec{\Delta}$ (for $\vec{\Delta}$ a euclidean quiver of the same type as A), while $\mathcal{T}_0$, $\mathcal{T}_1$ are $\mathbb{P}_1(k)$ - families of quasi-tubes, the stable part of each is a tubular $\mathbb{P}_1(k)$ - family of tubes of the same tubular type as A. Finally, $\mathrm{Hom}_{\hat{A}}(\mathcal{T}_p, \mathcal{X}_p) = 0$ and $\mathrm{Hom}_{\hat{A}}(\mathcal{X}_q \vee \mathcal{T}_q, \mathcal{X}_p \vee \mathcal{T}_p) = 0$ for $p, q \in \mathbb{Z}$, $q > p$.

(vii) We have $A = \mathrm{Cont}(\mathcal{X}_0 \vee \mathcal{T}_0 \vee \mathcal{X}_1 \vee \mathcal{T}_1)$, $D' = \mathrm{Cont}(\mathcal{T}_0 \vee \mathcal{X}_1 \vee \mathcal{T}_1 \vee \mathcal{X}_2)$, $A' = \mathrm{Cont}(\mathcal{X}_1 \vee \mathcal{T}_1 \vee \mathcal{X}_2 \vee \mathcal{T}_2)$, $D = \mathrm{Cont}(\mathcal{T}_1 \vee \mathcal{X}_2 \vee \mathcal{T}_2 \vee \mathcal{X}_3)$. It follows from this description and (iii) above that $C \subseteq \mathrm{Cont}(\mathcal{X}_0 \vee \mathcal{T}_0 \vee \mathcal{X}_1)$ while $C' \subseteq \mathrm{Cont}(\mathcal{X}_1 \vee \mathcal{T}_1 \vee \mathcal{X}_2)$.

(viii) Let $\mathcal{T}_0 = (T_{\lambda,0})_{\lambda \in \mathbb{P}_1(k)}$, $\mathcal{T}_1 = (T_{\lambda,1})_{\lambda \in \mathbb{P}_1(k)}$. Comparing A and D', one sees that $\mathrm{Cont}(\mathcal{X}_0)_0 = \{1, \dots a_0\}$. Similarly, it follows from (v) that we may assume $\mathrm{Cont}(T_{0,0})_0 = \{a_0 + 1, \dots a_1\}$, $\mathrm{Cont}(T_{1,0})_0 = \{a_1 + 1, \dots a_2\}$, $\mathrm{Cont}(T_{\infty,0})_0 = \{a_2 + 1, \dots a_3\}$. Also, $\mathrm{Cont}(\mathcal{X}_1)_0 = \{a_3 + 1, \dots a_4\}$, $\mathrm{Cont}(T_{0,1})_0 = \{a_4 + 1, \dots a_5\}$, $\mathrm{Cont}(T_{1,1})_0 = \{a_5 + 1, \dots a_6\}$ and $\mathrm{Cont}(T_{\infty,1})_0 = \{a_6 + 1, \dots a_7\}$.

(ix) The support of any indecomposable $\hat{A}$-module in $\mathcal{X}_0$ which is not projective-injective, lies in one of the categories $\nu^{-1}(S_r^+ \dots S_1^+ A)$ with $0 \leq r \leq a_0$. Similarly, the support of any indecomposable $\hat{A}$-module in $\mathcal{X}_1$, which is not projective-injective, lies in one of the categories $\nu^{-1}(S_s^+ \dots S_{a_3+1}^+ A')$ with $a_3 \leq s \leq a_4$.

(x) The support of any indecomposable lying in a stable tube of $\mathcal{T}_0$ (respectively, $\mathcal{T}_1$) lies in $\nu^{-1}C'$ (respectively, in C). The support of any indecomposable lying in one of the quasi-tubes $T_{\lambda,1}$ with $\lambda = 0, 1, \infty$ is contained in $\mathrm{Cont}(T_{\lambda,-1} \vee \mathcal{X}_0 \vee \mathcal{T}_0 \vee \mathcal{X}_1 \vee T_{\lambda,1})$. For, clearly, this support lies in $\mathrm{Cont}(\mathcal{T}_{-1} \vee \mathcal{X}_0 \vee \mathcal{T}_0 \vee \mathcal{X}_1 \vee \mathcal{T}_1)$. On the other hand, there exists no morphism $P(b) \to M$ such that $P(b) \in T_{\mu,-1}$ is projective and $M \in T_{\lambda,1}$ with $\lambda \neq \mu$ and $\lambda, \mu \in \{0, 1, \infty\}$, since the existence of such a morphism would imply the existence of a morphism $M \to I(b) = P(\nu b) \in T_{\mu,1}$, a contradiction to the orthogonality of the quasi-tubes $T_{\lambda,1}$ and $T_{\mu,1}$. Similarly, the support of any indecomposable lying in one of the quasi-tubes $T_{\lambda,0}$ with $\lambda = 0, 1, \infty$ is contained in $\mathrm{Cont}(T_{\lambda,-2} \vee \mathcal{X}_{-1} \vee \mathcal{T}_{-1} \vee \mathcal{X}_0 \vee T_{\lambda,0})$.

Observe that the sequence $1, \dots a_0, \dots a_1, \dots a_7$ is not unique. Actually, for each strong ν-reflection sequence $i_1, \dots i_t$ with $t \leq a_0$ in A_0, there exists a sequence as above with $i_1 = 1, \dots i_t = t$.

1.4 Description of ind $\hat{A}$, for A tubular (see [27])

It follows from [17] and [22] (3) that there exists an integer N, $3 \leq N \leq 10$, and a ν-reflection sequence of sinks which forms an enumeration of the objects of A and is of the form:

$$\begin{array}{lll} 1, \ldots\ldots\ldots & a_1, a_1+1, \ldots a_2, a_2+1, \ldots\ldots & a_r, \\ a_r+1, \ldots\ldots & a_{r+1}, \ldots\ldots\ldots\ldots\ldots\ldots & a_{2r}, \\ \vdots & \vdots & \vdots \\ a_{(N-1)r}+1, \ldots\ldots & a_{(N-1)r+1}, \ldots\ldots\ldots\ldots\ldots\ldots, & a_{Nr}. \end{array}$$

(where r = 3 if the tubular type of A is (3, 3, 3), (2, 4, 4) or (2, 3, 6) and r = 4 if it equals (2, 2, 2, 2)) such that:

(i) $S^+_{a_{Nr}} \ldots S^+_1 A = \nu A$, while $A_j = S^+_{jr} \ldots S^+_1 A$, for $1 \leq j \leq N$, is a tubular category of the same tubular type as A. The categories $\{\nu^q A_j\}_{q \in \mathbb{Z}}$ with $0 \leq j < N$ and $A_0 = A$ form a complete family of tubular subcategories of $\hat{A}$.

(ii) For each $0 \leq j < N$, A_j is a tubular extension of a tame concealed full convex subcategory C_j of $\hat{A}$ and a tubular coextension of a tame concealed full convex subcategory C_{j+1} (where $C_n = \nu C_0$). The categories $\{\nu^q C_j\}_{q \in \mathbb{Z},\, 0 \leq j < N}$ form a complete family of tame concealed full convex subcategories of $\hat{A}$.

(iii) The objects $a_{jr}+1, \ldots a_{jr+1}$ $(0 \leq j < N, a_0 = 0)$ belong to branches of A_j corresponding to one tube of mod A_j. Similarly, the other objects belong to branches of A_j corresponding to the other tubes of mod A_j.

(iv) ind $\hat{A} = \bigvee_{p \in \mathbb{Z}} (\mathcal{M}_{p-1,p} \vee \mathcal{T}_p)$ where $\nu^q(\mathcal{T}_p) = \mathcal{T}_{p+Nq}$ and $\nu^q(\mathcal{M}_{p-1,p}) = \mathcal{M}_{p+Nq-1,p+Nq}$ (for all $p, q \in \mathbb{Z}$). Here $\mathcal{T}_p = (T_{\lambda,p})_{\lambda \in \mathbb{P}_1(k)}$ is a $\mathbb{P}_1(k)$-family of quasi-tubes containing at least one projective-injective module, and whose stable Auslander-Reiten quiver is a tubular $\mathbb{P}_1(k)$-family of the same tubular type as A.

Also, for each $p \in \mathbb{Z}$, we have $\mathcal{M}_{p-1,p} = \bigvee_{q \in \mathbb{Q}_p^{p-1}} \mathcal{T}_q$, where $\mathbb{Q}_p^{p-1}$ is the set of all rationals q with $p-1 < q < p$, and each $\mathcal{T}_q = (T_{\lambda,q})_{\lambda \in \mathbb{P}_1(k)}$ is a stable tubular $\mathbb{P}_1(k)$-family of the same tubular type as A such that $\nu(\mathcal{T}_q) = \mathcal{T}_{q+N}$ for each $q \in \mathbb{Q}$. Finally, $\mathrm{Hom}_{\hat{A}}(\mathcal{T}_p, \mathcal{M}_{p-1,p}) = 0$ and $\mathrm{Hom}_{\hat{A}}(\mathcal{M}_{q-1,q} \vee \mathcal{T}_q, \mathcal{M}_{p-1,p} \vee \mathcal{T}_p) = 0$ for $p, q \in \mathbb{Z}$ with $p < q$.

(v) For each $0 \leq j < N$, $A_j = \mathrm{Cont}(\mathcal{T}_{j+1} \vee \ldots \vee \mathcal{T}_{j+N})$ while $C_j \subseteq \mathrm{Cont}(\mathcal{T}_{j+1} \vee \ldots \vee \mathcal{T}_{j+N-1})$.

(vi) It follows from (iii) that we can take Cont $(T_{0,j+1})_0 = \{a_{jr} + 1, \ldots a_{jr+1}\}$, Cont $(T_{1,j+1})_0 = \{a_{jr+1} + 1, \ldots a_{jr+2}\}$ and Cont $(T_{\infty,j+1})_0 = \{a_{jr+2} + 1, \ldots a_{jr+3}\}$ if $r = 3$, while if $r = 4$, then also Cont $(T_{\lambda,j+1})_0 = \{a_{jr+3} + 1, \ldots a_{jr+4}\}$ for some $\lambda \neq 0, 1, \infty$.

(vii) The support of any indecomposable $\hat{A}$-module lying in a tube in $\mathcal{M}_{p,p+1}$ lies in $\nu^{-1}A_p$. The support of any indecomposable $\hat{A}$-module lying in a stable tube of $\mathcal{T}_p$ ($p \in \mathbf{Z}$) lies in $\nu^{-1}C_p$. Finally, the support of any indecomposable $\hat{A}$-module lying in one of the tubes $T_{\lambda,p}(p \in \mathbf{Z}, \lambda \in \mathbb{P}_1(k))$ lies in Cont $(T_{\lambda,p-N} \vee \mathcal{T}_{p+1-N} \vee \ldots \vee \mathcal{T}_{p-1} \vee T_{\lambda,p})$.

Again the above enumeration is not unique.

1.5 For the remaining part of this article, we assume that A is either euclidean or tubular, and denote by F_λ: ind $\hat{A} \to$ ind T(A) the push-down functor. We shall need the following two easy lemmas.

LEMMA. Let A be a euclidean category, and $M_{\hat{A}} \in \mathcal{X}_p$ be such that $F_\lambda M$ is a sincere T(A)-module and M is a successor (respectively, predecessor) of every projective-injective in $\mathcal{X}_p$. Then Supp M = Cont $(\mathcal{T}_{p-2} \vee \mathcal{X}_{p-1} \vee \mathcal{T}_{p-1} \vee \mathcal{X}_p)$ (respectively, Cont $(\mathcal{X}_{p-2} \vee \mathcal{T}_{p-2} \vee \mathcal{X}_{p-1} \vee \mathcal{T}_{p-1})$.

<u>Proof</u>. Since $F_\lambda M$ is sincere, there exists for any $P \in \mathcal{T}_{p-2} \vee \mathcal{X}_{p-1} \vee \mathcal{T}_{p-1} \vee \mathcal{X}_p$ a non-zero morphism $\nu^q P \to M$ for some q. Now, if M is a successor of P, then $\mathrm{Hom}_{\hat{A}}(\nu^q P, M) = 0$ for all $q < 0$. Since, trivially, $\mathrm{Hom}_{\hat{A}}(\nu^q P, M) = 0$ for all $q > 0$, we must have $q = 0$. Thus Supp M $\supseteq$ Cont $(\mathcal{T}_{p-2} \vee \mathcal{X}_{p-1} \vee \mathcal{T}_{p-1} \vee \mathcal{X}_p)$. The converse is trivial.

LEMMA. Let A be a tubular category, and $M_{\hat{A}} \in \mathcal{M}_{p,p+1}$ be such that $F_\lambda M$ is a sincere T(A)-module. Then Supp M = Cont $(\mathcal{T}_{p-N+1} \vee \ldots \vee \mathcal{T}_p)$.

<u>Proof</u>. This follows immediately from (1.4) (vii).

2. <u>PERIODS</u>

2.1 Let A be a euclidean or a tubular category and G = <g> be an admissible group of automorphisms of $\hat{A}$. We denote by m the least integer such that ${}^g\mathcal{X}_p = \mathcal{X}_{p+m}$ and ${}^g\mathcal{T}_p = \mathcal{T}_{p+m}$ for A euclidean, or ${}^g\mathcal{M}_{p,p+1} = \mathcal{M}_{p+m,p+m+1}$ and ${}^g\mathcal{T}_p = \mathcal{T}_{p+m}$ for A tubular (see [27] (2.9) (3.6)). Our objective is to describe the periods of $\hat{A}$/G in ind $\hat{A}$.

DEFINITION. A <u>G-period</u> in ind $\hat{A}$ (or, briefly, a <u>period</u>) is a full convex and connected subcategory of ind $\hat{A}$ such that:

(i) $\mathcal{T} \cap {}^g\mathcal{T} = \emptyset$

(ii) The restriction to $\mathcal{T}$ of the push-down functor F_λ : ind $\hat{A} \to$ ind $(\hat{A}/G)$ is dense.

Assume now that A is euclidean. By a section Σ (see [10]) in the stable part of a transjective component $\mathcal{X}_p$ of ind $\hat{A}$ is meant a full convex subquiver of the stable part of $\mathcal{X}_p$ such that Σ_0 consists of a complete set of representatives of the τ-orbits. A full convex subquiver Σ of $\mathcal{X}_p$ is called an extended section if the full subquiver of Σ consisting of all points which are not projective-injective is a section in the stable part of $\mathcal{X}_p$. Let Σ be an extended section in a transjective component $\mathcal{X}_p$, and let $[\Sigma, {}^g\Sigma[$ be the full subcategory of ind $\hat{A}$ consisting of all modules $M \notin {}^g\Sigma_0$ such that there exists a path of the form $M_0 \to M_1 \to \dots M_s = M \to \dots \to M_t$ in ind $\hat{A}$, with $M_0 \in \Sigma_0$, $M_t \in {}^g\Sigma_0$ and $0 \le s < t$. Then it is clear that $[\Sigma, {}^g\Sigma[$ is a period in ind $\hat{A}$.

2.2 LEMMA. Let A be a euclidean category, and $\mathcal{T}$ be a period in ind $\hat{A}$. Then either

(i) there exists an extended section Σ in a transjective component $\mathcal{X}_p$ of ind $\hat{A}$ for some $p \in \mathbb{Z}$ such that $\mathcal{T} = [\Sigma, {}^g\Sigma[$, or

(ii) there exists a subset $\Lambda \subseteq \mathbb{P}_1(k)$ and $j \in \mathbb{Z}$ such that

$$\mathcal{T} = \left(\bigvee_{\lambda \in \Lambda} T_{\lambda,j}\right) \vee \mathcal{X}_{j+1} \vee \tau_{j+1} \vee \dots \vee \left(\bigvee_{\lambda \notin \Lambda} T_{\lambda,j+N}\right)$$

Proof. Clearly, if $\mathcal{T}$ is of one of these two forms, then $\mathcal{T}$ is a period. We must show that, conversely, any period $\mathcal{T}$ of ind $\hat{A}$ occurs in one of these two ways. Let thus Σ denote the full subquiver of $\Gamma_{\hat{A}}$ consisting of all $\hat{A}$-modules $M \in \mathcal{T}_0$ such that $\tau M \notin \mathcal{T}_0$, and Σ' denote the full subquiver of Σ consisting of all points which are not projective-injective. We have two cases to consider:

(i) $\Sigma' \neq \emptyset$. In this case, any $M \in \Sigma'_0$ must belong to a transjective component $\mathcal{X}_p$ of $\Gamma_{\hat{A}}$. For, otherwise, M lies in a tube and we have a path of the form $M \to \dots \to \tau M \to * \to M$ in ind $\hat{A}$. Since $M \in \mathcal{T}_0$, we deduce from the convexity of $\mathcal{T}$ that $\tau M \in \mathcal{T}_0$, a contradiction, since $M \in \Sigma'_0$. Next, observe that all elements of Σ' must lie in the same transjective component $\mathcal{X}_p$. For, otherwise, there exist $M, M' \in \Sigma'_0$ such that $M \in \mathcal{X}_p$ and $M' \in \mathcal{X}_{p'}$ with $p < p'$. But then there exists a path in ind $\hat{A}$ of the form $M \to \dots \to \tau M' \to * \to M'$. Since $M, M' \in \mathcal{T}_0$, we deduce from the convexity of $\mathcal{T}$ that $\tau M' \in \mathcal{T}_0$, a contradiction, since $M' \in \Sigma'_0$.

We now claim that Σ' is a section in the stable part of a transjective component $\mathcal{X}_p$ of $\Gamma_{\hat{A}}$. Firstly, the convexity of $\mathcal{T}$ and the definition of Σ' imply that no two vertices of Σ' lie in the same τ-orbit. Secondly, Σ' is convex. For, assume we have a path $M = M_0 \to M_1 \to \dots \to M_t = M'$ in ind $\hat{A}$ with $M, M' \in \Sigma'_0$. Since $\mathcal{T}$ is convex, we have $M_s \in \mathcal{T}_0$ for all $1 \le s < t$. Assume that for some $1 \le s < t$, we have $\tau M_s \in \mathcal{T}_0$, then the path $\tau M_s \to \ldots \to \tau M_t \to M_{t-1} \to M_t = M'$ shows that $\tau M_t = \tau M' \in \mathcal{C}_0$, a contradiction, since $M' \in \Sigma'_0$. In order to show that Σ' is a section, it suffices to show that Σ'_0 contains a vertex of each stable τ-orbit in $\mathcal{X}_p$. Let indeed $M \in \Sigma'_0$. Since $\tau M \notin \mathcal{C}_0$, we know by property (ii) of the definition of a period that ${}^{g^\ell}(\tau M) \in \mathcal{C}_0$ for some $\ell \in \mathbb{Z}$, and by the convexity of $\mathcal{C}$, we may assume $\ell = 1$. Let now N be any successor of M in $\mathcal{X}_p$. Then we have a path $M \to \ldots \to N \to \ldots \to {}^g(\tau M)$ in ind $\hat{A}$, thus $N \in \mathcal{C}_0$, since $\mathcal{C}$ is convex. Dually, any predecessor of τM in $\mathcal{X}_p$ does not lie in $\mathcal{C}$. Consequently, in each stable τ-orbit of $\mathcal{X}_p$, there exists an indecomposable module L such that $L \in \mathcal{C}_0$, but $\tau L \notin \mathcal{C}_0$. This completes the proof of our claim.

Finally, we show that $\mathcal{C} = [\Sigma, {}^g\Sigma[$. Indeed, for any $M \in \Sigma_0$, we have $\tau M \notin \mathcal{C}_0$ and consequently ${}^g(\tau M) \in \mathcal{C}_0$. But ${}^g(\tau M) = \tau({}^gM)$ and thus $\tau({}^g\Sigma) \subseteq \mathcal{C}_0$. We obtain $[\Sigma, {}^g\Sigma[\subseteq \mathcal{C}$, since by the convexity of $\mathcal{C}$, also any projective-injective with radical in $\tau({}^g\Sigma)$ lies in $\mathcal{C}$. On the other hand, assume there exists $M \in (\mathcal{C} \setminus [\Sigma, {}^g\Sigma[)_0$. Then M is either a predecessor of $\tau\Sigma$ or a successor of ${}^g\Sigma$, and the convexity of $\mathcal{C}$ yields a contradiction to the fact that $(\tau\Sigma)_0 \cap \mathcal{C}_0 = \emptyset$ or to the fact that $({}^g\Sigma)_0 \cap \mathcal{C}_0 = \emptyset$. This completes the proof of (i).

(ii) $\Sigma = \emptyset$. Then any $M \in \mathcal{C}_0$ is either a projective-injective $\hat{A}$-module or fulfills $\tau M \in \mathcal{C}_0$. Since $\mathcal{C}$ is a period, it must contain at least one module M which is not projective-injective. Now, if M lies in a transjective component $\mathcal{X}_p$, the above remark and the convexity of $\mathcal{C}$ imply that all predecessors of M lie in $\mathcal{C}$. Assume that a successor N of M does not belong to $\mathcal{C}$. Then there exists t such that $\tau^t N \notin \mathcal{C}_0$ but $\tau^{t+1} N \in \mathcal{C}_0$. But then by property (ii) of the definition of a period there exists $\ell \in \mathbb{Z}$ such that ${}^{g^\ell}(\tau^t N) \in \mathcal{C}_0$. But then by property (i) we have ${}^{g^\ell}(\tau^{t+1} N) = \tau\, {}^{g^\ell}(\tau^t N) \notin \mathcal{C}_0$. Thus ${}^{g^\ell}(\tau^t N) \in \Sigma_0$, a contradiction to our assumption that $\Sigma = \emptyset$. Hence $\mathcal{X}_p \subseteq \mathcal{C}$. Similarly, assume that an $\hat{A}$-module in $\mathcal{C}$, which is not projective-injective, lies in a quasi-tube. It follows again from the convexity of $\mathcal{C}$ that the whole quasi-tube lies in $\mathcal{C}$. The conclusion now follows from the convexity of $\mathcal{C}$.

2.3 Assume that A is a tubular algebra. For two rational numbers r, r' such that $p - 1 \le r < r' < p$ ($p \in \mathbb{Z}$) we shall let

$$\mathcal{N}_{r,r'} = \bigvee_{\substack{\gamma \in \mathbb{Q}_p^{p-1} \\ r \le \gamma < r'}} \mathcal{T}_\gamma$$

(Thus $\mathcal{N}_{p-1,p} = \mathcal{M}_{p-1,p}$)

LEMMA. let A be tubular and $\mathcal{C}$ be a period in ind $\hat{A}$. Then there exist $r \in \mathbb{Q}$ and $\Lambda \subseteq \mathbb{P}_1(k)$ such that

$$\mathcal{T} = (\bigvee_{\lambda\in\Lambda} T_{\lambda,r}) \vee \mathcal{N}_{r,[r]+1} \vee \mathcal{S}_{[r]+1} \vee \mathcal{M}_{[r]+1,[r]+2} \vee \cdots$$
$$\vee \mathcal{S}_{[r]+N} \vee \mathcal{N}_{[r]+N,r+N} \vee (\bigvee_{\lambda\notin\Lambda} T_{\lambda,r+N})$$

Proof. Follows from the proof above and [24] (5.2.2) with only the obvious modifications.

2.4 From now on, we shall only consider $<\nu>$ - periods in ind $\hat{A}$, which we call simply periods.

LEMMA. Let $\mathcal{T}$ be a period in ind $\hat{A}$. Then Supp $(\mathcal{T})$ = Cont $(\nu^{-1}\mathcal{T} \vee \mathcal{T})$.

Proof. Observe that, for every projective-injective $\hat{A}$-module $P(i)$ lying in $\mathcal{T}$, we have $i \in$ Supp $(\mathcal{T})_0$. Since moreover $P(i) = I(\nu^{-1}i)$, we have $\nu^{-1}i \in$ Supp $(\mathcal{T})_0$. Thus Cont $(\nu^{-1}\mathcal{T} \vee \mathcal{T}) \subseteq$ Supp $(\mathcal{T})$. On the other hand, let $j \in$ Supp $(\mathcal{T})_0$. By property (ii) of the definition of a period, we have $P(\nu^{\ell}j) \in \mathcal{T}_0$ for some $\ell \geq 0$. But if $\ell \geq 2$, then $P(\nu^{\ell-1}j) = I(\nu^{\ell-2}j)$ is in $\nu^{-1}\mathcal{T}$. Thus $I(j)$ is a predecessor of $\mathcal{T}$, a contradiction to the fact that $j \in$ Supp $(\mathcal{T})_0$. Hence $P(j) \in \mathcal{T}_0$ or $P(j) \in (\nu^{-1}\mathcal{T})_0$ so that $j \in$ Cont $(\nu^{-1}\mathcal{T} \vee \mathcal{T})_0$.

2.5 We are now able to prove the main result of this section.

PROPOSITION. Let A be euclidean or tubular. Then the support algebra B of a period $\mathcal{T}$ in ind $\hat{A}$ is isomorphic to a duplicated algebra $\bar{E}$, where E is a domain algebra such that $T(E) \xrightarrow{\sim} T(A)$.

Proof. We restrict ourselves to the euclidean case. Using the characterisation of periods in (2.2), we divide the proof into two cases. Then the proof in the tubular case is entirely similar to the second case.

(i) Assume that $\mathcal{T} = [\Sigma, {}^{\nu}\Sigma[$ for some extended section Σ in a transjective component $\mathcal{X}_p$ and such that at least one of the projective-injectives in $\mathcal{X}_p$ is a predecessor of Σ, and at least one is a successor of Σ.

It follows from (2.4) that B = Cont $(\nu^{-1}\mathcal{T} \vee \mathcal{T})$. Let E be the full convex subcategory of $\hat{A}$ such that E = Cont $(\nu^{-1}\mathcal{T})$. Then clearly $B \xrightarrow{\sim} \bar{E}$. We shall show that E is a complete ν-slice of $\hat{A}$ which is a representation-finite tilted algebra of euclidean type (thus E is a domain algebra such that $T(E) \xrightarrow{\sim} T(A)$). Since E contains exactly one object from each ν-orbit of $\hat{A}$, we have that E is indeed a complete ν-slice of $\hat{A}$. That E is a tilted algebra follows from the fact that Σ divides the projective-injectives in $\mathcal{X}_p$ into two non-empty sets, the set $\{P(1), \dots P(r)\}$ of projective-injectives lying on the left of Σ

and the set {P(r+1), ... P(a_0)} of projective-injectives lying on its right. From (1.3), we know that 1, ... a_0 is a strong ν-reflection sequence of sinks in some $\nu^q A$ or $\nu^q A'$ ($q \in \mathbb{Z}$). Without loss of generality, we may assume that 1, ... a_0 are in A. Then $E = \nu^{-1} (S_r^+ \ldots S_1^+ A)$ is a tilted algebra of the same type as A (see [7] (4.3)). In order to show that E is representation-finite, it suffices to show that E contains no tame concealed algebra as a full convex subcategory (1.3) (iii). But this follows from the fact that 1 is a strong sink in A and therefore belongs to C (by [7] (4.1)). Consequently C is not contained in E. Dually, C' is not contained in E. Thus E is as wanted.

(ii) Assume now that C is as in (2.2) (ii) or else is of the form $[\Sigma, {}^{\nu}\Sigma[$, where Σ is an extended section in a transjective component, and such that all projective-injectives in this component are either predecessors or successors of Σ. Then there exist a subset $\Lambda \subseteq \mathbb{P}_1(k)$ and an integer $j \in \mathbb{Z}$ such that

$$\text{Cont}(\mathcal{C}) = \text{Cont}\Big(\big(\bigvee_{\lambda \in \Lambda} T_{\lambda,j}\big) \vee \mathcal{X}_{j+1} \vee \mathcal{T}_{j+1} \vee \mathcal{X}_{j+2} \vee \big(\bigvee_{\lambda \notin \Lambda} T_{\lambda,j+2}\big)\Big).$$

We may assume that j = -1, thus we have

$$\text{Cont}(\mathcal{C}) = \text{Cont}\Big(\big(\bigvee_{\lambda \in \Lambda} T_{\lambda,-1}\big) \vee \mathcal{X}_0 \vee \mathcal{T}_0 \vee \mathcal{X}_1 \vee \big(\bigvee_{\lambda \notin \Lambda} T_{\lambda,1}\big)\Big).$$

Let now E be the full convex subcategory of $\hat{A}$ given by $E = \text{Cont}(\nu^{-1}\mathcal{C})$. Then again $B \xrightarrow{\sim} \overline{E}$ and E is a complete ν-slice of $\hat{A}$. We shall show that E is a representation-infinite domain algebra. There exists a sequence as in (1.3) such that $E_0 = \{\nu^{-1}(s+1), \ldots \nu^{-1}(a_7), 1, \ldots a_0, a_0 + 1, \ldots a_1, a_1 + 1, \ldots a_4, a_4 + 1, \ldots s\}$, where s equals a_4, a_5, a_6 or a_7. Then $A = S^+_{\nu^{-1}a_7} \ldots S^+_{\nu^{-1}(s+1)} E$. We know that E is a euclidean algebra, that is, a truncated branch extension of a tame concealed algebra C, and that the objects $\{s + 1, \ldots a_7\}$ are precisely the objects lying in the tubes $T_{\lambda,1}$ with $\lambda \in \Lambda$. Consequently, $E = S^-_{s+1} \ldots S^-_{a_7} A$ is obtained from A by replacing the extension branches containing these objects by coextension branches containing the objects $\nu^{-1}(s + 1), \ldots \nu^{-1}a_7$. This completes the proof.

2.6 We finish the section with a lemma which later proves to be useful in determining sections in the stable part of a transjective component.

LEMMA. Let A be euclidean, and $\mathcal{X}$ be a transjective component of ind $\hat{A}$. Assume that Cont ($\mathcal{X}$) is the disjoint union of a category $\mathcal{A}$ closed under successors and a category $\mathcal{B}$ closed under predecessors. Then there exists an extended section Σ in $\mathcal{X}$

such that $\mathfrak{A}$ (respectively, $\mathfrak{B}$) is the full subcategory consisting of all $i \in \mathrm{Cont}\,(\mathfrak{X})_0$ such that P(i) is a successor (respectively, predecessor) of Σ.

Proof. We have to show that, for any two projective-injective indecomposables P(i) and P(j) in $\mathfrak{X}$ which are not comparable in the natural partial order of $\mathfrak{X}$, their radicals too are not comparable. Namely, we have to show that the following situation cannot occur:

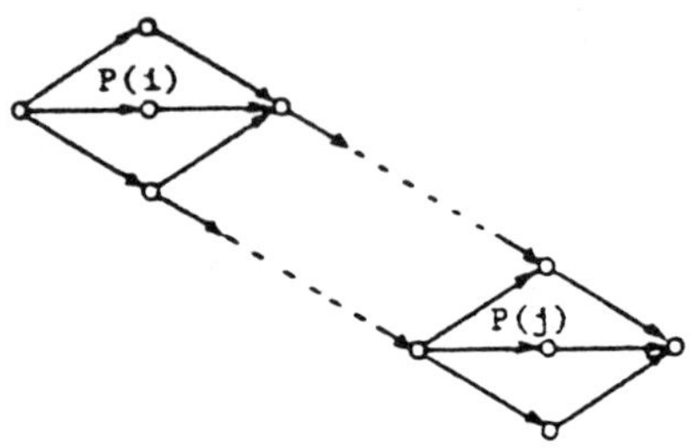

But, if this were the case, there would exist a non-zero morphism f: rad P(i) $\rightarrow$ rad P(j) $\rightarrow$ P(j) (indeed, the path from rad P(i) to rad P(j) is sectional [9]). Since P(j) is injective, the exact diagram

$$\begin{array}{ccccc} 0 & \rightarrow & \mathrm{rad}\,P(i) & \xrightarrow{g} & P(i) \\ & & f \downarrow & & \\ & & P(j) & & \end{array}$$

(where g denotes the canonical inclusion) can be completed by h: P(i) $\rightarrow$ P(j) such that f = hg, a contradiction.

3. DESCRIPTION OF THE DUPLICATED ALGEBRAS

3.1 We shall start the proof of our main theorem by showing the implication (iv) $\Rightarrow$ (i).

PROPOSITION. Let E be a complete ν-slice of $\hat{A}$ such that E is a domain algebra. Then $\overline{E}$ is a fundamental domain for T(A) in $\hat{A}$ and, for any $0 \neq e^2 = e \in A$, the quotient algebra $\overline{E}$ / <e> is not a fundamental domain for T(A) in $\hat{A}$.

Proof. We split the proof into two cases.

(i) Suppose that E is representation-finite tilted of euclidean type. There exists a strong ν-reflection sequence of sources $i_1, \dots i_t$ such that, for all $1 < s \leq t$, the algebra $S^-_{i_s} \dots S^-_{i_t} E$ is representation-finite, while $F = S^-_{i_1} \dots S^-_{i_t} E$ is a euclidean algebra of the same type as E. Indeed, if E is simply connected, that is, by [4], of type $\neq \tilde{\mathbf{A}}_m$, this follows from [7] (3.4), while if E is of type $\mathbf{A}_m$, this follows from [21] (or directly by using the classifications of the tilted and iterated tilted algebras given in [3]). Since $S^+_{\nu^{-1}i_t} \dots S^+_{\nu^{-1}i_1} F = E$ and $\nu^{-1}i_1, \dots \nu^{-1}i_t$ is a strong ν-reflection sequence of sinks in F, there exists an enumeration of the objects of F of the form $1, \dots a_0, a_0 + 1, \dots a_1, \dots a_7$ as in (1.3) such that $i_1 = 1, \dots i_t = t$ and $t < a_0$, since E is representation-finite. It follows from the definition of a duplicated algebra that $\overline{E}$ is the full convex subcategory of $\hat{A} = \hat{F}$ consisting of the objects $t + 1, \dots a_0, a_0 + 1, \dots a_7, \nu 1, \dots, \nu a_0, \dots \nu a_7, \nu^2 1, \dots \nu^2 t$. Thus, the indecomposable $\overline{E}$-modules can be identified with those indecomposable $\hat{A}$-modules whose support consists of objects inside this set.

It follows from (1.3) (ix) that the objects belonging to $S^+_r \dots S^+_1 F$ with $t \le r \le a_0$ or $a_3 \le r \le a_4$, and $\nu(S^+_r \dots S^+_1 F)$, with $0 \le r < t$, form a complete set of representatives of the ν-orbits of the supports of the indecomposable $\hat{F}$-modules, which are not projective-injective and lie in the transjective components. Observe that all these objects lie in $\overline{E}$.

Next, observe that the objects of C' and νC belong to $\overline{E}$ by (1.3) (vii) (here, C and C' are defined for F as in (1.3) (iii)). Moreover these are the only two tame concealed full convex subcategories of $\overline{E}$. (Indeed, by [7] (4.1), the vertex $1 \in F_0$ must belong to C, thus C does not lie in $\overline{E}$. Dually, νC' does not belong to $\overline{E}$.) By (1.3) (x), this implies that ind $\overline{E}$ contains a complete set of representatives of ν-orbits of all indecomposable $\hat{F}$-modules lying in a stable tube.

Similarly, since the sets of objects $\{a_{i-1} + 1, \dots a_i, a_3 + 1, \dots a_4, \dots a_7, \nu 1, \dots \nu a_0, \nu(a_{i-1} + 1), \dots \nu a_i\}$ for i = 1, 2, 3 and $\{a_{j-1} + 1, \dots a_j, \nu 1, \dots \nu a_0, \dots \nu a_3, \dots \nu a_4, \nu(a_{j-1} + 1), \dots \nu a_j\}$ for j = 5, 6, 7 belong to $\overline{E}$, we deduce from (1.3) (x) that ind $\overline{E}$ contains a complete set of representatives of ν-orbits of indecomposable $\hat{F}$-modules lying in non-stable quasi-tubes.

Finally, observe that for any $b \in E_0$, both b and νb belong to $\overline{E}_0$. Since $\overline{E}$ is convex in $\hat{A}$, the convex hull of b and νb lies inside $\overline{E}_0$. But this convex hull is precisely the support of the indecomposable projective-injective $\hat{F}$-module P(νb). This shows that $\overline{E}$ contains one representative from each ν-orbit of indecomposable projective-injective $\hat{F}$-modules and completes the proof that $\overline{E}$ is a fundamental domain for T(A) in $\hat{A}$.

In order to show the minimality of $\overline{E}$, let $b \in E_0$. We claim that the ν-orbit of the indecomposable projective-injective $\hat{F}$-module P(b) has no representative in ind $\overline{E}_b$, where $\overline{E}_b$ denotes the full subcategory of $\overline{E}$ consisting of all objects of $\overline{E}$ except b. For, if this were the case, such an indecomposable projective-injective must have $S(\nu^p b)$ as its top, for some $p \in \mathbb{Z}$. Thus it must equal $P(\nu b)$. However, soc $P(\nu b) = S(b)$ and $b \notin (\overline{E}_b)_0$. Dually, let $b \in (\nu E)_0$. Then the indecomposable projective-injective $\hat{F}$-module $I(b)$ has no representative in ind $\overline{E}_b$. This completes the proof.

(ii) Suppose now that E is a representation-infinite domain algebra. Then we have

$$E = {}_{j=1}^{q}[Y_j, L_j] \; C \; [X_i, K_i]_{i=1}^{p}$$

where X_i and Y_j belong to different tubes in mod C, for each pair (i, j). Since E is a full convex subcategory of $\hat{A}$, the tubular type of E is necessarily either domestic or tubular. It follows from [7] (2.4) (2.5) and (2.6) that there exists a strong ν-reflection sequence of sinks $i_1, \dots i_t$ such that $F = S_{i_t}^{+} \dots S_{i_1}^{+} E$ is a euclidean category or a tubular category of the same tubular type as E. We shall only write the proof in the euclidean case, leaving to the reader the straightforward modifications to cover the tubular case. Since $E = S_{\nu i_1}^{-} \dots S_{\nu i_t}^{-} F$, we may assume that the sequence $i_1, \dots i_t$ coincides with one of the sequences $\{a_4 + 1, \dots a_7\}$, $\{a_5 + 1, \dots a_7\}$ or $\{a_6 + 1, \dots a_7\}$. Thus the set of objects of E equals $\{\nu^{-1}i_1, \dots \nu^{-1}a_7, 1, \dots a_0, a_0 + 1, \dots a_4, a_4 + 1, \dots i_1 - 1\}$. As before, the duplicated algebra $\overline{E}$ of E can be identified with the full convex subcategory of $\hat{A}$ consisting of the objects $\{\nu^{-1}i_1, \dots \nu^{-1}a_7, 1, \dots a_0, \dots a_4, a_4 + 1, \dots i_1 - 1, i_1, \dots a_7, \nu 1, \dots \nu a_0, \dots \nu a_4, \nu(a_4 + 1), \dots \nu(i_1 - 1)\}$. Thus the indecomposable $\overline{E}$-modules are precisely those indecomposable $\hat{A}$-modules whose support has objects in this set.

It is easily seen that $\overline{E}$ contains precisely three tame concealed full convex subcategories, namely C, C' and νC. As in case (i) above, it is easy to see that ind $\overline{E}$ contains a complete set of representatives of ν-orbits of all indecomposable $\hat{F}$-modules, and that $\overline{E}$ is minimal with this property. This finishes the proof.

REMARK. Observe that, if A is euclidean and E is representation-finite, then $\overline{E}$ is 2-parametric, while, if E is representation-infinite, then $\overline{E}$ is 3-parametric.

3.2 It is easy to derive from (3.1), [7] (4.3) and [20] (3) that ind B for B as in condition (iv) of our main theorem fulfills condition (v). We are now going to show that B with ind B as in condition (v) is the support algebra of a period $\mathcal{C}$, hence, by (2.5), of the form described in condition (iv).

PROPOSITION. Let B be a full subcategory of $\hat{A}$ such that ind B contains a full convex subcategory $\mathcal{C}$ isomorphic to a period of $\hat{A}$ such that:

(1) All indecomposable projective B-modules which are not injective are predecessors of $\mathcal{C}$.

(2) All indecomposable projective-injective B-modules lie in $\mathcal{C}$.

(3) All indecomposable injective B-modules which are not projective are successors of $\mathfrak{T}$.

Then B is the support algebra of $\mathfrak{T}$.

Proof. Assume ind B satisfies the stated properties. We claim that B = Supp ($\mathfrak{T}$). By (2.4), it suffices to show that B = Cont ($\nu^{-1}\mathfrak{T} \vee \mathfrak{T}$). Denote by $\mathfrak{P}\backslash\mathfrak{I}$ the set of indecomposable projective B-modules which are not injective, and by $\mathfrak{P} \cap \mathfrak{I}$ the set of indecomposable projective-injective B-modules. By hypothesis, the former precede $\mathfrak{T}$ while the latter lie in $\mathfrak{T}$. We need to show that there exists a bijection $\mathfrak{P} \cap \mathfrak{I} \to \mathfrak{P}\backslash\mathfrak{I}$ given by P(i) $\to$ P(ν^{-1}i).

We first observe that, if $P(i)_B$ lies in $\mathfrak{T}$, then P(i) = I (ν^{-1}i). Thus P(ν^{-1}i) is a predecessor of $\mathfrak{T}$ in ind Â and therefore a predecessor of $\mathfrak{T}$ in ind B. Hence we may define a map as above.

Assume now that this map is not surjective. Then there exists j $\in B_0$ such that $P(j)_B \in (\mathfrak{P}\backslash\mathfrak{I})_0$ and P(νj) $\notin \mathfrak{T}_0$. Without loss of generality, we may assume that j is a neighbour (in B) of i $\in B_0$ such that $I(i)_B = P(\nu i)_B \in \mathfrak{T}_0$. If there exists an arrow i $\to$ j in B, then there exists a non-zero morphism $I(j)_B \to I(i)_B = P(\nu i)_B$. Thus I(j) is a predecessor of I(i) $\in \mathfrak{T}_0$, a contradiction to assumption (3). Thus there exists an arrow j $\to$ i in B, hence S(j) is a composition factor of I(i) $\in \mathfrak{T}$. However, this implies j $\in$ Supp $(\mathfrak{T})_0$ = Cont ($\nu^{-1}\mathfrak{T} \vee \mathfrak{T}$). Since j $\notin$ Cont $(\mathfrak{T})_0$, we have j $\in$ Cont $(\nu^{-1}\mathfrak{T})_0$, that is, νj $\in$ Cont $(\mathfrak{T})_0$. We have reached another contradiction, which completes our proof.

4. PROOF OF THE MAIN THEOREM

We shall now complete the proof of our main theorem. The implication (iii) $\Rightarrow$ (iv) was shown in (2.5), the implication (iv) $\Rightarrow$ (i) in (3.1) and the implications (iv) $\Rightarrow$ (v) and (v) $\Rightarrow$ (iii) in (3.2). Since the implication (i) $\Rightarrow$ (ii) is trivial, it only remains to show (ii) $\Rightarrow$ (iii), that is, that if B is a fundamental domain for T(A) in Â such that no proper full convex subcategory of B is a fundamental domain for T(A) in Â, then B is the support algebra of a $\langle\nu\rangle$ - period $\mathfrak{T}$ of ind Â, or, equivalently, B = Cont ($\nu^{-1}\mathfrak{T} \vee \mathfrak{T}$) for some period $\mathfrak{T}$.

We first assume that A is euclidean. Recall that we have a canonical inclusion of mod B into mod Â. We shall fix one such inclusion, and identify each indecomposable B-module with the corresponding Â-module. Let M $\in \mathfrak{X}_0$ be an indecomposable Â-module such that $F_\lambda M$ is a sincere T(A)-module. We claim that M can be chosen in such a way that for any indecomposable projective-injective Â-module P $\in \mathfrak{I}_{-2} \vee \mathfrak{X}_{-1} \vee \mathfrak{I}_{-1} \vee \mathfrak{X}_0$ there exists a non-zero morphism P $\to$ M. First observe that $\mathfrak{X}_0$ contains only finitely many indecomposable projective-injectives. Let thus M be a successor of all these projective-injectives which lies in $\mathfrak{X}_0$ and such that $F_\lambda M$ is sincere. Our claim then follows from (1.5).

By hypothesis, there exists an indecomposable B-module M' such that $F_\lambda M' \xrightarrow{\sim} F_\lambda M$. Since M' lies in the ν-orbit of M, we have that $M' \in \mathfrak{X}_t$ for some even integer t. We may assume that M' is chosen so that t is maximal. We conclude from (1.5) that Supp M' = Cont $(\mathfrak{T}_{t-2} \vee \mathfrak{X}_{t-1} \vee \mathfrak{T}_{t-1} \vee \mathfrak{X}_t)$ is contained in B.

Similarly, by taking a successor N of all the indecomposable projective-injectives in $\mathfrak{X}_1$ such that $F_\lambda N$ is sincere, we conclude that Cont $(\mathfrak{T}_{s-2} \vee \mathfrak{X}_{s-1} \vee \mathfrak{T}_{s-1} \vee \mathfrak{X}_s)$ is contained in B for some odd integer s. Without loss of generality, we may assume that $t > s$. We claim that $s = t - 1$. For, if this is not the case, then $s \leq t - 3$ and hence, by the convexity of B in $\hat{A}$, we have that Cont $(\mathfrak{T}_{t-5} \vee \mathfrak{X}_{t-4} \vee \mathfrak{T}_{t-4} \vee \ldots \vee \mathfrak{X}_{t-1} \vee \mathfrak{T}_{t-1} \vee \mathfrak{X}_t)$ is contained in B. But $\mathfrak{C} = \mathfrak{T}_{t-2} \vee \mathfrak{X}_{t-1} \vee \mathfrak{T}_{t-1} \vee \mathfrak{X}_t$ is a period in ind $\hat{A}$, and with the help of (2.4), (2.5) and (3.1), we conclude that Supp $(\mathfrak{C})$ = Cont $(\nu^{-1}\mathfrak{C} \vee \mathfrak{C})$ = Cont $(\mathfrak{T}_{t-4} \vee \ldots \vee \mathfrak{X}_t)$ is a fundamental domain for T(A). This contradicts the minimality of B. Thus $s = t - 1$ and consequently Cont $(\mathfrak{T}_{t-3} \vee \mathfrak{X}_{t-2} \vee \mathfrak{T}_{t-2} \vee \mathfrak{X}_{t-1} \vee \mathfrak{T}_{t-1} \vee \mathfrak{X}_t)$ is contained in B.

Let now $L \in \mathfrak{X}_1$ be such that $F_\lambda L$ is a sincere T(A)-module and L is a predecessor of all indecomposable projective-injectives in $\mathfrak{X}_1$. Again by (1.5), we have that Supp L = Cont $(\mathfrak{X}_{-1} \vee \mathfrak{T}_{-1} \vee \mathfrak{X}_0 \vee \mathfrak{T}_0)$. Thus there exists an odd integer r such that Cont $(\mathfrak{X}_{r-2} \vee \mathfrak{T}_{r-2} \vee \mathfrak{X}_{r-1} \vee \mathfrak{T}_{r-1})$ is contained in B. We claim that either $r = t - 1$ or $r = t + 1$. Assume that $r < t$. If $r \leq t - 3$, then by the convexity of B we have that Cont $(\mathfrak{X}_{t-5} \vee \mathfrak{T}_{t-5} \vee \ldots \vee \mathfrak{X}_t)$ is contained in B. As before, this yields a contradiction to the minimality of B. Therefore $r = t - 1$ in this case. Next assume that $r > t$. But then we conclude by the same argument that $r = t + 1$.

We thus split the remaining part of the proof into two cases:

(1) Assume that $r = t + 1$. Then the full subcategory Cont $(\mathfrak{T}_{t-3} \vee \mathfrak{X}_{t-2} \vee \ldots \vee \mathfrak{T}_t)$ of $\hat{A}$ lies in B. We shall show that there exists an extended section Σ in $\mathfrak{X}_{t-1}$ such that B = Cont $[\nu^{-1}\Sigma, \nu\Sigma[$. For this, we first observe that for any $a \in \text{Cont}(\mathfrak{X}_{t-1})_0$, we have $\nu a \in B_0$ or $\nu^{-1}a \in B_0$. Indeed, if every indecomposable $\hat{A}$-module M such that $M(a) \neq 0$ is a B-module, then $\text{Hom}_{\hat{A}}(P(a), P(\nu a)) \neq 0$ implies that $P(\nu a)$ is a B-module and thus $\nu a \in B_0$. Otherwise, let M be an indecomposable $\hat{A}$-module such that $M(a) \neq 0$ and which is not a B-module. There exists an indecomposable B-module M' in the ν-orbit of M such that $F_\lambda M' \xrightarrow{\sim} F_\lambda M$ and $M' \xrightarrow{\sim} \nu^m M$ for some $m \in \mathbb{Z}$, $m \neq 0$. If $m > 0$, it follows from the convexity of B in $\hat{A}$ and the fact that $a \in B_0$ (because

Cont($\mathfrak{X}_{t-1}$) is contained in B) that a, νa, ...ν^ma all belong to B. In particular, νa $\in$ B_0. Similarly, m < 0 yields $\nu^{-1}a \in B_0$.

Next, let i, j $\in$ Cont $(\mathfrak{X}_{t-1})_0$ be such that P(i) is a proper predecessor of P(j). We claim that there exists an indecomposable module M $\in$ $(\mathfrak{X}_{t-1})_0$ and a sequence of non-zero morphisms P(i) $\rightarrow$ M $\rightarrow$ P(j). Indeed, $\mathrm{Hom}_{\hat{A}}$ (P(i), P(νi)) $\neq$ 0, hence the morphism mapping the top of P(i) to the socle of P(νi) factors through a module M lying on the maximal subsection ending in P(j), that is, the set of all modules N such that there exists a sectional path from N to P(j) (see [9]). Thus $\mathrm{Hom}_{\hat{A}}$ (P(i), M) $\neq 0 \neq$ $\mathrm{Hom}_{\hat{A}}$ (M, P(j)). We claim that this implies that ν^{-1}j or νi lies in B. Indeed, M(i) $\neq$ 0 and M(ν^{-1}j) $\neq$ 0. Then there exists a B-module M' in the ν-orbit of M such that M' = ν^mM and $F_\lambda M' \xrightarrow{\sim} F_\lambda M$. But then M'($\nu^m$i) $\neq$ 0 and M'(ν^{m-1}j) $\neq$ 0, that is, $\nu^m i \in B_0$ and $\nu^{m-1} j \in B_0$ for some m $\in$ $\mathbb{Z}$. Assume m > 0. Since i $\in$ B_0 and B is convex in $\hat{A}$, we have that i, νi, ... ν^mi all belong to B. In particular, $\nu i \in B_0$. On the other hand, if m $\leq$ 0 (or equivalently m - 1 < 0) then ν^{m-1}j, ... ν^{-1}j, j all belong to B_0. In particular $\nu^{-1}j \in B_0$. Hence for i, j $\in$ Cont $(\mathfrak{X}_{t-1})_0$ such that P(i) is a proper predecessor of P(j), we have $\nu^{-1}j \in B_0$ or $\nu i \in B_0$.

Observe now that, if Cont ($\mathfrak{X}_{t-3}$) is contained totally in B (respectively, Cont ($\mathfrak{X}_{t+1}$) is contained totally in B), then the full subcategory Cont ($\mathfrak{X}_{t-3} \vee \ldots \vee \mathfrak{Y}_t$) (respectively, Cont ($\mathfrak{Y}_{t-3} \vee \ldots \vee \mathfrak{X}_{t+1}$)) is a fundamental domain for T(A) in $\hat{A}$ and hence, by the minimality of B, it equals B. Therefore B is the support algebra of the period $\mathfrak{X}_{t-1} \vee \mathfrak{Y}_{t-1} \vee \mathfrak{X}_t \vee \mathfrak{Y}_t$ (respectively, $\mathfrak{Y}_{t-1} \vee \mathfrak{X}_t \vee \mathfrak{Y}_t \vee \mathfrak{X}_{t+1}$). So assume that this is not the case. Then there exists i $\in$ Cont $(\mathfrak{X}_{t-1})_0$ such that $\nu i \in B_0$ but $\nu^{-1} i \notin B_0$ and j $\in$ Cont $(\mathfrak{X}_{t-1})_0$ such that $\nu^{-1} j \in B_0$ but $\nu j \notin B_0$.

Let a $\in$ Cont $(\mathfrak{X}_{t-1})_0$ be such that P(a) is a predecessor of P(i), then $\nu a \in B_0$. (Indeed, we have seen that either $\nu a \in B_0$ or $\nu^{-1} i \in B_0$, and we have $\nu^{-1} i \notin B_0$ by hypothesis). We shall now show that $\nu^{-1} a \notin B_0$. Suppose that $\nu^{-1} a \in B_0$. Let b $\in$ B_0 be a sink such that P(b) is a predecessor of P(ν^{-1}a) (such a sink must exist since the quiver of B has no oriented cycles), and let M be any B-module such that M(b) $\neq$ 0. We claim that ${}^\nu M$ is also a B-module, and thus is a B_b-module, where B_b is the full convex subcategory of B consisting of all objects except b. This would show that B_b is a fundamental domain, a contradiction to minimality. Suppose M(c) $\neq$ 0 for some c $\in$ B_0. Then P(c) is a predecessor of P(a). Indeed, there exists a chain of non-zero morphisms P(b) $\rightarrow$ M $\rightarrow$ P(νb) $\rightarrow \ldots \rightarrow$ P(a)$\rightarrow$ P(i) (for, P(b) is a predecessor of P(ν^{-1}a), which is a predecessor of P(ν^{-1}i)), and also a morphism P(c) $\rightarrow$ M. As we did for a, we deduce that $\nu c \in B_0$. Since M(c) $\neq$ 0 and $\nu c \in B_0$, this shows that ${}^\nu M$ is indeed a B-module. Clearly ${}^\nu M(b) = 0$. Therefore ${}^\nu M$ is a B_b-module such that $F_\lambda(M) \xrightarrow{\sim} F_\lambda({}^\nu M)$, and we reach a contradiction to the minimality. Dually, if $b \in \mathrm{Cont}(\mathcal{X}_{t-1})_0$ is such that $P(b)$ is a successor of $P(j)$, then $\nu^{-1}b \in B_0$ and $\nu b \notin B_0$.

Next, we want to show that $B \subseteq \mathrm{Cont}\,(\mathcal{X}_{t-3} \vee \mathcal{T}_{t-3} \vee \ldots \vee \mathcal{X}_{t+1})$. Suppose $b \in B_0$ belongs to $\mathrm{Cont}\,(\mathcal{T}_{t+1} \vee \mathcal{X}_{t+2} \vee \mathcal{T}_{t+2} \vee \ldots)$. We clearly may assume that b is a source. We shall show that for any B-module M such that $M(b) \neq 0$, then $\nu^{-1}M$ is a B_b-module (where, as usual, B_b denotes the full convex subcategory of B consisting of all objects except b) such that $F_\lambda(M) \xrightarrow{\sim} F_\lambda(\nu^{-1}M)$. It suffices to show that $\nu^{-1}M$ is a B-module. Again, assume that $M(c) \neq 0$, we must show that $\nu^{-1}c \in B_0$. Since there exist non-zero morphisms $P(b) \to M$ and $P(c) \to M$, then $c \in \mathrm{Cont}\,(\mathcal{T}_{t-1} \vee \mathcal{X}_t \vee \mathcal{T}_t \vee \ldots)_0$. Thus there exists $\ell > 0$ such that $\nu^{-\ell}c \in \mathrm{Cont}\,(\mathcal{T}_{t-3} \vee \mathcal{X}_{t-2} \vee \mathcal{T}_{t-2} \vee \mathcal{X}_{t-1})$. Since this content lies in B, it follows by convexity that $\nu^{-1}c \in B_0$, which proves our claim. Dually, the intersection of B with $\mathrm{Cont}\,(\ldots \vee \mathcal{X}_{t-4} \vee \mathcal{T}_{t-4})$ is empty, which proves that B is contained in $\mathrm{Cont}\,(\mathcal{X}_{t-3} \vee \mathcal{T}_{t-3} \vee \ldots \vee \mathcal{X}_{t+1})$.

We now want to show that the set $S = \{a \in \mathrm{Cont}\,(\mathcal{X}_{t-1})_0 \mid \nu a, \nu^{-1}a \in B_0\}$ is empty. For, if this is not the case, then the full subcategory of B consisting of all its objects except those of the form $\nu^{-1}a$, $a \in S$, is convex and is isomorphic to the duplicated algebra of a domain algebra which is a complete ν-slice of $\hat{A}$ (by (2.6)). The minimality of B then yields a contradiction. Consequently, $S = \emptyset$.

It now follows from (2.6) that there exists an extended sectionΣ in $\mathcal{X}_{t-1}$ such that

$$B = \mathrm{Cont}\,[\nu^{-1}\Sigma, \nu\Sigma[$$

This finishes the proof in the first case.

(2) Assume now that $r = t - 1$. In this case, the subcategory $\mathrm{Cont}\,(\mathcal{X}_{t-3} \vee \mathcal{T}_{t-3} \vee \mathcal{X}_{t-2} \vee \mathcal{T}_{t-2} \vee \mathcal{X}_{t-1} \vee \mathcal{T}_{t-1} \vee \mathcal{X}_t)$ is contained in B. We clearly may assume that $\mathrm{Cont}\,(\mathcal{T}_{t-2}) \neq \emptyset$ (for, if $\mathrm{Cont}\,(\mathcal{T}_{t-2}) = \emptyset$, then $\mathrm{Cont}\,(\mathcal{T}_t) = \emptyset$ so that, if $\mathcal{C} = \mathcal{X}_{t-3} \vee \mathcal{T}_{t-3} \vee \mathcal{X}_{t-2} \vee \mathcal{T}_{t-2}$, then $\mathrm{Cont}\,(\mathcal{C} \vee \nu\mathcal{C})$ is contained in B, so that, by minimality, it is equal to it and we are done).

Observe that, if $P(a_1), \ldots P(a_p)$ are all the projective-injectives lying in one tube $T_{\lambda,t-2} \in \mathcal{T}_{t-2}$, then there exists an indecomposable module M in this tube such that $M(a_\ell) \neq 0$, $M(\nu^{-1}a_\ell) \neq 0$ for all $1 \leq \ell \leq p$. Thus there exists a B-module M' such that $F_\lambda(M) \xrightarrow{\sim} F_\lambda(M')$. For each such tube $T_{\lambda,t-2}$, we fix one such module M'. But then there exists $m_\lambda \in \mathbf{Z}$ such that $M'(\nu^{m_\lambda} a_\ell) \neq 0$, $M'(\nu^{m_\lambda - 1} a_\ell) \neq 0$ for all $1 \leq \ell \leq p$. We claim that if $m_\lambda \leq 0$, then $\mathrm{Cont}\,(T_{\lambda,t-4})$ is contained in B, while if $m_\lambda > 0$, then $\mathrm{Cont}\,(T_{\lambda,t})$ is contained in B.

Suppose first that $m_\lambda \leq 0$. By hypothesis, we have Cont $(T_{\lambda,t-2})$ contained in B. It follows that $\nu^{m_\lambda -1}$ Cont $(T_{\lambda,t-2})$ is contained in B. By convexity, ν^{-1} Cont $(T_{\lambda,t-2})$ = Cont $(T_{\lambda,t-4})$ is contained in B. Thus we are done. The proof is the same if $m_\lambda > 0$.

Let $\Lambda = \{\lambda \in \mathbb{P}_1(k) \mid m_\lambda \leq 0\}$. Then the full convex subcategory of $\hat{A}$ defined by

Cont $((\bigvee_{\lambda\in\Lambda} T_{\lambda,t-4}) \vee \mathcal{X}_{t-3} \vee \mathcal{T}_{t-3} \vee \mathcal{X}_{t-2} \vee \mathcal{T}_{t-2} \vee \mathcal{X}_{t-1} \vee \mathcal{T}_{t-1} \vee \mathcal{X}_t \vee (\bigvee_{\lambda\notin\Lambda} T_{\lambda,t}))$ is a fundamental domain. Thus, by minimality, it is equal to B. Since it is of the required form, we are done.

There only remains to consider the case where A is tubular.

Recall first that, for each M lying in a homogeneous tube of a component of the form $\mathcal{M}_{i,i+1}$, then $F_\lambda(M)$ is a sincere T(A)-module [24] (5.1) [22] (3). Thus, let M_1 be a B-module such that M_1 lies in a homogeneous tube of a component of the form $\mathcal{M}_{t,t+1}$ with t maximal (such an M_1 exists because B is finite dimensional). Let N be the integer as in (1.4), and choose B-modules $M_2, \ldots M_N$ in homogeneous tubes such that $F_\lambda(M_2)$, ... $F_\lambda(M_N)$ belong respectively to homogeneous tubes of the respective images of the components $\mathcal{M}_{t-1,t}, \ldots \mathcal{M}_{t-N+1,t-N+2}$. Since M_1 lies in $\mathcal{M}_{t,t+1}$, then, by (1.5), Cont $(\mathcal{T}_{t-N+1} \vee \ldots \vee \mathcal{T}_{t-1} \vee \mathcal{T}_t)$ is contained in B. The B-module M_2 belongs to a component of the form $\mathcal{M}_{t-1-\ell N,t-\ell N}$ with $\ell \geq 0$. Using (1.5) and the convexity of B, we deduce that Cont $(\mathcal{T}_{t-N})$ is contained in B. The same argument using $M_3, \ldots M_N$ implies that the full subcategory Cont $(\mathcal{T}_{t-2N+2} \vee \ldots \vee \mathcal{T}_{t-N} \vee \ldots \vee \mathcal{T}_t)$ is contained in B.

Observe that, since A is tubular, then Cont $(\mathcal{T}_{t+1}) \neq \emptyset$. Let $P(a_1), \ldots P(a_p)$ be all the projective-injectives lying in one tube $T_{\lambda,t+1} \in \mathcal{T}_{t+1}$, then there exists an indecomposable module M in this tube such that $M(a_\ell) \neq 0$, $M(\nu^{-1}a_\ell) \neq 0$ for all $1 \leq \ell \leq p$. Thus there exists a corresponding B-module M' with $F_\lambda(M) \xrightarrow{\sim} F_\lambda(M')$. For each such tube $T_{\lambda,t+1}$, we fix one such module M'. Observe that, for each λ, there exists $m_\lambda \in \mathbb{Z}$ with $M'(\nu^{m_\lambda}a_\ell) \neq 0$, $M'(\nu^{m_\lambda -1}a_\ell) \neq 0$ for all $1 \leq \ell \leq p$. We claim that if $m_\lambda \leq 0$ then Cont $(T_{\lambda,t-2N+1}) \subseteq$ B, while if $m_\lambda > 0$, then Cont $(T_{\lambda,t+1}) \subseteq$ B. Suppose first $m_\lambda \leq 0$. By hypothesis, we have Cont $(T_{\lambda,t-N+1}) \subseteq$ B. It follows that $\nu^{m_\lambda -1}$ Cont $(T_{\lambda,t-N+1}) \subseteq$ B. By convexity, ν^{-1} Cont $(T_{\lambda,t-N+1})$ = Cont $(T_{\lambda,t-2N+1})$ is contained in B and we are done. The proof is the same if $m_\lambda > 0$.

Letting $\Lambda = \{\lambda \in \mathbb{P}_1(k) \mid m_\lambda \leq 0\}$, then the full convex subcategory of $\hat{A}$ defined

by Cont $((\bigvee_{\lambda\in\Lambda} T_{\lambda,t-2N+1}) \vee \mathcal{T}_{t-2N+2} \vee \ldots \vee \mathcal{T}_t \vee (\bigvee_{\lambda\notin\Lambda} T_{\lambda,t+1}))$ is a fundamental domain in $\hat{A}$. Thus by minimality, it is equal to B. Since it is of the required form, we are done.

5. REMARKS AND EXAMPLES

5.1 PROPOSITION. Let E be a domain algebra, then gl. dim. $E \leq 2$.

Proof. The statement is well-known if E is a tilted algebra [16] (5). So assume it is not and let $E = {}_{j=1}^{q}[X_j, L_j]\, C\, [Y_i, K_i]_{i=1}^{p}$ where for each pair (i, j), X_j and Y_i lie in different tubes and also $p \geq 1$, $q \geq 1$. It suffices to show that for any $a \in E_0$, we have pd $S(a) \leq 2$. If $a \in ({}_{j=1}^{q}[X_j, L_j]\, C)_0$ then the statement is clear since in a projective resolution for S(a) in mod E, no point from the branches K_i ($1 \leq i \leq p$) may appear (because there is no path from a to these branches). Consequently, such a projective resolution is the same as in ${}_{j=1}^{q}[X_j, L_j]\, C$ which is a tilted algebra and thus of global dimension ≤ 2. There remains to consider the case where $a \in (K_i)_0$ for some i. Consider the short exact sequence of E-modules

$$0 \to R(a) \to P(a) \to S(a) \to 0$$

with $R(a) = \text{rad}\, P(a)_E$. We thus need to show that pd $R(a)_E \leq 1$, or, equivalently, that $\text{Hom}_E(DE, \tau R(a)) = 0$. However, indecomposable injective E-modules either lie in the preinjective component, or in tubes orthogonal to the one containing a. Then, clearly, $\text{Hom}_E(DE, \tau R(a)) = 0$ and we are done.

COROLLARY. Let B be a minimal fundamental domain for T(A) in $\hat{A}$, where A is euclidean or tubular. Then

$$2 \leq \text{gl. dim. } B \leq 5$$

Proof. This follows from the proposition and [2] (1.1).

REMARK. It is not hard to show that in fact gl. dim. $B \neq 2$. We do not know of any example with gl. dim. $B = 5$.

5.2 EXAMPLE. let A be given by the quiver

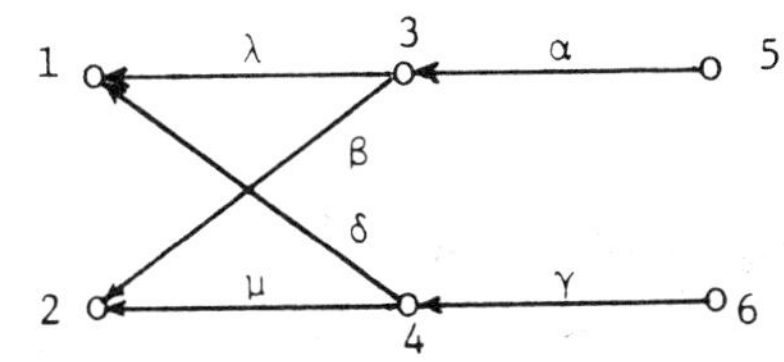

bound by $\alpha\beta = 0$, $\gamma\mu = 0$. Clearly A is a euclidean algebra. A ν-reflection sequence of sinks as in (1.3) is given by 1, 2, 3, 4, 5, 6. The representation-finite domain algebras E such that $\bar{E}$ is a minimal fundamental domain for T(A) in $\hat{A}$ are given by

$S_1^+ A$

$\alpha\beta = 0, \gamma\mu = 0$

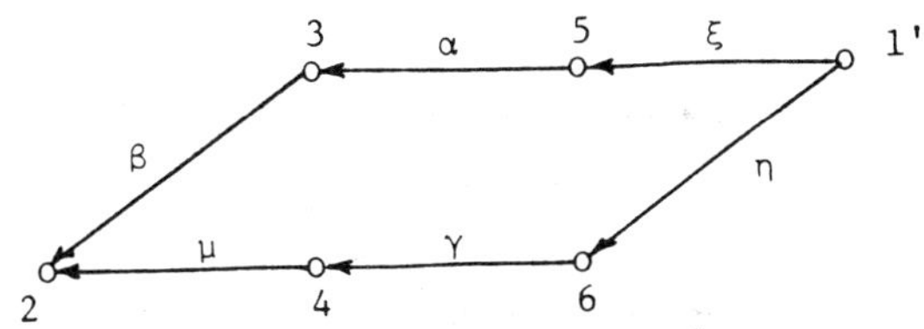

$S_2^+ A$

$\zeta\lambda = 0, \sigma\delta = 0$

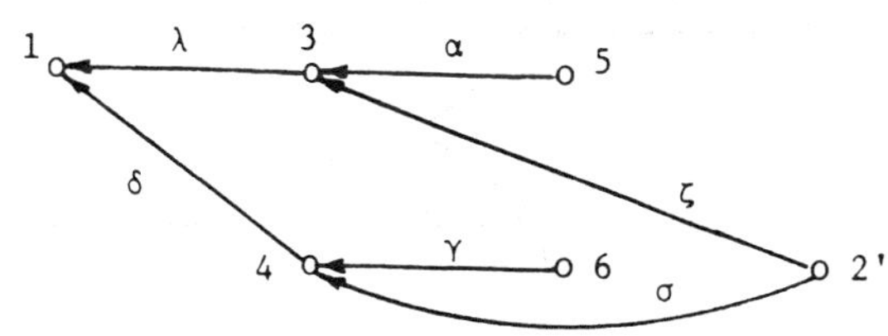

$S_3^+ S_2^+ S_1^+ A$

$\lambda\eta = 0, \beta\sigma = 0$

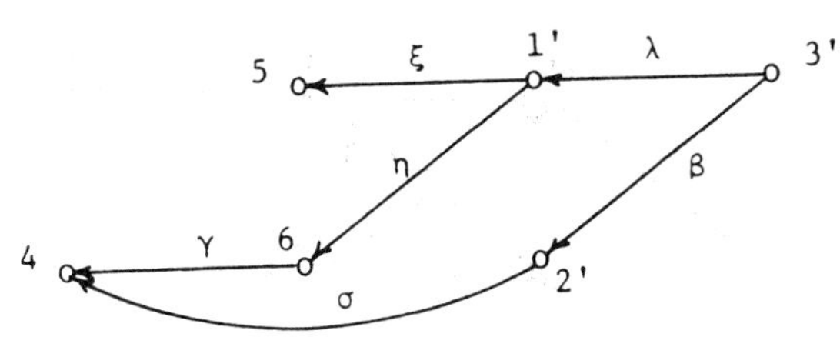

$S_4^+ S_2^+ S_1^+ A$

$\delta\xi = 0$, $\mu\zeta = 0$

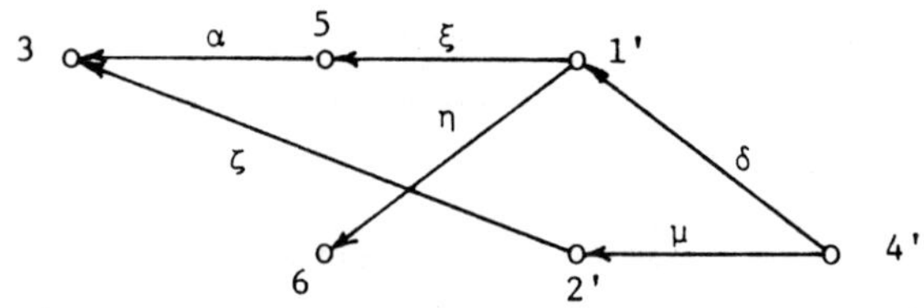

The representation-infinite domain algebras E such that $\overline{E}$ is a minimal fundamental domain for T(A) in $\hat{A}$ are given by A, and

$S_2^+ S_1^+ A$

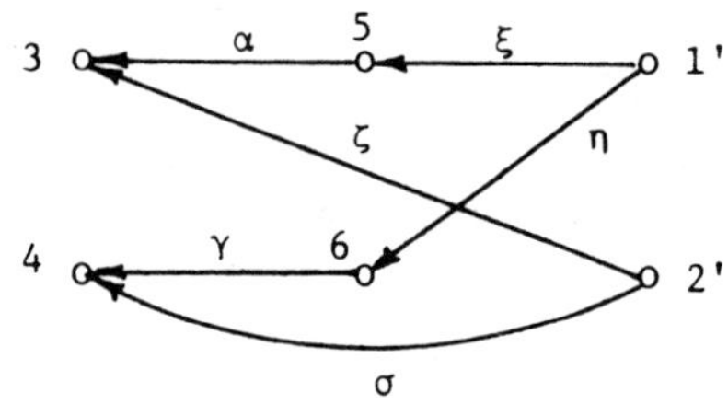

$S_6^+ S_4^+ S_3^+ S_2^+ S_1^+ A$

$\delta\xi = 0$, $\gamma\mu = 0$

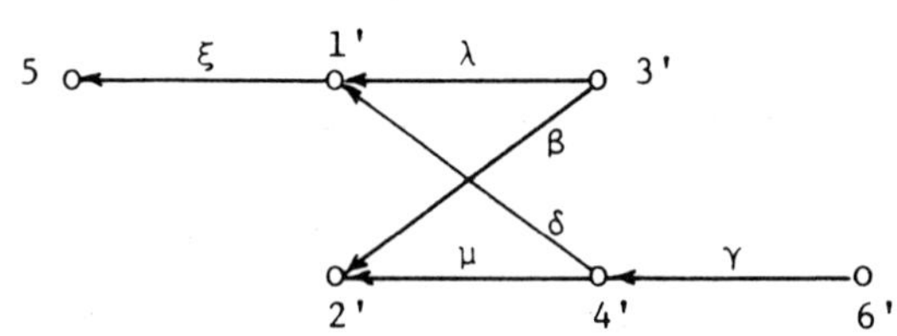

$S_5^+ S_4^+ S_3^+ S_2^+ S_1^+ A$

$\lambda\eta = 0$, $\alpha\beta = 0$

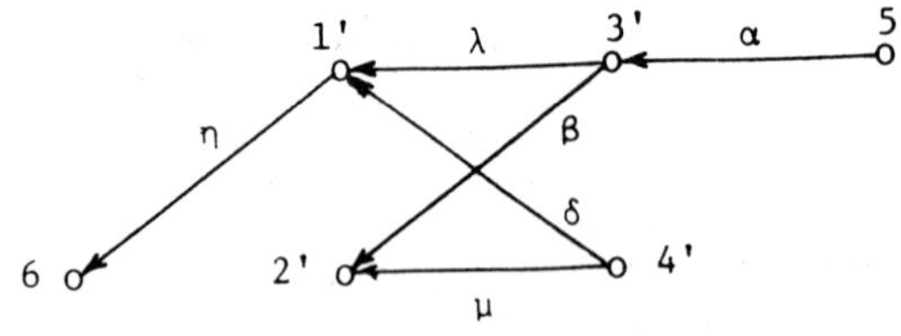

$S_4^+ S_3^+ S_2^+ S_1^+ A$ $\lambda\eta = 0,\ \delta\xi = 0$

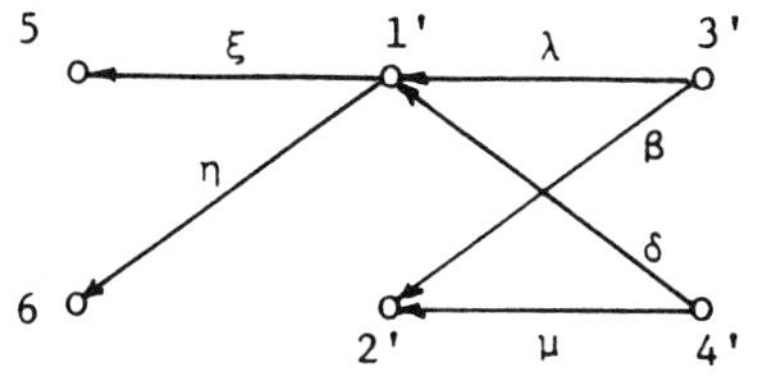

Observe that the 2nd and 3rd domain algebras are not tilted algebras.

ACKNOWLEDGEMENT.

This article was written while the third author was a Feodor Lynen fellow, partially supported by the Alexander von Humboldt Foundation and the University of Sherbrooke. The first and second authors gratefully acknowledge the hospitality of each other's university and the supports of the NSERC of Canada and the Polish Ministry of Education, project R.P.I. 10.

REFERENCES

[1] ASSEM, I.: On representations of the bimodule DA, Tsukuba J. Math., Vol. 9, No. 2 (1985), 217-232.

[2] ASSEM, I. and IWANAGA, Y.: On a class of representation-finite QF-3 algebras, Tsukuba J. Math., Vol. 11, No. 2 (1987), 199-217.

[3] ASSEM, I. and SKOWROŃSKI, A.: Iterated tilted algebras of type $\widetilde{\mathbf{A}}_n$, Math. Z. 195 (1987), 269-290.

[4] ASSEM, I. and SKOWROŃSKI, A.: On some classes of simply connected algebras, Proc. London Math. Soc. (3) 56 (1988), 417-450.

[5] ASSEM, I. and SKOWROŃSKI, A.: Algebras with cycle-finite derived categories, Math. Ann. 280 (1988), 441-463.

[6] ASSEM, I and SKOWROŃSKI, A.: On tame repetitive algebras, to appear, preprint No. 49 (1989), Université de Sherbrooke.

[7] ASSEM, I., NEHRING, J. and SKOWROŃSKI, A.: Domestic trivial extensions of simply connected algebras, Tsukuba J. Math., Vol 13, No. 1 (1989), 31-72.

[8] AUSLANDER, M. and REITEN, I.: Representation theory of artin algebras III and IV, Comm. Algebra 3 (1975), 239-294 and 5 (1977), 443-518.

[9] BAUTISTA, R.: Sections in Auslander-Reiten quivers, Proc. ICRA II (Ottawa, 1979), Lecture Notes in Math. 832, Springer, Berlin (1980), 74-96.

[10] BONGARTZ, K.: Tilted algebras, Proc. ICRA III (Puebla, 1980), Lecture Notes in Math. 903, Springer, Berlin (1980), 26-38.

[11] BONGARTZ, K. and GABRIEL, P.: Covering spaces in representation theory, Invent. Math. 65 (1981 / 82), No. 3, 331-378.

[12] BRETSCHER, O., LÄSER, C. and RIEDTMANN, C.: Selfinjective and simply connected algebras, Manuscripta Math. 36 (3) (1981), 253-308.

[13] DOWBOR, P. and SKOWROŃSKI, A.: Galois coverings of representation-infinite algebras, Comment. Math. Helv. 62 (1987), 311-337.

[14] GABRIEL, P.: The universal cover of a representation-finite algebra, Proc. ICRA III (Puebla, 1980), Lecture Notes in Math. 903, Springer, Berlin (1980), 68-105.

[15] HAPPEL, D.: On the derived category of a finite dimensional algebra, Comment. Math. Helv. 62 (1987), 339-389.

[16] HAPPEL, D. and RINGEL, C.M.: Tilted algebras, Trans. Amer. Math. Soc 274 (1982), 399-443.

[17] HAPPEL, D. and RINGEL, C.M.: The derived category of a tubular algebra, Proc. ICRA IV (Ottawa, 1984), Lecture Notes in Math. 1177, Springer, Berlin (1986), 156-180.

[18] HUGHES, D. and WASCHBÜSCH, J.: Trivial extensions of tilted algebras, Proc. London Math. Soc. 46 (1983), 347-364.

[19] LARRIÓN, F. and SALMERÓN, L.: On Auslander-Reiten quivers without oriented cycles, Bull. London Math. Soc. 16 (1984), 47-51.

[20] NEHRING, J.: Polynomial growth trivial extensions of non-simply connected algebras, Bull. Pol. Acad. Sci., Vol. 36, No. 7-8 (1988) 441-445.

[21] NEHRING, J.: Trywialne rozszerzenia wielomianowego wzrostu, Ph.D. Thesis, Nicolas Copernicus University (1989).

[22] NEHRING, J. and SKOWROŃSKI, A.: Polynomial growth trivial extensions of simply connected algebras, Fund. Math. 132 (1989), 117-134.

[23] OKUNO, H.: Isomorphisms between coverings of trivial extension algebras, Comm. Algebra 15 (4), 791-812 (1987).

[24] RINGEL, C.M.: Tame algebras and integral quadratic forms, Lecture Notes in Math. 1099, Springer, Berlin (1984).

[25] SCHEWE, W.: Fundamental domains for representation-finite algebras, to appear, preprint No. 74 (1990), Université de Sherbrooke.

[26] SKOWROŃSKI, A.: Group algebras of polynomial growth, Manuscripta Math. 59 (1987), 499-516.

[27] SKOWROŃSKI, A.: Selfinjective algebras of polynomial growth, Math. Ann. 285 (1989), 177-199.

[28] WAKAMATSU, T.: Note on trivial extensions of artin algebras, Comm. Algebra 12 (1984), 33-41.

Département de mathématiques et d'informatique
Université de Sherbrooke
Sherbrooke (Québec)
Canada, J1K 2R1

Institute of Mathematics
Nicholas Copernicus University
Chopina 12 / 18
87-100, Toruń
Poland

Canadian Mathematical Society
Conference Proceedings
Volume 11 (1991)

On a theorem of E. Green on the dual of the transpose

MAURICE AUSLANDER AND IDUN REITEN

Let Λ be an artin algebra and denote by Tr the transpose and by D the ordinary duality. We denote by $mod\,\Lambda$ the category of finitely generated Λ-modules. Then $DTr : \underline{mod}\Lambda \to \overline{mod}\Lambda$ is a functor, even an equivalence of categories, from $mod\,\Lambda$ modulo projectives to $mod\,\Lambda$ modulo injectives. It was proved about 10 years ago by E. Green (unpublished [**3**]) that there is a left exact functor $F : mod\,\Lambda \to mod\,\Lambda$ which induces $DTr : \underline{mod}\Lambda \to \overline{mod}\Lambda$. Also he described a bimodule I with the property that $F = Hom_\Lambda(I,\ )$.

In general it is interesting to decide whether functors which are defined on factor categories of module categories can be lifted to a functor between the module categories. Also it is useful to decide if the new functor has any exactness properties. Another example of such a functor is the first syzygy functor $\Omega^1 : \underline{mod}\Lambda \to \underline{mod}\Lambda$. We shall see that Ω^1 can be lifted to an exact functor from $mod\,\Lambda$ to $mod\,\Lambda$.

The main aim of this note is to give a short proof of Green's theorem. In section 1 we discuss what the problems are in lifting Tr, TrD and DTr. In section 2 we give some basic properties of enveloping algebras Λ^e of finite dimensional algebras over a field and use them to show that TrD can be lifted to a right exact functor. In section 3 we give some illustrative examples, and in section 4 we show that $Hom_\Lambda(Tr_{\Lambda^e}\Lambda,\ )$ provides a lifting of DTr. We end the paper with a discussion of how the lifting of DTr we give can be used to define preprojective algebras for all finite dimensional algebras.

1980 *Mathematics Subject Classification* (1985 *Revision*). Primary 16A46; Secondary 16A64.

The first author was partially supported in part by NSF Grant 4-50325..

1. The transpose

We first recall the reason why, as usually defined, Tr is not a functor from $mod\,\Lambda$ to $mod\,\Lambda^{op}$. Let B and C be in $mod\,\Lambda$, and let $P_1 \to P_0 \to B \to 0$ and $Q_1 \to Q_0 \to C \to 0$ be minimal projective presentations. When $h : B \to C$ is a map in $mod\,\Lambda$, we have an exact commutative diagram

$$\begin{array}{ccccccc} P_1 & \to & P_0 & \to & B & \to & 0 \\ \downarrow h_1 & & \downarrow h_0 & & \downarrow h & & \\ Q_1 & \to & Q_0 & \to & C & \to & 0 \end{array}$$

which gives rise to the exact commutative diagram

$$\begin{array}{ccccccc} Q_0^* & \to & Q_1^* & \to & TrC & \to & 0 \\ \downarrow h_0^* & & \downarrow h_1^* & & \downarrow g & & \\ P_0^* & \to & P_1^* & \to & TrB & \to & 0. \end{array}$$

Here $A^* = Hom_\Lambda(A, \Lambda)$ and $f^* = Hom_\Lambda(f, \Lambda)$ for A an object and f a map in $mod\,\Lambda$. But here $g : TrC \to TrB$ is not in general well defined since the liftings h_0 and h_1 are not unique.

It is well known and easily seen that if Λ is hereditary, then $Tr : mod\,\Lambda \to mod\,\Lambda^{op}$ is a functor. In this case we have $Tr = Ext^1_\Lambda(\ , \Lambda)$, which is clearly a right exact functor, so that DTr is left exact and TrD is right exact. We also note that if Λ is not hereditary, it is impossible to have a right exact functor $T : mod\,\Lambda \to mod\,\Lambda^{op}$ such that $DT(C) \simeq DTrC$ for all C in $mod\,\Lambda$. For choose a projective module P having a nonprojective submodule X. If DT was left exact, $DT(X) \neq 0$ would be a submodule of $DT(P) = 0$, a contradiction. But to have a lifting $T : mod\,\Lambda \to mod\,\Lambda^{op}$ of $Tr : \underline{mod}\Lambda \to \underline{mod}\Lambda^{op}$ we only need $T(C) \simeq TrC \amalg P$, where P is projective.

Analyzing what the problem is with lifting Tr, we see that we have to look for functorial projective presentations of Λ-modules. Denote by $\mathcal{C}_\Lambda$ the category whose objects are the complexes $C^\bullet$ of Λ-modules with $C_i = 0$ for $i < 0$ and whose morphisms are the ordinary maps between complexes. Let $\mathcal{P}_\Lambda$ be the full subcategory of $\mathcal{C}_\Lambda$ whose objects are the complexes $P^\bullet$ of projective modules, with $H_i(P^\bullet) = 0$ for $i \neq 0$. A sequence in $\mathcal{P}_\Lambda$ is exact, right exact or left exact if it has the corresponding property in the abelian category $\mathcal{C}_\Lambda$.

Associated with a covariant functor $F : mod\,\Lambda \to \mathcal{P}_\Lambda$ is a contravariant functor $T : mod\,\Lambda \to mod\,\Lambda^{op}$ defined as follows. For B in $mod\,\Lambda$, $T(B) = Coker(F(B)_0^* \to F(B)_1^*)$. If $g : B \to C$ is a map in $mod\,\Lambda$, then $T(g) : T(C) \to T(B)$ is defined to be the unique map making the following diagram commute

$$\begin{array}{ccccccc} F(C)_0^* & \to & F(C)_1^* & \to & T(C) & \to & 0 \\ \downarrow F(h)_0^* & & \downarrow F(h)_1^* & & \downarrow T(g) & & \\ F(B)_0^* & \to & F(B)_1^* & \to & T(B) & \to & 0. \end{array}$$

We then have the following.

PROPOSITION 1.1. *Let $F : \operatorname{mod}\Lambda \to \mathcal{P}_\Lambda$ be a covariant functor satisfying $H_0(F(C)) = C$ for C in $\operatorname{mod}\Lambda$ and $H_0(F(h)) = h$ for $h : B \to C$ a map in $\operatorname{mod}\Lambda$. Then the functor $T : \operatorname{mod}\Lambda \to \operatorname{mod}\Lambda^{op}$ defined above is a lifting of $Tr : \underline{\operatorname{mod}}\Lambda \to \underline{\operatorname{mod}}\Lambda^{op}$. Further T is right exact if F is right exact.*

PROOF. If $h : B \to C$ is in $\operatorname{mod}\Lambda$, we have by our assumptions not necessarily minimal projective presentations $F(B)_1 \to F(B)_0 \to B \to 0$ and $F(C)_1 \to F(C)_0 \to C \to 0$ and an exact commutative diagram

$$\begin{array}{ccccccc} F(B)_1 & \to & F(B)_0 & \to & B & \to & 0 \\ \downarrow {\scriptstyle F(h)_1} & & \downarrow {\scriptstyle F(h)_0} & & \downarrow {\scriptstyle h} & & \\ F(C)_1 & \to & F(C)_0 & \to & C & \to & 0. \end{array}$$

Then clearly $T(C) \simeq TrC \amalg P$, where P is projective, and $T(\underline{h}) = Tr(\underline{h})$ in $\underline{\operatorname{mod}}\Lambda$.

Let $0 \to A \to B \to C \to 0$ be an exact sequence. If F is exact, we have the exact commutative diagrams

$$\begin{array}{ccccccccc} 0 & \to & F_1(A) & \to & F_1(B) & \to & F_1(C) & \to & 0 \\ & & \downarrow & & \downarrow & & \downarrow & & \\ 0 & \to & F_0(A) & \to & F_0(B) & \to & F_0(C) & \to & 0 \\ & & \downarrow & & \downarrow & & \downarrow & & \\ 0 & \to & A & \to & B & \to & C & \to & 0 \\ & & \downarrow & & \downarrow & & \downarrow & & \\ & & 0 & & 0 & & 0 & & \end{array}$$

and

$$\begin{array}{ccccccccc} 0 & \to & F(C)_0^* & \to & F(B)_0^* & \to & F(A)_0^* & \to & 0 \\ & & \downarrow & & \downarrow & & \downarrow & & \\ 0 & \to & F(C)_1^* & \to & F(B)_1^* & \to & F(A)_1^* & \to & 0 \\ & & \downarrow & & \downarrow & & \downarrow & & \\ & & T(C) & \to & T(B) & \to & T(A) & \to & 0 \\ & & \downarrow & & \downarrow & & \downarrow & & \\ & & 0 & & 0 & & 0. & & \end{array}$$

□

Similarly we can associate with a covariant functor $F : \operatorname{mod}\Lambda \to \mathcal{P}_\Lambda$ a covariant functor $U : \operatorname{mod}\Lambda \to \operatorname{mod}\Lambda$ defined as follows. For C in $\operatorname{mod}\Lambda$, $U(C) = \operatorname{Im}(F(C)_1 \to F(C)_0)$. If $h : B \to C$ is a map in $\operatorname{mod}\Lambda$, then $F(f)_0(U(B)) \subseteq U(C)$ and we define $U(h) = F(f)_0 \mid_{U(B)}: U(B) \to U(C)$. Then we have the following.

PROPOSITION 1.2. *Assume that we have a covariant functor $F : \operatorname{mod}\Lambda \to \mathcal{P}_\Lambda$ satisfying $H_0(F(B)) = B$ for B in $\operatorname{mod}\Lambda$ and $H_0(F(h)) = h$ for $h : B \to C$ a map in $\operatorname{mod}\Lambda$. Then the associated covariant functor $U : \operatorname{mod}\Lambda \to \operatorname{mod}\Lambda$ defined as above is a lifting of $\Omega^1 : \underline{\operatorname{mod}}\Lambda \to \underline{\operatorname{mod}}\Lambda$, which is exact if F is exact.*

PROOF. If $h : B \to C$ is in $mod\,\Lambda$, we have a commutative diagram

$$\begin{array}{ccccccccc} 0 & \to & U(B) & \to & F_0(B) & \to & B & \to & C \\ & & \downarrow U(h) & & \downarrow F_0(h) & & \downarrow h & & \\ 0 & \to & U(C) & \to & F_0(C) & \to & C & \to & 0 \end{array}$$

so that $U : mod\,\Lambda \to mod\,\Lambda$ is clearly a lifting of $\Omega^1 : \underline{mod}\Lambda \to \underline{mod}\Lambda$.

If F is exact, we see that U is exact. □

2. The enveloping algebra

In this section we shall construct functorial projective resolutions of Λ-modules. This is accomplished by considering the enveloping algebra $\Lambda^e = \Lambda \otimes_k \Lambda^{op}$ of a finite dimensional algebra Λ over a field k. Recall that the category $mod\,\Lambda^e$ of finitely generated (left) Λ^e-modules is equivalent to the category of Λ-bimodules with k acting centrally. If M is in $mod\,\Lambda^e$, we define $xmy = (x \otimes y)m$ for x, y in Λ and m in M. We then have the following preliminary result.

LEMMA 2.1. *Let ${}_\Lambda P_\Lambda$ be a Λ-bimodule which is a projective Λ^e-module. Then we have the following.*

a) *${}_\Lambda P$ and P_Λ are projective over Λ.*
b) *$P \otimes_\Lambda C$ is projective for all C in $\mathrm{mod}\,\Lambda$ and $B \otimes_\Lambda P$ is projective in $\mathrm{mod}\,\Lambda^{op}$ for all B in $\mathrm{mod}\,\Lambda^{op}$.*
c) *$Hom_\Lambda({}_\Lambda P, {}_\Lambda C)$ is an injective Λ-module for all C in $\mathrm{mod}\,\Lambda$ and $Hom_\Lambda(P_\Lambda, B_\Lambda)$ is injective in $\mathrm{mod}\,\Lambda^{op}$ for all B in $\mathrm{mod}\,\Lambda^{op}$.*

PROOF. We have $\Lambda^e \otimes_\Lambda C = \Lambda \otimes_k \Lambda^{op} \otimes_\Lambda C \simeq \Lambda \otimes_k C$, which clearly is projective in $mod\,\Lambda$. Since P is a direct summand of $(\Lambda^e)^n$ for some $n > 0$, b) and a) follow.

c) For a Λ-module X we have the natural isomorphism $Hom_\Lambda(X, Hom_\Lambda(P, C)) \simeq Hom_\Lambda(P \otimes_\Lambda X, C)$. Here $Hom_\Lambda(P, C)$ is a left Λ-module using the right action on P and $P \otimes_\Lambda X$ is a left Λ-module by the left action on P. Let $0 \to X \to Y \to Z \to 0$ be an exact sequence of Λ-modules. Then $0 \to P \otimes_\Lambda X \to P \otimes_\Lambda Y \to P \otimes_\Lambda Z \to 0$ is an exact sequence, which splits since the terms are projective Λ-modules by b). Hence $0 \to Hom_\Lambda(P \otimes_\Lambda Z, C) \to Hom_\Lambda(P \otimes_\Lambda Y, C) \to Hom_\Lambda(P \otimes_\Lambda X, C) \to 0$ is exact, and consequently $0 \to Hom_\Lambda(Z, Hom_\Lambda(P, C)) \to Hom_\Lambda(Y, Hom_\Lambda(P, C)) \to Hom_\Lambda(X, Hom_\Lambda(P, C)) \to 0$ is exact. This shows that $Hom_\Lambda(P, C)$ is an injective Λ-module. □

Let $\cdots \to P_i \to \cdots P_1 \to P_0 \to \Lambda \to 0$ be a minimal projective Λ^e-resolution of Λ. Since by Lemma 2.1 a) the P_i are projective in $mod\,\Lambda$, we have for C in $mod\,\Lambda$ an exact sequence $\cdots \to P_i \otimes_\Lambda C \to \cdots \to P_1 \otimes_\Lambda C \to P_0 \otimes_\Lambda C \to \Lambda \otimes_\Lambda C \to 0$. This is a projective resolution of C by Lemma 2.1 b). It also follows that we have an exact sequence $0 \to Hom_\Lambda(\Lambda, C) \to Hom_\Lambda(P_0, C) \to Hom_\Lambda(P_1, C) \to \cdots$, which is an injective resolution of C by Lemma 2.1 c).

Denote by $P^\bullet$ the complex $\cdots \to P_i \to \cdots \to P_1 \to P_0 \to 0$. Before we go on we mention the following curious application of the above resolutions: For B and C in $mod\,\Lambda$ we have $Ext^i_\Lambda(\ ,B)(C) = H^i(P^\bullet \otimes_\Lambda C, B) \simeq H^i(Hom_\Lambda(C, Hom_\Lambda(P^\bullet, B)) = Ext^i(C,\ )(B)$.

We now define an exact covariant functor $F : mod\,\Lambda \to \mathcal{P}_\Lambda$ in the following way. For B in $mod\,\Lambda$ we define $F(B) = P^\bullet \otimes_\Lambda B$ and for a map $h : B \to C$ in $mod\,\Lambda$, we define $F(h)$ by $F(h)_i = P_i \otimes h : F(B)_i \to F(C)_i$. Then we get by section 1 a lifting $T_l : mod\,\Lambda \to mod\,\Lambda^{op}$ of $Tr : \underline{mod}\Lambda \to \underline{mod}\Lambda^{op}$. By associating to the right module A the complex $A \otimes_\Lambda P^\bullet$, we get similarly a lifting $T_r : mod\,\Lambda^{op} \to mod\,\Lambda$ of $Tr : \underline{mod}\Lambda^{op} \to \underline{mod}\Lambda$. We then have the following consequence of Proposition 1.1.

COROLLARY 2.2. *The functors $T_l : mod\,\Lambda \to mod\,\Lambda^{op}$ and $T_r : mod\,\Lambda^{op} \to mod\,\Lambda$ are liftings of $Tr : \underline{mod}\Lambda \to \underline{mod}\Lambda^{op}$ and $Tr : \underline{mod}\Lambda^{op} \to \underline{mod}\Lambda$ respectively, and they are right exact functors.*

We also note that we have the following description of T_l and T_r, which could alternatively have been taken as definitions. That T_l and T_r are right exact would then be an immediate consequence of $Hom_\Lambda({}_\Lambda P_i, {}_\Lambda\Lambda)$ being injective in $mod\,\Lambda$ and $Hom_\Lambda(P_{i\Lambda}, \Lambda_\Lambda)$ being injective in $mod\,\Lambda^{op}$ when P_i is a projective Λ^e-module.

PROPOSITION 2.3. *We have $T_l \simeq Coker((\ , Hom_\Lambda({}_\Lambda P_0, {}_\Lambda\Lambda)) \to (\ , Hom_\Lambda({}_\Lambda P_1, {}_\Lambda\Lambda))$ and $T_r \simeq Coker((\ , Hom_\Lambda({}_\Lambda P_0, {}_\Lambda\Lambda)) \to (\ , Hom_\Lambda({}_\Lambda P_1, {}_\Lambda\Lambda))$.*

PROOF. Let $\cdots \to P_1 \to P_0 \to \Lambda \to 0$ be a minimal projective Λ^e-resolution of Λ, and let C be in $mod\,\Lambda$. We have the exact commutative diagram

$$\begin{array}{ccccccc}
(P_0 \otimes_\Lambda C, \Lambda) & \to & (P_1 \otimes_\Lambda C, \Lambda) & \to & T_l(C) & \to & 0 \\
\wr\| & & \wr\| & & \| & & \\
(C, ({}_\Lambda P_0, {}_\Lambda\Lambda)) & \to & (C, ({}_\Lambda P_1, {}_\Lambda\Lambda)) & \to & T_l(C) & \to & 0
\end{array}$$

□

We note that if Λ is hereditary and k is a perfect field, then $T_l(C) \simeq TrC$ for all C in $mod\,\Lambda$. For then we know that $0 \to P_1 \to P_0 \to \Lambda \to 0$ is exact since $pd_{\Lambda^e}\,\Lambda = gl.dim.\,\Lambda = 1$.

3. Projective modules

We have seen in Lemma 2.1 that if a two-sided Λ-module ${}_\Lambda C_\Lambda$ is a projective Λ^e-module, then ${}_\Lambda C$ and C_Λ are projective Λ and Λ^{op}-modules respectively. This suggests studying the subcategory $\mathcal{B}$ of Λ^e-modules which are projective in both

$mod\,\Lambda$ and $mod\,\Lambda^{op}$. We begin by pointing out that in general $\mathcal{B}$ is quite different from the subcategory of $mod\,\Lambda^e$ consisting of the projective Λ^e-modules.

Clearly the Λ^e-module Λ is in $\mathcal{B}$. However, it is well known that Λ is a projective Λ^e-module if and only if Λ^e is a semisimple algebra. Therefore $\mathcal{B}$ consists precisely of the projective Λ^e-modules if and only if Λ^e is a semisimple algebra. In particular, if Λ is not semisimple, then Λ in $\mathcal{B}$ is not a projective Λ^e-module. Moreover, it is easily seen that $\mathcal{B}$ is a resolving subcategory of $mod\,\Lambda^e$, i.e. $\mathcal{B}$ contains the projective Λ^e-modules, is closed under extensions and kernels of epimorphisms. From this it follows that $\mathcal{B}$ also contains $\Omega^i_{\Lambda^e}(\Lambda)$ for all $i \geq 0$. To further illustrate the point that the category $\mathcal{B}$ can be quite large and complex, we describe $\mathcal{B}$ in the case $\Lambda = k[x]/(x^2)$ with k an algebraically closed field.

Example. Let $\Lambda = k[x]/(x^2)$ where k is an algebraically closed field. Then $\Lambda^e \simeq k[x,y]/(x^2,y^2)$, and Λ^e is the only indecomposable projective Λ^e-module. On the other hand we claim that the indecomposable objects in the category $\mathcal{B}$ are exactly Λ^e and the modules in the regular components indexed by some $\lambda \neq 0$ in k.

For if an indecomposable Λ-bimodule ${}_\Lambda X_\Lambda$ is a projective Λ-module it must have even length and is hence regular or projective. We know that Λ^e is in $\mathcal{B}$. If $X = \Lambda^e/(x+\lambda y)$ for $\lambda \neq 0$, then X is not annihilated by x or y, so X is not semisimple over $k[x]/(x^2)$ or $k[y]/(y^2)$. Since X has length 2, it must then be free. Using that $\mathcal{B}$ is closed under extensions, it follows that the objects in the regular components indexed by $\lambda \neq 0$ are in $\mathcal{B}$. Finally it is easy to see that none of the objects in the remaining two regular components are in $\mathcal{B}$.

It is also worth noting that if B and C are in $\mathcal{B}$, then $B\otimes_\Lambda C$ is also in $\mathcal{B}$. This follows from the fact that the functors $C\otimes_\Lambda\ $, $\ \underset{\Lambda}{\otimes} C$, $B\otimes_\Lambda$ and $\underset{\Lambda}{\otimes} B$ are all exact and hence the compositions $B\otimes_\Lambda(C \underset{\Lambda}{\otimes}\ )$ and $(\ \otimes_\Lambda B) \underset{\Lambda}{\otimes} C$ are exact.

It is sometimes useful to be able to tell which modules in $\mathcal{B}$ are projective Λ^e-modules. In Lemma 2.1 we saw that if ${}_\Lambda P_\Lambda$ is a projective Λ^e-module, then $P \underset{\Lambda}{\otimes} C$ is a projective left Λ-module for all left Λ-modules C and $D \underset{\Lambda}{\otimes} P$ is a projective right Λ-module for all right Λ-modules D. The rest of this section is devoted to proving the converse of this statement when k is a perfect field. Specifically we prove the following more general result.

THEOREM 3.1. *Let Λ and Γ be finite dimensional algebras over a perfect field k and ${}_\Gamma X_\Lambda$ a Γ-Λ-bimodule with the property that ${}_\Gamma X$ is projective in* $\mathrm{mod}\,\Gamma$ *and $(\Gamma/ rad\,\Gamma)\otimes_\Gamma X$ is projective in* $\mathrm{mod}\,\Lambda^{op}$. *Then ${}_\Gamma X_\Lambda$ is a projective $\Gamma\otimes_k \Lambda^{op}$-module.*

For this we need the following preliminary result. Note that $(\Gamma\otimes_k\Lambda^{op})^{op} \simeq \Gamma^{op}\otimes_k\Lambda$.

LEMMA 3.2. *Let X be a Γ-Λ-bimodule and let C be in* $\operatorname{mod}\Lambda$ *and B in* $\operatorname{mod}\Gamma^{op}$. *Then we have the following.*

a) $(B\otimes_k C)\otimes_{\Gamma\otimes_k\Lambda^{op}} X \simeq B\otimes_\Gamma X\otimes_\Lambda C$, *and the isomorphism is functorial in B and C.*
b) *If ${}_\Gamma X$ is projective, we have an isomorphism* $\operatorname{Tor}_i^{\Gamma\otimes_k\Lambda^{op}}(B\otimes_k C, X)\simeq \operatorname{Tor}_i^\Lambda(B\otimes_\Gamma X, C)$ *which is functorial in B, C and X.*

PROOF. a) It is easy to show that $\alpha : (B\otimes_k C)\otimes_{\Gamma\otimes_k\Lambda^{op}} X \to B\otimes_\Gamma X\otimes_\Lambda C$ given by $\alpha((b\otimes c)\otimes x) = b\otimes x\otimes c$ is well defined, and is an isomorphism functorial in B and C.

b) Let $\cdots \to Q_i \to Q_{i-1} \to \cdots \to Q_0 \to X \to 0$ be a projective $\Gamma\otimes_k\Lambda^{op}$-resolution of X and denote by $Q^\bullet$ the complex $\cdots \to Q_i \to Q_{i-1} \to \cdots \to Q_0 \to 0$. Since ${}_\Gamma X$ is projective, the sequence splits in $\operatorname{mod}\Gamma$. Hence we have an exact sequence $\cdots B\otimes_\Gamma Q_i \to B\otimes_\Gamma Q_{i-1} \to \cdots \to B\otimes_\Gamma Q_0 \to B\otimes_\Gamma X \to 0$ which is a projective resolution in $\operatorname{mod}\Lambda^{op}$. Hence $H_i((B\otimes_\Gamma Q^\bullet)\otimes_\Lambda C) = \operatorname{Tor}_i(B\otimes_\Gamma X, C)$. Since by a) $B\otimes_\Gamma Q^\bullet\otimes_\Lambda C \simeq (B\otimes_k C)\otimes_{\Gamma\otimes_k\Lambda^{op}}\otimes Q^\bullet$, we have $\operatorname{Tor}_i^\Lambda(B\otimes_\Gamma X, C)\simeq \operatorname{Tor}_i^{\Gamma\otimes_k\Lambda^{op}}(B\otimes_k C, X)$. □

Proof of Theorem 3.1.Since k is a perfect field, $(\Gamma/\operatorname{rad}\Gamma)\otimes_k(\Lambda/\operatorname{rad}\Lambda)^{op}$ is semisimple and hence $(\Gamma\otimes_k\Lambda^{op})/\operatorname{rad}(\Gamma\otimes_k\Lambda^{op}) \simeq \Gamma/\operatorname{rad}\Gamma\otimes_k(\Lambda/\operatorname{rad}\Lambda)^{op}$. Then by Lemma 3.2 $\operatorname{Tor}_i^{\Gamma\otimes_k\Lambda^{op}}((\Gamma/\operatorname{rad}\Gamma)\otimes_k(\Lambda/\operatorname{rad}\Lambda)^{op}, X) \simeq \operatorname{Tor}_i^\Lambda((\Gamma/\operatorname{rad}\Gamma)\otimes_\Gamma X, \Lambda/\operatorname{rad}\Lambda)$, which is 0 since $(\Gamma/\operatorname{rad}\Gamma)\otimes_\Gamma X$ is projective in $\operatorname{mod}\Lambda^{op}$ by assumption. This shows that X is a projective $\Gamma\otimes_k\Lambda^{op}$-module.

COROLLARY 3.3. *Let Λ and Γ be finite dimensional algebras over a perfect field k, and let X be in* $\operatorname{mod}(\Gamma\otimes_k\Lambda^{op})$. *Then the following are equivalent.*

a) *X is projective in* $\operatorname{mod}(\Gamma\otimes_k\Lambda^{op})$.
b) *X_Λ is projective and $C\otimes_\Gamma X$ is projective in* $\operatorname{mod}\Lambda^{op}$ *for all C in* $\operatorname{mod}\Gamma^{op}$.

We point out that if k is not a perfect field, then Corollary 3.3 does not hold in general for finite dimensional k-algebras.

Example. Let k be a nonperfect field and let $K = k[x]$ be a purely inseparable field extension of k. Then $K\otimes_k K$ is a local ring which is not a field. Since K is a field, all bimodules ${}_K M_K$ are free over K as left and right module. Choose a nonfree $K\otimes_k K$-module M. Then $M\otimes_K X$ is a free K^{op}-module for all X in $\operatorname{mod}K$. Hence M satisfies condition b) in Corollary 3.3, but not condition a).

4. Representation by bimodules

Let, as in section 2, T_l and T_r denote the right exact functors lifting Tr which are obtained via the minimal projective Λ^e- resolution of Λ. Then $T_r D : \operatorname{mod}\Lambda \to \operatorname{mod}\Lambda$ and $T_l D : \operatorname{mod}\Lambda^{op} \to \operatorname{mod}\Lambda^{op}$ are also right exact and DT_l :

$mod\,\Lambda \to mod\,\Lambda$ and $DT_r : mod\,\Lambda^{op} \to mod\,\Lambda^{op}$ are left exact. We shall show that $T_rD = Tr_{\Lambda^e}\Lambda \otimes\ ,T_lD = \ \otimes_\Lambda Tr_{\Lambda^e}\Lambda\ , DT_l = \mathrm{Hom}_\Lambda({}_\Lambda Tr_{\Lambda^e}\Lambda,\ )$ and $DT_r = Hom_\Lambda((Tr_{\Lambda^e}\Lambda)_\Lambda,\ )$. Note that we have immediately that since T_rD and T_lD are right exact, then $T_rD = T_rD(\Lambda)\otimes_\Lambda$ and $T_lD = \ \otimes_\Lambda T_lD(\Lambda)$. We start out with some preliminary results.

LEMMA 4.1. *Let Λ be an artin algebra, ${}_\Lambda X_\Lambda$ a Λ-bimodule and U_Λ, V_Λ right Λ-modules.*

a) *We have a group isomorphism α : $\mathrm{Hom}_{\Lambda^e}(X, \mathrm{Hom}_k(U_\Lambda, V_\Lambda)) \to Hom_\Lambda(U\otimes_\Lambda X, V_\Lambda)$ functorial in X, U, V given by $\alpha(t)(u\otimes x) = t(x)(u)$ for $t : X \to \mathrm{Hom}_k(U_\Lambda, V_\Lambda)$.*
b) *If ${}_\Lambda U_\Lambda$ and ${}_\Lambda V_\Lambda$ are in addition Λ-bimodules, the left structures induce a Λ-bimodule structure on both modules in a), and α is a Λ-bimodule isomorphism.*

PROOF. a) Define β : $Hom_\Lambda(U\otimes_\Lambda X, V) \to Hom_{\Lambda^e}(X, \mathrm{Hom}_k(U_\Lambda, V_\Lambda))$ by $\beta(s)(x)(u) = s(u\otimes x)$, for $s : U\otimes_\Lambda X \to V$. It is easy to check that the map α is well defined and a group homomorphism and that $\alpha\beta = \mathrm{id}$ and $\beta\alpha = \mathrm{id}$. Hence it follows that α is a group isomorphism which is clearly functorial in X, U, V.

b) It is also routine to check that α is a Λ-bimodule map. □

As a special case we get the following.

COROLLARY 4.2. *Let Λ be a finite dimensional k-algebra and ${}_\Lambda X_\Lambda$ a Λ- bimodule. Then we have a Λ-bimodule isomorphism α : $\mathrm{Hom}_{\Lambda^e}(X, \mathrm{Hom}_k(D\Lambda, \Lambda)) \xrightarrow{\sim} \mathrm{Hom}_\Lambda(D\Lambda\otimes_\Lambda X, \Lambda_\Lambda)$.*

PROOF. Let ${}_\Lambda U_\Lambda = {}_\Lambda D\Lambda_\Lambda$ and ${}_\Lambda V_\Lambda = {}_\Lambda\Lambda_\Lambda$ in Lemma 4.1. □

To investigate further we shall need an appropriate reformulation of $Hom_k(D\Lambda, \Lambda)$. This will be a special case of the following, which is easy to check.

LEMMA 4.3. *Let R and S be artin algebras and ${}_SP_R$ and ${}_RX_S$ bimodules, where ${}_RP$ is projective. Then the natural isomorphism δ : ${}_SP_R \otimes_R ({}_RX_S) \to \mathrm{Hom}_R({}_RP^*_S, {}_RX_S)$, where $P^* = \mathrm{Hom}_R(P, R)$ and $\delta(p\otimes x)(f) = f(p)x$ for $p \in P, x \in X$ and $f \in P^*$ is an isomorphism of S-bimodules.*

COROLLARY 4.4. *For a finite dimensional k-algebra Λ we have a natural Λ-bimodule isomorphism $\delta : \Lambda \otimes_k \Lambda \to Hom_k(D\Lambda, \Lambda)$.*

PROOF. We use Lemma 4.3 with $R = k, S = \Lambda, {}_SR_R = {}_\Lambda\Lambda_k$ and ${}_RX_S = {}_k\Lambda_\Lambda$. □

We are now in the position to prove our desired bimodule isomorphism. Observe that we have a ring isomorphism $h : \Lambda^e \to (\Lambda^e)^{op}$ given by $h(x\otimes y) = y\otimes x$, inducing an equivalence of categories between $mod\,\Lambda^e$ and $mod(\Lambda^e)^{op}$. We view a left Λ^e-module as a right Λ^e-module via this isomorphism.

THEOREM 4.5. *We have isomorphisms of bimodules* $T_rD({}_\Lambda\Lambda) \simeq T_lD(\Lambda_\Lambda) \simeq Tr_{\Lambda^e}({}_{\Lambda^e}\Lambda) \simeq Tr_{\Lambda^e}(\Lambda_{\Lambda^e})$.

PROOF. Let $P_1 \to P_0 \to \Lambda \to 0$ be a minimal projective Λ^e-presentation of Λ, and consider the induced projective presentation $D\Lambda\otimes_\Lambda P_1 \to D\Lambda\otimes_\Lambda P_0 \to D\Lambda \to 0$ of $(D\Lambda)_\Lambda$ in $mod\,\Lambda^{op}$. Using Corollaries 4.2 and 4.4 we get the following exact commutative diagram

$$\begin{array}{ccccccc}
Hom_\Lambda((D\Lambda\otimes_\Lambda P_0)_\Lambda, \Lambda_\Lambda) & \to & Hom_\Lambda((D\Lambda\otimes_\Lambda P_1)_\Lambda, \Lambda_\Lambda) & \to & T_rD\Lambda & \to 0 \\
\wr\| & & \wr\| & & \| & \\
Hom_{\Lambda^e}(P_0, Hom_k(D\Lambda, \Lambda) & \to & Hom_{\Lambda^e}(P_1, Hom_k(D\Lambda, \Lambda)) & \to & T_rD\Lambda & \to 0 \\
\wr\| & & \wr\| & & \wr\| & \\
Hom_{\Lambda^e}(P_0, \Lambda^e) & \to & Hom_{\Lambda^e}(P_1, \Lambda^e) & \to & Tr_{\Lambda^e}(\Lambda) & \to 0
\end{array}$$

Viewing P_0, P_1 and Λ^e as right Λ^e-modules in the last row, it is easy to see that $Tr_{\Lambda^e}({}_{\Lambda^e}\Lambda) \simeq Tr_{\Lambda^e}(\Lambda_{\Lambda^e})$. We here use that the modules ${}_{\Lambda^e}\Lambda$ and Λ_{Λ^e} with the natural Λ^e-actions correspond under the equivalence between $mod\,\Lambda^e$ and $mod(\Lambda^e)^{op}$ induced by the ring isomorphism $h : \Lambda^e \to (\Lambda^e)^{op}$.

Similarly we get $T_lD(\Lambda) \simeq Tr_{\Lambda^e}\Lambda$. □

This provides the following information on the functors T_rD and DT_l. Here we write $Tr_{\Lambda^e}(\Lambda)$ for $Tr_{\Lambda^e}({}_{\Lambda^e}\Lambda) \simeq Tr_{\Lambda^e}(\Lambda_{\Lambda^e})$.

COROLLARY 4.6. *For a finite dimensional k-algebra* Λ, $T_rD : \mathrm{mod}\,\Lambda \to \mathrm{mod}\,\Lambda$ *is isomorphic to* $Tr_{\Lambda^e}\Lambda \otimes \;\;$, *and* $DT_l : \mathrm{mod}\,\Lambda \to \mathrm{mod}\,\Lambda$ *is isomorphic to* $Hom_\Lambda(Tr_{\Lambda^e}\Lambda, \;\;)$.

PROOF. Since T_rD is right exact, we have $T_rD = T_rD(\Lambda) \otimes \;\;$. Then we use Theorem 4.5 for the first claim.

We have for C in $mod\,\Lambda$ the exact sequences $P_1\otimes_\Lambda C \to P_0\otimes_\Lambda C \to C \to 0$ and

$$\begin{array}{ccccccc}
0 & \to & DT_lC & \to & D(Hom_\Lambda(P_1\otimes_\Lambda C, \Lambda)) & \to & D(Hom_\Lambda(P_0\otimes_\Lambda C, \Lambda)) \\
& & & & \wr\| & & \wr\| \\
& & & & (D\Lambda\otimes_\Lambda P_1)\otimes_\Lambda C & \to & (D\Lambda\otimes_\Lambda P_0)\otimes_\Lambda C.
\end{array}$$

The exact sequence $Hom_\Lambda(D\Lambda\otimes_\Lambda P_0, \Lambda) \to Hom_\Lambda(D\Lambda\otimes_\Lambda P_0, \Lambda) \to T_rD\Lambda \to 0$ gives rise to the exact sequence $0 \to Hom_\Lambda(T_rD\Lambda, C) \to$

$Hom_\Lambda(Hom_\Lambda(D\Lambda \otimes_\Lambda P_1, \Lambda), C) \to Hom_\Lambda(Hom_\Lambda(D\Lambda \otimes_\Lambda P_0, \Lambda), C)$. Since we have natural isomorphisms $(D\Lambda \otimes_\Lambda P_i) \otimes_\Lambda C \to Hom_\Lambda(Hom_\Lambda(D\Lambda \otimes_\Lambda P_i, \Lambda), C)$ because $D\Lambda \otimes_\Lambda P_i$ is a projective Λ^{op}-module, we conclude that $DT_l \simeq Hom_\Lambda(T_r D\Lambda, \ ) \simeq Hom_\Lambda(Tr_{\Lambda^e}\Lambda, \ )$. □

5. Preprojective algebras

We illustrate the advantage of being able to lift TrD to a functor $T_r D$: $mod\,\Lambda \to mod\,\Lambda$ by studying generalizations of preprojective algebras. We here write TD istead of $T_r D$.

Let Λ be as usual a finite dimensional algebra over a field k. If Λ is hereditary, there is an associated preprojective algebra $\Pi(\Lambda)$ introduced by Gelfand and Riedtmann and generalized by Dlab-Ringel. From the work of Baer-Geigle-Lenzing, $\Pi(\Lambda)$ can be defined as the Z-graded algebra $\Pi(\Lambda) = \Lambda \amalg TrD\Lambda \amalg \cdots \amalg (TrD)^i\Lambda \amalg \cdots$ where multiplication is given as follows. Viewing the natural isomorphism $(TrD)^i\Lambda \to Hom_\Lambda(\Lambda, (TrD)^i\Lambda)$ as an identification, given $f : \Lambda \to TrD^i\Lambda$ and $g : \Lambda \to TrD^j\Lambda$, we have since TrD : $mod\,\Lambda \to mod\,\Lambda$ is a functor for hereditary algebras, an induced map $TrD^j(f) : TrD^j\Lambda \to TrD^{i+j}\Lambda$. Then we define fg to be $TrD^j(f)g$ [**2**].

We recall how ring theoretic properties of $\Pi(\Lambda)$ reflect the representation theory of Λ.

PROPOSITION 5.1. *Let Λ be a hereditary k-algebra.*

a) *Λ is of finite representation type if and only if $\Pi(\Lambda)$ is an artin algebra.*

b) *Λ is of finite or tame representation type if and only if $\Pi(\Lambda)$ is noetherian.*

Part a) follows directly from the definition in [**2**] (see also [**2**] for earlier references), and part b) from [**4**].

As pointed out to us by Lenzing, we can use the functor TD : $mod\,\Lambda \to mod\,\Lambda$ discussed in this paper to give the following definition of a preprojective algebra of an arbitrary finite dimensional k-algebra:

$\Pi(\Lambda) = \Lambda \amalg (TD)\Lambda \amalg \cdots \amalg (TD)^i\Lambda \amalg \cdots$, and the multiplication is given as follows. We view the natural isomorphism $(TD)^i\Lambda \to Hom_\Lambda(\Lambda, (TD)^i\Lambda)$ as an identification. For $f : \Lambda \to (TD)^i\Lambda$ and $g : \Lambda \to (TD)^j\Lambda$ we define $f \cdot g = (TD)^j(f)g$.

Note that since we have seen that $TD(C) \simeq TrD(C)$ for all C in $mod\,\Lambda$ when Λ is hereditary, this gives a generalization of the ordinary preprojective algebra.

The following result shows that Proposition 5.1 does not generalize to arbitrary artin algebras.

PROPOSITION 5.2. *Let Λ be an indecomposable Nakayama algebra over an algebraically closed field k. Then $\Pi(\Lambda)$ is noetherian, and it is finite dimensional if and only if Λ has a simple projective module.*

For this we first need the following preliminary result.

LEMMA 5.3. *Let Λ be an indecomposable Nakayama algebra over an algebraically closed field k. Let $P_1, \dots, P_n$ be the indecomposable projective Λ-modules, ordered in such a way that $P_{i+1}/\underline{r}P_{i+1} \simeq \underline{r}P_i/\underline{r}^2 P_i$ for $i = 1, \dots, n-1$ and $P_1/\underline{r}P_1 \simeq \underline{r}P_n/\underline{r}^2 P_n$ if P_n is not simple. Let A_j be an indecomposable Λ-module with $A_j/\underline{r}A_j \simeq P_j/\underline{r}P_j$.*

a) *$TDA_j \simeq TrDA_j$ if and only if A_j is not injective or A_j is injective and $A_j/\underline{r}A_j$ is injective.*
b) *If A_j is injective and $A_j/\underline{r}A_j$ is not injective, then $TD(A_j) \simeq P_{i-1}$.*

PROOF. Let $S_i = P_i/\underline{r}P_i$ and $S' = D(S_i)$. We write $A_j = \begin{pmatrix} S_j \\ \vdots \\ S_{j+t} \end{pmatrix}$ and $D(A_j) = \begin{pmatrix} S'_{j+t} \\ \vdots \\ S'_j \end{pmatrix}$. Then one can check that any artin algebra Λ has a minimal projective Λ^e-presentation of the form $\coprod_{i\cdot \xrightarrow{\alpha} t} \Lambda e_t \otimes_k e_i\Lambda \xrightarrow{h} \coprod_i \Lambda e_i \otimes_k e_i\Lambda \to \Lambda \to 0$, as pointed out to us by E. Green. Here $i\cdot \xrightarrow{\alpha} t$ is an arrow in the ordinary quiver for Λ and h is defined by $h(e_t \otimes e_i) = e_t \otimes \alpha e_i - e_t\alpha \otimes e_i = e_t \otimes e_t\alpha - \alpha e_i \otimes e_i$. Then consider the projective presentation $P_1 = D(A_j)\otimes_\Lambda(\coprod_{i\xrightarrow{\alpha}t} \Lambda e_t \otimes_k e_i\Lambda) \to P_0 = (\coprod_i \Lambda e_i \otimes_k e_i\Lambda) \to D(A_j) \to 0$ of $D(A_j)$ in $\operatorname{mod}\Lambda^{op}$. We then have $D(A_j)\otimes_\Lambda \Lambda e_i$ copies of $e_i\Lambda$ in P_0, that is, the number is the number of composition factors of S'_i in $D(A_j)$. Hence we have $P_0 = e_j\Lambda \amalg \cdots \amalg e_{j+t}\Lambda$. Further, for $i \xrightarrow{\alpha} t$, $Dim_k(DA_j\otimes_\Lambda \Lambda e_t)$ gives the number of copies of $e_i\Lambda$ in P_1. Hence we get $P_1 = e_{j-1}\Lambda \amalg \cdots \amalg e_{j+t-1}\Lambda$ if S_j is not simple injective and $P_1 = e_j\Lambda \amalg \cdots \amalg e_{j+t-1}\Lambda$ if S_j is simple injective. So in the first case we can use the presentation $e_{j-1}\Lambda \to e_{j+t}\Lambda \to DA \to 0$ to compute TDA_j and in the second case we can use $0 \to e_{j+t}\Lambda \to DA_j \to 0$. If S_j is simple injective, we then get $TDA_j = 0$. If S_j is not simple injective, we get $TDA_j = TrDA_j$ if A_j is injective. This finishes the proof of the lemma. $\square$

Returning to the proof of Proposition 5.2 assume that Λ has a simple projective module, that is, we have the quiver $\underset{1}{\cdot} \to \underset{2}{\cdot} \ \cdots \ \cdot \to \underset{n}{\cdot}$. Then S_1 is the only simple injective Λ-module. If for an indecomposable module A_i we have $A_i/\underline{r}A_i \simeq S_i$, then if $i > 1$ it follows that TDA_i is indecomposable and a factor of P_{i-1} and $TDA_1 = 0$. This shows that for every indecomposable Λ-module A

there is some n such that $(TD)^n A = 0$. It follows that $\Pi(\Lambda)$ is finite dimensional in this case.

Assume now that Λ has no simple projective module, that is, we have the quiver

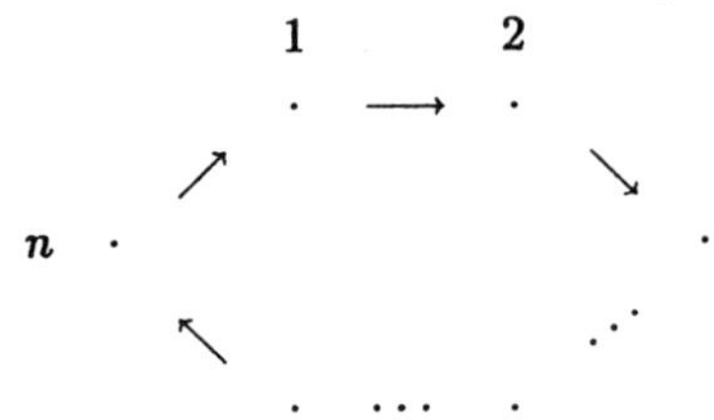

. Then there are no simple injective modules. For the indecomposable projective module P_i we have $TD(P_i) = P_{i-1}/K_{i-1}$ for some Λ-module K_{i-1}, and hence $TD(\Lambda) = \Lambda/K$ for some Λ-module K. This shows that $\Pi(\Lambda)$ is a factor of the polynomial ring $\Lambda[x]$ and is hence noetherian.

Since for any $A \neq 0$ we have $TDA \neq 0$, it follows that $\Pi(\Lambda)$ is not artin in this case.

The following result shows that $\Pi(\Lambda)$ is not necessarily noetherian when Λ is tame.

PROPOSITION 5.4. *Let Λ be a local selfinjective algebra over an algebraically closed field k. Then $\Pi(\Lambda)$ is noetherian if and only if Λ is of finite representation type.*

PROOF. If Λ is of finite representation type, it is Nakayama, and hence $\Pi(\Lambda)$ is noetherian by Proposition 5.2.

If Λ is not of finite representation type, Λ is not Nakayama, and hence $\dim_k \underline{r}/\underline{r}^2 = n \geq 2$ where $\underline{r} = rad\,\Lambda$. We then have a minimal projective Λ^e-presentation of Λ:

$$n(\Lambda \otimes_k \Lambda) \to \Lambda \otimes_k \Lambda \to \Lambda \to 0.$$

Letting t be the length of Λ, this is as a sequence of Λ-modules: $nt\Lambda \to t\Lambda \to \Lambda \to 0$. Hence $TD\Lambda = T\Lambda = (nt - t + 1)\Lambda = (t(n-1)+1)\Lambda$, so that $\Pi(\Lambda)$ is the free algebra $k < x_1, \dots, x_r >$, where $r = t(n-1)+1 \geq 2$, and consequently $\Pi(\Lambda)$ is not noetherian. □

It would be interesting to know if $\Pi(\Lambda)$ is noetherian when Λ is of finite representation type, and if $\Pi(\Lambda)$ being noetherian implies that Λ is of finite or tame representation type.

It is not difficult to check that the graded subgroup $\underline{a} = \tau(\Lambda)\amalg\tau(TD\Lambda)\amalg\dots$ of $\Pi(\Lambda)$ is a two-sided ideal of $\Pi(\Lambda)$ where $\tau(X)$ denotes the trace of the nonpreprojective modules in X for all Λ-modules X (see [1] for definition of preprojective modules). S.Smalø has suggested that $\Pi_1(\Lambda) = \Pi(\Lambda)/\underline{a}$ should be an interesting ring to study in connection with representation theory. For example $\underline{a} = 0$, that is $\Pi_1(\Lambda) = \Pi(\Lambda)$, when Λ is hereditary or of finite representation type.

References

1. M. Auslander and S.O. Smalø, *Preprojective modules over artin algebras*, J. Algebra 66 (1980) 61-122
2. D. Baer, W. Geigle and H. Lenzing, *The preprojective algebra of a tame hereditary artin algebra*, Comm. in Alg. 15 (1987) 425-457
3. E. Green, *Coxeter functors revisited*, preprint 1979 (unpublished)
4. H. Lenzing, *Homological transfer from finitely presented to infinite modules*, Abelian Group Theory, Lecture Notes in Math, Springer-Verlag 1006 (1983) 734-761

Maurice Auslander, Department of Mathematics, Brandeis University, Waltham, Mass. 02254, USA

E-mail: Auslander@brandeis.bitnet

Idun Reiten, Institutt for matematikk og statistikk, Universitetet i Trondheim, AVH, 7055 Dragvoll, Norway

E-mail: Idun.Reiten@avh.unit.no

Canadian Mathematical Society
Conference Proceedings
Volume 11 (1991)

POSETS WITH AN EQUIVALENCE RELATION OF TAME TYPE AND OF FINITE GROWTH

V.M.Bondarenko, A.G.Zavadskij

ABSTRACT. Two theorems for posets with an equivalence relation are proved: a tameness criterion and a finite-growthness criterion.

0. INTRODUCTION. In the present paper representations of posets with an equivalence relation are considered. These representations (called also representations of weakly completed posets) were introduced in 1973 by L.A.Nazarova and A.V.Roiter [5].

Posets with an equivalence relation of finite representation type were described in [7], and these of tame type and finite growth, in the shurian case, in [4] (for ordinary posets, which obviously are shurian, it has been done earlier respectively in [2,3,11]).

Our aim here is to present a tameness criterion (Theorem 1) and a finite-growthness criterion (Theorem 2) for arbitrary posets with an equivalence relation */. The theorems generalize the results obtained in [3,4,11] for the shurian case. The first criterion is formulated (with help of a suitable terminology) in the same form as in the shurian case [4], and the second one in a new form.

In order to simplify considerations we put some natural restrictions on posets (see the conditions 1) and 2) in the definition of a reduced poset before Theorem 1). In fact they do not decrease a generality, as one can see in the section 3.

*/A part of this contents (concerning Theorem 1) was also submitted to Proceedings of Novosibirsk-89 Conference on Algebra.

1980 Mathematics Subject Classification (1985 Revision). 16A64

The algorithm of differentiation, used for the proofs, is described in the paper [8] being submitted to these proceedings too.

1. FORMULATION OF THE RESULTS AND EXAMPLES. A poset with an equivalence relation is a finite partially ordered set $\mathfrak{M}$ together with some equivalence relation defined on the set of all its points (in a general case the order relation and the equivalence relation are arbitrary and don't depend from each other).

Definitions of representations of these posets, their dimensions, etc. are contained in [8], and we shall use them. Denote by $r(x)$ the number of elements in the equivalence class of a point $x \in \mathfrak{M}$. A point x will be called small if $r(x)=1$, and big if $r(x) \geq 2$ (small and big points are pictured respectively by o and ●).

Posets with an equivalence relation of tame type and of finite growth are introduced naturally. Namely, in the case of infinite field k denote by $m(d)$ the smallest number of representations of $\mathfrak{M}$ over the polynomial ring $k[X]$ generating (see [8,11]) almost all indecomposable representations of $\mathfrak{M}$ over k of a given dimension d. Then $\mathfrak{M}$ is called tame (of finite growth) over k if $m(d) < \infty$ for all d (if $\sup_d m(d) < \infty$). In the case of a finite field k one has to pass to some its infinite extension k'.

A poset $\mathfrak{M}$ will be called reduced if two conditions hold:

1) $r(x) \leq 2$ for all $x \in \mathfrak{M}$;

2) every big point is contained in a subset of the form

● ● or ● o o .

Denote by $\widetilde{\mathfrak{M}}$ the ordinary poset (i.e. without any equivalence relation) obtained from a reduced poset $\mathfrak{M}$ by replacing every big point x by a pair of small incomparable points x', x'' inheriting all the relations of the point x in $\mathfrak{M}$.

<u>THEOREM 1</u>. For a reduced poset $\mathfrak{M}$ with an equivalence relation the following statements are equivalent:

a) $\mathfrak{M}$ is tame (over an arbitrary field k);

b) $\mathfrak{M}$ doesn't contain subsets of the form $N_1=(1,1,1,1,1)^{\circ}$, $N_2=(1,1,1,2)$, $N_3=(2,2,3)$, $N_4=(1,3,4)$, $N_5=(N,5)$, $N_6=(1,2,6)$ and

$N_7=$ ● o o o , $N_8=$ ● o 8 , $N_9=$ ● ● o , $N_{10}=$ ● ● ● ;

c) $\widetilde{\mathfrak{M}}$ doesn't contain subsets $N_1,\dots,N_6$.

REMARK. A similar criterion was proved in [4] for the shurian case, i.e. for posets satisfying the above condition 1) and (instead of 2)) the following more strong condition:

2') equivalent points aren't comparable.

Now we pass to a formulation of a finite-growthness criterion. We'll write $x \sim y$ if x,y are equivalent and $x \neq y$. We'll write $x-y$ if x,y are arbitrary incomparable points. For subsets $X,Y \subset \mathfrak{M}$, a designation $X-Y$ means that $x-y$ for all $x \in X$, $y \in Y$ (if $X=\{x\}$ we write simply $x-Y$).

We shall consider nonoriented graphs with edges of two types: $\sim$ and $-$. Such a graph will be called a cycle (in $\mathfrak{M}$) if it has the form $x_1 \sim x_2 - x_3 \sim x_4 - \dots - x_{2n-1} \sim x_{2n}$ (with an edge $-$ joining x_{2n} and x_1) $(n \geq 1)$

A graph will be called a chain (in $\mathfrak{M}$) if it has the form $A-B$ or $A-x_1 \sim x_2 - \dots - x_{2n-1} \sim x_{2n} - B$ $(n \geq 1)$, where the vertices $A=\{a_1,a_2\}$ and $B=\{b_1,b_2\}$ (the ends of a chain) are two-element small-point subsets satisfying the conditions a_1-a_2 and b_1-b_2.

Note that, generally speaking, in chains and cycles some vertices can be repeated (for example, chains can have the same both inner vertices and ends). If a graph (chain or cycle) has at least twice the same vertices, we say that it has repetitions. A cycle is called periodical if its vertices are repeated periodically.

THEOREM 2. A tame poset $\mathfrak{M}$ with an equivalence relation satisfying the condition $r(x) \leq 2$ for all x is of finite growth if and only if all chains and nonperiodical cycles in $\mathfrak{M}$ have no repetitions and don't intersect mutually (i.e. have no common vertices as graphs).

EXAMPLES. Denote by $G(\mathfrak{M})$ a graph with the set of vertices $\{v\}$ either of the form $v=\{a\}$, where a is big, or of the form $v=\{a_1,a_2\}$, where a_1,a_2 are both small and a_1-a_2 ; and the set of arrows of the types $\sim$ and $-$ defined above. In the following examples the graph $G(\mathfrak{M})$ (or its full subgraph) is pictured, with help of which one can easily verify whether or not the conditions of Theorem 2 are satisfied:

Poset | Graph $G(\mathfrak{M})$ (or its subgraph) | Growth

1) 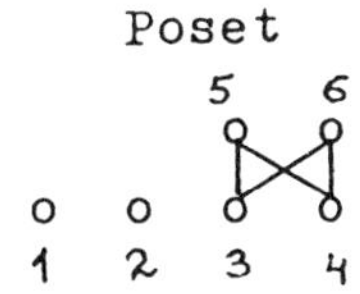

(subgraph)

∞

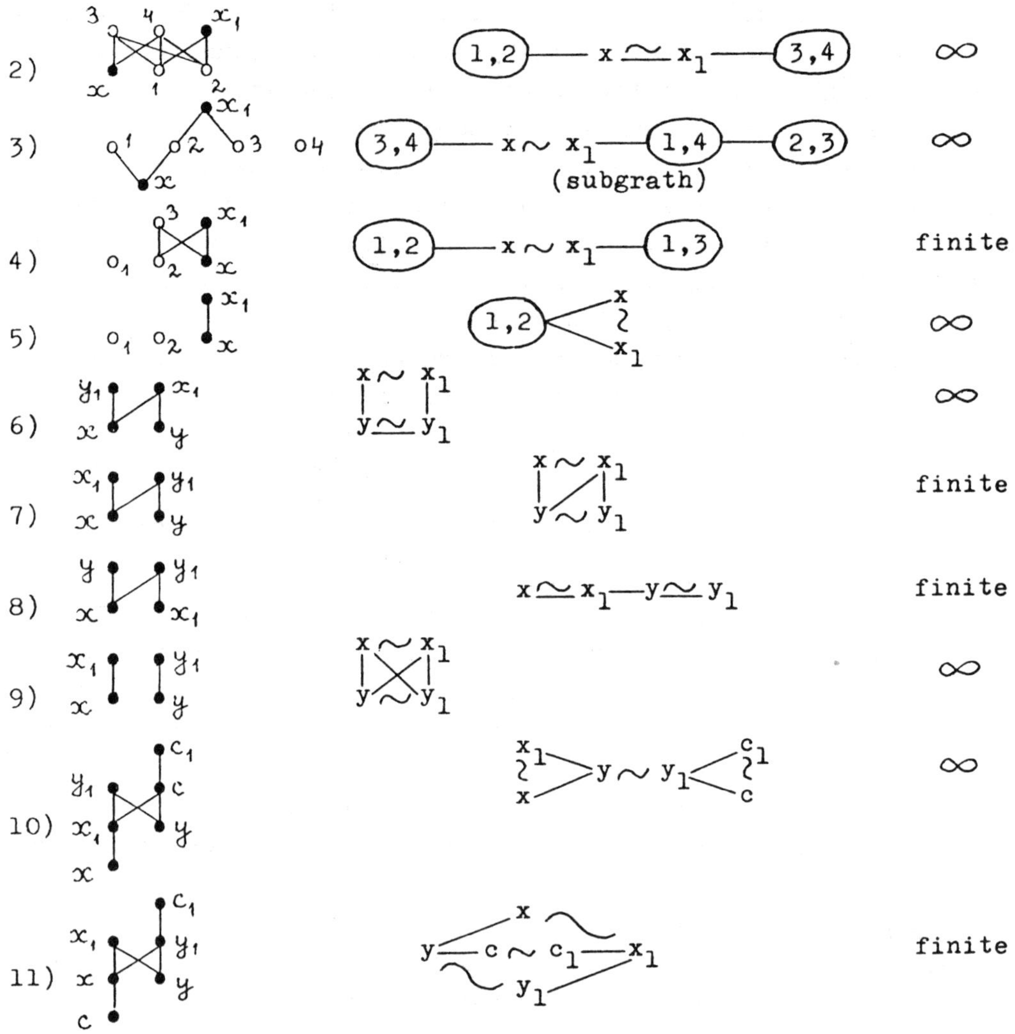

Obviously, in the cases 1) and 3) we have two chains with a common end; in the case 2) the chain (1,2) — $x \sim x_1$ — (3,4) intersects a cycle $x \simeq x_1$; in the case 5) the chain (1,2) — $x \sim x_1$ — (1,2) has the same ends; in the case 6) two cycles intersect evidently; in the case 9) the cycles $x \sim x_1 - y \sim y_1$ and $x \sim x_1 - y_1 \sim y$ intersect; in the case 10) we have a nonperiodical cycle with repetitions

$$y \sim y_1 - c \sim c_1 - y_1 \sim y \quad (x \sim x_1)$$

REMARK 1. The formulation of Theorem 2 can be modified in several ways (in particular, it can be done more laconic if we consider some generalized cycles embraced both ordinary cycles and chains). But the given form is the most convenient for practical applications in the considered class of problems.

REMARK 2. In fact Theorem 2 is true for arbitrary posets with an equivalence relation (without restrictions for r(x)) if we consider the ends of chains not only in the form $A=\{a_1,a_2\}$, but also in the form $A=\{a\}$, where r(a)=3. But, for a simplicity (and taking into account Section 3), we restrict our considerations to the case $r(x)\leqslant 2$.

2. PROOF OF THEOREM 1. Obviously, the statements b) and c) are equivalent for an arbitrary poset. Hence we have to prove an equivalence of a) and b).

b)$\Rightarrow$ a). THE SUFFICIENCY OF THE CRITERION. Assume $\mathfrak{M} \not\supset N_1, N_2,\dots,N_{10}$. Denote by $\delta(\mathfrak{M})$ the width of a poset $\mathfrak{M}$, i.e. the maximal number of its mutually incomparable points. One can suppose $\delta(\mathfrak{M})\geqslant 2$ (in the contrary case the tameness is obvious). We shall use natural properties of the differentiation algorithm [8] consisting of the main operations I,...,V and the additional ones $\alpha, \bar{\alpha}, \beta, \gamma$ (see the consequences 1 and 2 in [8]). Taking into account these properties (and also the placed below lemma 4), it is sufficient to show that there exists a pair (or a group) of points which is suitable for one of the differentiations I,...,V or $\bar{\alpha}$ and, moreover, that , for any of the differentiations I,...,V or $\alpha, \bar{\alpha}, \beta, \gamma$, the derivative poset $\mathfrak{M}'$ doesn't contain subsets $N_1,\dots,N_{10}$.

A poset with an equivalence relation will be called garland if it is an arbitrary ordinal sum of any chains (consisting both of big and small points) and two-point subsets consisting of small points.

Denote by $\min\mathfrak{M}$ ($\max\mathfrak{M}$) the set of all minimal (maximal) points of $\mathfrak{M}$.

LEMMA 1. Let $\mathfrak{N}$ be an ordinary poset, $\delta(\min\mathfrak{N})\geqslant 2$ and $\mathfrak{N}\not\supset N_1,\dots,N_6$. Then there exists a I-suitable pair of points (a,b) or a V-suitable quadruple of points (a_1,a_2,b_1,b_2), where $a_1,a_2\in \min\mathfrak{N}$.

PROOF. First recall that , according to Dilworth's theorem any poset $\mathfrak{N}$ with $\delta(\mathfrak{N})=n$ can be presented as a sum of n chains, and if $\min\mathfrak{N}=\{a_1,\dots,a_m\}$, where $m\leqslant n-1$, then one of the cones a_i^{Δ} contains at least two of Dilworth's chains of the subset $\mathfrak{N}\setminus\min\mathfrak{N}$, i.e. $\delta(\mathfrak{N}\setminus a_i^{\Delta})\leqslant n-2$.

Assume $\min\mathfrak{N}=\{a_1,\dots,a_m\}$, where $m\geqslant 2$. Clearly $\delta(\mathfrak{M})\leqslant 4$.

If $m \leq 3$, then $\delta(\mathfrak{N} \setminus a_i^\Delta) \leq 2$ for some i, and therefore (a_i, b) is a I-suitable pair of points where $b \in \max(\mathfrak{N} \setminus a_i^\Delta)$. Hence m=4 and $\min \mathfrak{N} = \{a_1, a_2, a_3, a_4\}$. By a symmetry $\delta(\max \mathfrak{N})=4$. Denote $A_{ij}=a_i^\Delta \cap a_j^\Delta + \{a_i, a_j\}$ and $B_{ij} = \mathfrak{N} \setminus A_{ij}$.

Suppose that $\delta(B_{ij}) \leq 2$ for some pair i,j, for example $\delta(B_{12}) \leq 2$. If $a_1^\Delta \cap B_{12} = a_2^\Delta \cap B_{12} = \emptyset$ then B_{12} is a garland and therefore either (a_1, b) is a I-suitable pair, where $\max B_{12} = \{b\}$, or (a_1, a_2, b_1, b_2) is a V-suitable quadruple, where $\max B_{12} = \{b_1, b_2\}$. If, for example, $x \in a_2^\Delta \cap B_{12}$ and $x < b$, where $b \in \max B_{12}$, then (a_1, b) is I-suitable. Now the proof can be finished by an application of the following fact:

(A) Let an ordinary poset $\mathfrak{N}$ be a sum of four chains $\mathfrak{N} = A_1 + \ldots + A_4$ and has exactly four minimal points $a_1, \ldots, a_4$ $(a_i \in A_i)$ and exactly four maximal points. Assume $\mathfrak{N}$ doesn't contain I-suitable pairs of points. Then there exist minimal points a_i, a_j $(i \neq j)$ such that $A_i + A_j \subset A_{ij}$, and, in particular, $\delta(B_{ij}) \leq 2$.

A proof of the statement (A) is exactly Step 1 of the proof of Lemma 1.2 from [10] (p.125-126), where the points a_i, a_j with the mentioned property were called main points (note that in [10] another letters were used and maximal points were considered instead of minimal ones). This finishes the proof of Lemma 1. The corresponding statement for posets with an equivalence relation will be obtained as a consequence of it.

LEMMA 2. Let $\mathfrak{M}$ be a reduced poset with an equivalence relation, $\delta(\min \mathfrak{M}) \geq 2$ and $\mathfrak{M} \not\supset N_1, \ldots, N_{10}$. Then there exists a pair (or a group) of points of $\mathfrak{M}$ which is suitable for one of the differentiations I,...,V or $\overline{\mathcal{L}}$.

For the proof, one can assume that $\min \mathfrak{M}$ doesn't contain big points. Really, if $a \in \min \mathfrak{M}$ is a big point, then $B = \mathfrak{M} \setminus a^\Delta$ is a garland, and therefore either (a,b) is $\overline{\mathcal{L}}$-suitable, II-suitable, III-suitable pair in the case $\max B = \{b\}$, or (a, b_1, b_2) is a IV-suitable triple in the case $\max B = \{b_1, b_2\}$.

Consider now an ordinary poset $\widetilde{\mathfrak{M}}$ which doesn't contain subsets $N_1, \ldots, N_6$ but contains (according to Lemma 1) a I-suitable pair (a,b) or a V-suitable quadruple (a_1, a_2, b_1, b_2), where $a_1, a_2 \in \min \widetilde{\mathfrak{M}}$. Looking on the converse transition from $\widetilde{\mathfrak{M}}$ to $\mathfrak{M}$ it is easy to see that $\mathfrak{M}$ contains the mentioned in Lemma 2 group of points. This finishes the proof of the lemma.

The following combinatorial fact is very simple:

(B) Assume $\mathfrak{M} \not\supset N_1,\dots,N_{10}$. Then

1) replacing any big point of $\mathfrak{M}$ by any garland we obtain again a poset not containing subsets $N_1,\dots,N_{10}$;

2) if $\mathfrak{M}$ contains a pair of incomparable small points a_1,a_2 with property $a_1^{\Delta}\setminus a_1=a_2^{\Delta}\setminus a_2=A$, then, inserting any garland between $\{a_1,a_2\}$ and A, we obtain again a poset not containing subsets $N_1,\dots,N_{10}$.

LEMMA 3. Let $\mathfrak{M}$ be a poset with an equivalence relation, $\mathfrak{M} \not\supset N_1,\dots,N_{10}$ and $r(x) \leqslant 2$ for all $x \in \mathfrak{M}$. Then the derivative poset $\mathfrak{M}'$ doesn't contain subsets $N_1,\dots,N_{10}$ for any of the differentiations I,...,V or $\alpha, \bar{\alpha}, \beta, \gamma$.

Really, this statement follows immediatly from (B) for all differentiations except of I. But for Differentiation I it has been proved in [10], Proposition 2.1, in the main case of subsets $N_1,\dots,N_6$ (considerations for the case $N_7,\dots,N_{10}$ are trivial). Note that in [9], Sectin 2.4, there is contained a more simple proof of Proposition 2.1, using the Tits form.

As the condition 2) in the definition of a reduced poset can be broken by the differentiation procedure, it's necessary to explain how it can be restored. Assume that a poset $\mathfrak{M}$ contains a big point a belonging to a two-point class $\{a,a_1\}$. Let a chain $\{c_1,\dots, c_n\}$ be the set of all points which aren't comparable with the point a (possibly empty), and assume that all c_i are small. Denote by $\mathfrak{M}'_\alpha$ the poset with an equivalence relation which is obtained from $\mathfrak{M}$ by rejection the point a and replacing the point a_1 by a small-point chain of length n+2. Then the following statement takes place.

LEMMA 4. There exists a natural one-to-one correspondence between the classes of indecomposable representations of the posets $\mathfrak{M}$ and $\mathfrak{M}'_\alpha$, in particular, these posets are simultaneously either tame or wild.

An obvious proof of Lemma 4, in the matrix language, is based on reducing of the matrix corresponding to the point a (this procedure doesn't increase dimensions of representations). Note that the combinatorial fact (B) can be applied to the operation $\mathfrak{M} \to \mathfrak{M}'_\alpha$.

Applying to some poset $\mathfrak{N}$ with the condition $r(x) \leqslant 2$ for all x the operation $\mathfrak{M} \to \mathfrak{M}'_\alpha$ as many times as possible, we obtain a reduced poset $\hat{\mathfrak{N}}$. A transition from $\mathfrak{N}$ to $\hat{\mathfrak{N}}$ will be called a reducing of a poset $\mathfrak{N}$.

It is clear now that Consequences 1,2 from [8] and just obtained Lemmas 2,3,4 prove the sufficiency of the criterion.

a) $\Rightarrow$ b). THE NECESSITY OF THE CRITERION. First consider some additional fact. Let $\mathfrak{M}$ be an arbitrary poset with an equivalence relation, and the condition $r(x) \leqslant 2$ for all x holds. For any point $a \in \mathfrak{M}$ denote $U(a)=\{x \in \mathfrak{M} \mid x-a\}$. If a is big, then a_1 will denote the point which is equivalent to a.

For every big point $a \in \mathfrak{M}$ we consider a poset $\bar{U}(a)= =\{\bar{u} \mid u \in U(a)\} \cup \{u', u''\}$ consisting of some new points $\bar{u}, u', u''$, where u', u'' are small and $u' < \bar{u} < u''$ for all $\bar{u}$ (suppose $\bar{U}(a)= =\{u' < u''\}$ if $U(a)=\emptyset$). Moreover, we assume that the order relation on $\bar{U}(a) \setminus \{u', u''\}$ completely corresponds to the order relation on $U(a)$.

Let X be any big-point subset in $U(a) \setminus \{a_1\}$ satisfying the following condition: if $x \in X$ and $x_1 \in U(a)$ then $x_1 \in X$. Denote by $\mathfrak{M}_\alpha(X)$ the poset with an equivalence relation which is obtained from $\mathfrak{M}$ by rejection of the point a and replacing of the equivalent point a_1 by the subset $\bar{U}(a)$, and together with the following conditions determining the equivalence relation:

1) $\bar{u}$ is small if $u \in U(a) \setminus X$;
2) $\bar{x} \sim x_1$ if $x \in X$, $x_1 \notin U(a)$ (here the point x becomes small);
3) $\bar{x} \sim x_1$, $\bar{x}_1 \sim x$ if $x, x_1 \in X$;
4) a relation $x \sim x_1$ isn't changed if $x, x_1 \notin X$ $(x, x_1 \in \mathfrak{M} \setminus \{a_1, a_2\})$.

The following lemma will be used frequently.

LEMMA 5. If $\mathfrak{M}$ is tame (has finite growth) then $\mathfrak{M}_\alpha(X)$ also is tame (has finite growth).

One can easily prove this lemma considering some special representations of $\mathfrak{M}$ and reducing the corresponding matrices (see, in particular, [7], Section 5).

For any consisting of big points subset $X \subset U(a) \setminus \{a_1\}$ suppose $\mathfrak{M}_\alpha(X)=\mathfrak{M}_\alpha(\tilde{X})$, where $\tilde{X}$ is the smallest subset containing X and such that $x_1 \in \tilde{X}$ if $x \in X$ and $x_1 \in U(a)$. Moreover, suppose by induction (later lower indices simply denote numbers of big points a_i and subsets $X_i \subset U(a_i) \setminus \{(a_i)_1\}$):

$$\mathfrak{M}_{a_1,\ldots,a_k}(X_1,\ldots,X_k)=(\mathfrak{M}_{a_1,\ldots,a_{k-1}}(X_1,\ldots,X_{k-1}))_{a_k}(X_k).$$

Also put $\mathfrak{M}_{a_1,\ldots,a_k}(x_1,\ldots,x_k)=\mathfrak{M}_{a_1,\ldots,a_k}(X_1,\ldots,X_k)$, where $X_i = \{x_i\}$ are one-point sets and $\mathfrak{M}^{o}_{a_1,\dots,a_k} = \mathfrak{M}_{a_1,\dots,a_k}(\emptyset,\dots,\emptyset)$. For $X \subset U(a)$, let $\bar{X} = \{\bar{x} \mid x \in X\}$; clearly $\bar{X} \subset \bar{U}(a)$.

Now we shall prove that a reduced poset containing a subset of the form $A = N_i$, $6 < i \leqslant 10$, is wild (for $i \leqslant 6$ it follows from [3]). Note that in the case $x - x_1$ for any big point $x \in N_i$ it has been proved in [4] (a poset N_i with this property will be denoted by N_i^{sh} - "shurian hypercritical"). That's why it is sufficient to show (taking into account Lemma 5) that there exists a poset $M = \mathfrak{M}_{x_1,\dots,x_k}(X_1,\dots,X_k)$ containing a subset N_i for $i \leqslant 6$ or N_i^{sh} for $i > 6$.

CASE $A = N_7 = \underset{a}{\bullet}\ \circ\ \circ\ \circ$. Here take $M = \mathfrak{M}^{o}_{a_1} \supset N_2$.

CASE $A = N_8 = \underset{a}{\bullet}\ \circ\ \overset{\circ}{\circ}$. According to reducibility of $\mathfrak{M}$, there exists one of the following graphs (consisting of different elements):

a) $a \simeq a_1$; b) $a \sim \underbrace{a_1 - b' - b''}$; c) $a \sim a_1 - b \simeq b_1$; d) $a \sim \underbrace{a_1 - b \sim b_1}$;

e) $\underbrace{a \sim a_1 - b \sim b_1}$; f) $a \sim a_1 - b \sim \underbrace{b_1 - c' - c''}$; g) $a \sim a_1 - b \sim b_1 - c \sim c_1$.

Then take M in the following form correspondently:

a) $\mathfrak{M} \supset N_8^{sh}$; b) $\mathfrak{M}^{o}_{a_1} \supset N_2$; c) $\mathfrak{M}^{o}_{a} \supset N_8^{sh}$; d) $\mathfrak{M}^{o}_{a,b} \supset N_2$;

e) $\mathfrak{M}_{a_1}(b) \supset N_8^{sh}$; f) $\mathfrak{M}^{o}_{b_1,a_1} \supset N_2$; g) $\mathfrak{M}^{o}_{c_1,b_1,a_1} \supset N_6$.

CASE $A = N_9 = \underset{a}{\bullet}\ \underset{b}{\bullet}\ \circ$. Assume that the points a and a_1 are comparable (if $a - a_1$ and $b - b_1$, then $\mathfrak{M} \supset N_9^{sh}$). Then take M in the form:

a) $\mathfrak{M}^{o}_{b_1,a_1} \supset N_4$ if $a_1 - b$; b) $\mathfrak{M}_{a_1}(b_1) \supset N_9^{sh}$ if $a_1 - b_1$;

c) $\mathfrak{M}^{o}_{a_1,b_1} \supset N_2$ if $\underbrace{a_1 - c' - c''}$; d) $\mathfrak{M}^{o}_{b_1,c_1,a_1} \supset N_4$ if $a_1 - c$ for a big point $c \neq a, b, b_1$.

CASE $A = N_{10} = \underset{a}{\bullet}\ \underset{b}{\bullet}\ \underset{c}{\bullet}$. Take M in the form $\mathfrak{M}^{o}_{a_1,b_1,c_1} \supset N_3$.

This finishes the proof of Theorem 1.

3. SUBSTANTIATION OF THE REDUCIBILITY CONDITION IN THEOREM 1.

Let $\mathfrak{M}$ be a poset with an equivalence relation. It is easy to show that if $\mathfrak{M}$ is tame then $r(x) \leqslant 4$ for any $x \in \mathfrak{M}$ and, moreover, any point x with $r(x)=4$ is comparable with all other points of $\mathfrak{M}$. Hence it is sufficient to consider the case $r(x)\leqslant 3$ for any $x\in\mathfrak{M}$.

Let $a\in\mathfrak{M}$ be a point with $r(a)\geqslant 2$ and A be some ordinary poset. Denote by $\mathfrak{M}_a^*(A)$ the poset obtained from $\mathfrak{M}$ by rejection of all points $x\sim a$ and replacement the point a by the poset A.

Let Z_n denote a small-point chain consisting of $n\geqslant 0$ points and $Z_n\times Z_m$ be a naturally defined direct product of chains Z_n and Z_m. Let again $U(x)=\{y\in\mathfrak{M} \mid y-x\}$.

LEMMA 6. Assume $r(a)\geqslant 2$ and $U(x)$ is a small-point chain for any $x\sim a$. Let $\mathfrak{M}_a'=\mathfrak{M}_a^*(Z_{m+2})$, if $r(a)=2$, $a\sim b$, $U(b)=Z_m$, and $\mathfrak{M}_a'=\mathfrak{M}_a^*(Z_{m+2}\times Z_{n+2})$, if $r(a)=3$, $a\sim b\sim c$, $U(b)=Z_m$, $U(c)=Z_n$. Then there exists a natural one-to-one correspondence between the classes of indecomposable representations of the posets $\mathfrak{M}$ and $\mathfrak{M}_a'$.

The statement can be proved easily by reduction of the matrices corresponding to the points $x\sim a$ (note that the cases $r(a)=2$ and $n=0$, $r(a)=3$ were considered in [7], Lemmas 9,10).

One can prove analogously as in the proof of Proposition 3 in [7] that, if $r(a)=3$, $a\sim b\sim c$ and the points a,b,c don't satisfy the condition of Lemma 6, then $\mathfrak{M}$ is wild. But if every triple $a\sim b\sim c$ contains a point satisfying the condition of Lemma 6, then after several applications of this lemma we obtain a poset $\hat{\mathfrak{M}}$ which satisfy both conditions 1) and 2) from definitions of a reduced poset; and $\hat{\mathfrak{M}}$ is tame iff $\mathfrak{M}$ is tame.

There is no difficulties to describe, analogously to [7], an evident shape of the poset $\hat{\mathfrak{M}}$. Namely, choose in the set of all points $x\in\mathfrak{M}$ with $2\leqslant r(x)\leqslant 3$ some subsets $A_1, A_2, \ldots, A_i, \ldots$, where $A_i\subset A_{i+1}$ for any $i>0$. Set $a\in A_1$ iff $U(a)$ is a small-point chain. Then suppose by induction that $a\in A_k$ iff $U(a)$ is a chain consisting of points x with $1\leqslant r(x)\leqslant 2$ and, moreover, for $b\in U(a)$, $r(b)=2$ and $b\sim b_1$ the points b and b_1 are comparable and $b_1\in A_{k-1}\setminus U(a)$. A point a with $r(a)\geqslant 2$ will be called singular if $a\in\bigcup_{i=1}^{\infty} A_i$. Denote by $h(a)$ the smallest i for which $a\in A_i$. Define a weight $m(a)$ of a singular point a in the following way. If $h(a)=1$ then let $m(a)=|U(a)|+2$. If $h(a)>1$ then define $m(a)$ by induction via $h(a)$, namely $m(a)=2+\sum_{b\in U(a)} m'(b)$, where $m'(b)=1$ if $r(b)=1$; and $m'(b)=m(c)$ if $r(b)=2$, $b\sim c$ (by definition of a singular point $h(c)<h(b)$).

Since from any triple of equivalent points one can choose at least one satisfying the condition of Lemma 6, then there exists in $\mathfrak{M}$ a subset $\mathfrak{M}_0$ (generally speaking, not uniquely defined) with the following properties:

1) $\mathfrak{M}_0$ contains at least one point from every equivalence class;

2) any point $a \notin \mathfrak{M}_0$ is singular;

3) if a is a singular point from $\mathfrak{M}_0$ and $a \sim b$, then $b \notin \mathfrak{M}_0$ and $h(a) \geq h(b)$.

Denote $\mathfrak{M}_{0,2} = \{x \in \mathfrak{M}_0 \mid r(x)=2;\ x \sim y;\ y \notin \mathfrak{M}_0\}$, $\mathfrak{M}_{0,3} = \{x \in \mathfrak{M}_0 \mid r(x)=3\}$.

If one transits from $\mathfrak{M}$ to $\hat{\mathfrak{M}}$ by an application of Lemma 6 to all such points $a \in \mathfrak{M}_0$, for which there exists $b \sim a$, $b \notin \mathfrak{M}_0$ (according to the order of increasing of $h(b)$), then $\hat{\mathfrak{M}}$ is obtained from $\mathfrak{M}_0$ by replacing of any point $a \in \mathfrak{M}_{0,2}$ by a chain $Z_{m(b)}$, where $a \sim b$, and any point $a \in \mathfrak{M}_{0,3}$ by a direct product $Z_{m(b)} \times Z_{m(c)}$, where $a \sim b \sim c$.

4. PROOF OF THEOREM 2. It isn't difficult to verify that the statement of Theorem 2 is equivalent to the following one:

(C) Chains and nonperiodical cycles have no repetitions, and chains have no common ends.

SUFFICIENCY OF THE CRITERION. We'll show that, if a poset $\mathfrak{M}$ with the condition $r(x) \leq 2$ for all x satisfies the condition (C), then it has finite growth. First we'll establish a heredity of the condition (C) with respect to the algorithm.

LEMMA 7. Let $\mathfrak{M}$ be tame reduced poset satisfying the condition (C). Then for any of the differentiations I,...,V or α, $\bar{\alpha}$, β, γ the derivative poset $\mathfrak{M}'$ also satisfies the condition (C).

PROOF. The lemma is obvious for the additional differentiations α, $\bar{\alpha}$, β, γ. Really, according to Theorem 1, a poset $\mathfrak{M}$ doesn't contain subsets N_7 and N_9, hence new small points (arising by additional differentiations) can not belong to ends of chains. So, the additional differentiations don't lead to arising of new chains (and don't change the old ones). Now we have to consider the main differentiations I,...,V. In each case we'll keep to notations from [8].

CASE OF DIFFERENTIATION I. Here $\mathfrak{M}=a^{\vartriangle}+b^{\triangledown}+C$, where $C=\{c_1<\dots<c_n\}$, and $\mathfrak{M}'=a^{\vartriangle}\cup b^{\triangledown}+C^-+C^+$, where $C^-=\{c_1^-<\dots<c_n^-\}$, $C^+=\{c_1^+<\dots<c_n^+\}$. Obviously, cycles in $\mathfrak{M}$ and $\mathfrak{M}'$ are the same, hence one has to consider chains.

Let $A-x\sim x_1-\dots-B$ be some chain in $\mathfrak{M}'$ containing a big point x where $A=\{a',a''\}$. If $a'=c_i^+$, $a''=c_j^-$ $(i<j)$ and, for example, $x\in b^{\triangledown}$, then $x-a$ and $\mathfrak{M}\supset\{x,a,c_i<c_j\}=N_8$, but it's impossible. Therefore, one of the point a', a'' doesn't belong to the subset $C^-\cup C^+$ and, considering the "pre-images" $\bar{t}\in\mathfrak{M}$ of points $t\in\mathfrak{M}'$ (i.e. taking $\bar{t}=t$ for $t\notin C^-\cup C^+$ and $\bar{t}=c_i$ for $t=c_i^{\pm}$), we obtain a chain $\bar{A}-x\sim x_1-\dots-B$ (where $\bar{X}=\{\bar{x}\}$, $x\in X$) being "primitive" for the initial one. The primitive chains for different chains (containing big points) can not coincide because, if $\bar{A}=\bar{A}_1$ for $A\neq A_1$ and $A_1=\{a_1',a_1''\}$, A_1-x, then it must be (up to a permutation of points) $a'=a_1'$ and $a''=c_i^-$, $a_1''=c_i^+$, i.e. $\mathfrak{M}'\supset\{x,a',c_i^-<c_j^+\}=N_8$, and this is in a contradiction with Lemma 3. Hence the chains in $\mathfrak{M}'$ (containing a big point) have no repetitions and don't intersect.

Moreover, $\mathfrak{M}'$ doesn't contain intersecting chains of type $A-B$ (without big points), i.e. it doesn't contain a subset $G=\{a,b,x_1,x_2,y_1,y_2\}$, where $x_i<y_j$ for all i,j. It follows from a combinatorial property of Differentiation I for ordinary posets proved in [1], Proposition 1.1, that, if $\mathfrak{M}$ is an ordinary tame poset and $\mathfrak{M}\not\supset G$, then $\mathfrak{M}'\not\supset G$.

In conclusion, supposing that a chain $A-x\sim x_1-\dots-B$ intersects a chain $A-E$, where $A=\{a',a''\}$, $E=\{e',e''\}$, we obtain from the condition $\mathfrak{M}'\not\supset N_2,N_8$ (considering different variants) the following fact: the intersection $(C^-\cup C^+)\cap(A\cup E)$ contains no more then one point. Hence $\bar{A}\cup\bar{E}$ is a four-point set, i.e. we have in $\mathfrak{M}$ chains with a common end: $A-x\sim x_1-\dots-B$ and $A-E$. This is a contradiction.

Hence, $\mathfrak{M}'$ satisfies the condition (C), and every chain in $\mathfrak{M}'$ with the ends A,B has a primitive chain in $\mathfrak{M}$ with the ends $\bar{A},\bar{B}$. Further, one can easily verify that any chain in $\mathfrak{M}$ is primitive for some chain in $\mathfrak{M}'$. Moreover, if $c_i\in\bar{A}$, for a primitive chain with a big point $\bar{A}-x\sim x_1-\dots-\bar{B}$, then $c_i^-\in A$ for $x\in b^{\triangledown}$ and $c_i^+\in A$ for $x\in a^{\vartriangle}$; but, if $c_i\in\bar{A}\cup\bar{B}$ for the chain $\bar{A}-\bar{B}$, then $c_i^-\in A\cup B$ if $|b^{\triangledown}\cap(A\cup B)|\geqslant 2$, and $c_i^+\in A\cup B$ if $|a^{\vartriangle}\cap(A\cup B)|\geqslant 2$.

CASE OF DIFFERENTIATION II. Here for any chain (or nonperiodical cycle) L in $\mathfrak{M}'$ we define the primitive chain (cycle) $L^{\uparrow}$ in $\mathfrak{M}$ substituting in L every subgraph $a_1^o \sim b_1^o$ by the subgraph $a_1 \sim a - b \sim b_1$ (and preserving all other vertices and arrows in L). It is assumed that dashes, which were incident to the vertexes a_1^o, b_1^o before, become incident to the vertices a_1, b_1. It is easy to see that, if L is a nonperiodical cycle, then $L^{\uparrow}$ also is. Moreover, repeated vertices in L lead to repeated vertices in $L^{\uparrow}$. Therefore $\mathfrak{M}'$ satisfies the condition (C). Note that the correspondence $L \mapsto L^{\uparrow}$ defines a bijection of the set of all chains (nonperiodical cycles) in $\mathfrak{M}'$ onto the set of all chains (nonperiodical cycles) in $\mathfrak{M}$.

CASE OF DIFFERENTIATION III. If there exists a chain (cycle) in $\mathfrak{M}'$ containing at least one pair $a_n \sim b_n$, then, substituting every subgraph $a_n \sim b_n$ $(n \geqslant 1)$ by the subgraph $a \sim b$, we obtain a chain (cycle) in $\mathfrak{M}$ containing at least one pair $a \sim b$ (and also some other points of $\mathfrak{M}$). Substituting after some pair $a \sim b$ by a sufficiently long chain $a \sim b - a \sim b - \ldots - a \sim b$, we obtain a chain (nonperiodical cycle) with repetitions; this is a contradiction. Hence, the chains and cycles in $\mathfrak{M}'$ don't contain the points a_n, b_n $(n \geqslant 1)$ and lie in $\mathfrak{M}$, i.e. $\mathfrak{M}'$ satisfies the condition (C).

CASE OF DIFFERENTIATION IV. If there exists a cycle in $\mathfrak{M}'$ containing the pair $p \sim p_1$, then, substituting in it every such pair $p \sim p_1$ by the subgraph $B - a \sim a_1$, where $B = \{b_1, b_2\}$, we obtain a graph containing as a subgraph a chain with the same ends: B and B; this is a contradiction (direct arrows of the cycle, which were incident to p and p_1 before, will become incident to B and a_1 correspondently). Therefore, the cycles in $\mathfrak{M}'$ don't contain the pairs $p \sim p_1$, i.e. they are cycles in $\mathfrak{M}$ (as a consequence they have no repetitions).

Assume that there exists a chain in $\mathfrak{M}'$ containing the pair $p \sim p_1$. Then, substituting again any such a pair $p \sim p_1$ by the subgraph $B - a \sim a_1$ and also substituting every end $\Sigma = \{\xi_1, \xi_2\}$ by $B - a \sim a_1$ (joining the direct arrow leading to Σ with a_1), we obtain among subgraphs two different chains with a common end (or one chain with the same ends) in $\mathfrak{M}$, however it is impossible. Hence the following statements take place:

1) any chain in $\mathfrak{M}'$ doesn't contain the pair $p \sim p_1$ and has no the same ends;

2) different chains in $\mathfrak{M}'$ have no common ends;

3) chains in $\mathfrak{M}'$ with the end Σ don't contain the vertices a, a_1 and B.

So, the poset $\mathfrak{M}'$ satisfies the condition (C); cycles in $\mathfrak{M}$ and $\mathfrak{M}'$ are the same; and the correspondence, given by the substitution the end Σ by $B-a\sim a_1$, defines a bijectin of the set of all chains in $\mathfrak{M}'$ onto the set of all chains in $\mathfrak{M}$.

CASE OF DIFFERENTIATION V. We don't consider Differentiation V_1 because geometrically it is dual to the additional differentiation γ.

In the case of Differentiation V_2 suppose $\bar{x}=a_i$ for $x=\overset{+}{\bar{a}}_i$ and $\bar{x}=x$ for other $x\in\mathfrak{M}'$. For any chain L in $\mathfrak{M}'$ of the shape $M-x\sim x_1-\ldots-H$ (in particular M-H) we define the primitive chain $L^{\uparrow}$ in $\mathfrak{M}$ of the shape $\bar{M}-x\sim x_1-\ldots-\bar{H}$, where $\bar{X}=\{\bar{x}\}$, $x\in X$. Taking into account that, according to Lemma 3, $\mathfrak{M}'\not\supset N_2, N_8$, we conclude $L=L_1$ for $L^{\uparrow}=L_1^{\uparrow}$, i.e. the correspondence $L\mapsto L^{\uparrow}$ is an injection. That's why the poset $\mathfrak{M}'$ satisfies the condition (C). Moreover, the mentioned correspondence is a bijection. In particular, during the transition from $L^{\uparrow}$ to L (differentiation of a chain) the end $\{a_1,a_2\}$ go into $\{a_1^+,a_2^+\}$; the end $\{a_i,b_j\}$ go into $\{a_i^+,b_j\}$ $(i\neq j)$; the end $\{a_i,x\}$, where $x\in B$, go into a_i^-,x; and other two-element vertices aren't changed.

In the case of Differentiation V_3 suppose that the pair $a_o\sim b_o$ belongs to a chain (or to a nonperiodical cycle differed from $a_o\simeq b_o$) in $\mathfrak{M}'$. Then after a substitution every such a pair $a_o\sim b_o$ by the chain A-B, where $A=\{a_1,a_2\}$, $B=\{b_1,b_2\}$, we obtain a graph containing (as subgraphs) some different chains in $\mathfrak{M}$ with a common end. However it's impossible. Therefore, the chains (nonperiodical cycles) in $\mathfrak{M}'$ ($\mathfrak{M}$) coincide with those in $\mathfrak{M}$ ($\mathfrak{M}'$) which differ from A-B (from $a_o\simeq b_o$). In particular, $\mathfrak{M}'$ satisfies the condition (C). It finishes the proof of Lemma 7.

Now we turn immediatly to the proof of the finite-growthness. First we shall introduce two types of standard representations of posets with an equivalence relation over the polynomial ring $k[t]$ (corresponding to cycles and chains).

According to notations and definitions from [8], let$(\mathfrak{M},\Theta)$ be a poset with an equivalence relation and $(U,U_K)_{K\in\Theta}$ some its representation. Denote by $\widetilde{U}_K$ an arbitrary generator system of the space U_K taken modulo the subspace $\underline{U}_K$ (note that in a matrix realization of U the columns of matrices are just vectors consisting of coordinates of elements from $\tilde{U}_K$, $K\in\Theta$). In the case of k[t]-representations the set U_K has an analogous sense.

In the sequel we shall use lower indices simply for a numeration of points, i.e. not necessary x_1 will be considered as a point equivalent to x.

For any cycle L of the shape

$$(1)\qquad \underbrace{x_1\sim y_1-x_2\sim y_2-\dots-x_n\sim y_n}\qquad (n\geqslant 1)$$

we consider an indecomposable representation $[L]=(U,U_K)_{K\in\Theta}$ over k, where U is a k-space with a basis $e_1,\dots,e_n$, supposing

$$(2)\qquad \begin{aligned}&\tilde{U}_{(x_i,y_i)}=(e_{i-1},e_i)\quad \text{for all } 1\leqslant i\leqslant n;\\ &\tilde{U}_K=\emptyset \text{ for other } K\in\Theta\ ,\end{aligned}$$

(consider lower indices modulo n). With a help of this representation[L]we'll construct a representation over k[t] choosing any fixed index $1\leqslant j\leqslant n$ and taking $\tilde{U}_{(x_j,y_j)}=\{(te_{j-1},e_j)\}$ or $\tilde{U}_{(x_j,y_j)}=\{(e_{j-1},te_j)$, but for other indices $i\neq j$ let the formulas (2) be unchangable. The obtained k[t]-representation will be called a representation of type [L,t] . It is defined not uniquele and depends upon a choice of the point x_j or y_j. Strictly speaking, it has to be written in the form $[L,t,x_j]$ or $[L,t,y_j]$ but the series of indecomposable representations over k generated by it (see [8,11]) doesn't depend upon x_j,y_j (i.e. up to a finite number it gives the same representations in each fixed dimension for different x_j,y_j).

Anologously, for any chain L of the shape

$$(3)\qquad \{a',a''\}-x_1\sim y_1-\dots-x_n\sim y_n-\{b',b''\}$$

we first construct a representation $[L]=(U,U_K)_{K\in\Theta}$ assuming U to be a vector space of dimension 2n+1 with a basis $e_1,f_1,\dots,e_{n+1}$, f_{n+1} and also taking

$$(4)\qquad \begin{aligned}&\tilde{U}_{(x_i,y_i)}=\{(e_i,e_{i+1}),(f_i,f_{i+1})\}\ \text{for all } 1\leqslant i\leqslant n;\\ &\tilde{U}_{a'}=\{e_1\}\ ;\ \ \tilde{U}_{a''}=\{f_1\}\ ;\\ &\tilde{U}_{b'}=\{e_{n+1}+f_{n+1}\}\ ;\ \ \tilde{U}_{b''}=\{e_{n+1}+f_{n+1}\}\ ;\\ &\tilde{U}_K=\emptyset \text{ for others } K.\end{aligned}$$

Then define k[t]-representation of type [L,t] preserving in

(4) the formulas for $\widetilde{U}_{(x_i,y_i)}$, $\widetilde{U}_{a'}$, $\widetilde{U}_{a''}$ and substituting one coefficient at either e_{n+1} or f_{n+1} in the formulas for $\widetilde{U}_{b'}$ or $\widetilde{U}_{b''}$ by the variable t (three other coefficients remain equal to 1). For example, $\widetilde{U}_{b'}=\{te_{n+1}+f_{n+1}\}$, $\widetilde{U}_{b''}=\{e_{n+1}+f_{n+1}\}$. Here also the series, generated by [L,t], doesn't depend from a way of putting t.

Note that a representation over the field of rational functions k(t) dual to [L,t] is again a representation of the same type (for a cycle instead of t one has to take $\pm t$) under an assumption that $\mathfrak{M}$ is tame and satisfies the condition (C).

It turns out that all those series of indecomposables of posets of finite growth, which are exact at least in one big point, are generated by representations of type [L,t]. For a proof, it is sufficiently to verify that any integration maps a representation of type [L,t] into a representation of a similar type. Clearly, we may not consider the additional differentiations and the differentiation V_1, because arising after their applications points can't belong to the ends of any chain. Moreover, one can suppose that the initial (primitive) poset $\mathfrak{M}$ is reduced since the reduction-procedure (see Section 3) also leads to small points not belonging to the ends of chains.

A verification of the fact, that the class of representations of a type [L,t] is closed with respect to integration, is very simple. It is performed for any case separatly by taking in account the proved above Lemma 7 and the described in [8] construction of integration.

Namely, in the case of Differentiation II, consider a representation of type [L,t] of a derivative poset $\mathfrak{M}'$, where L is a cycle in $\mathfrak{M}'$ of a shape (1). Without loss of generality we can assume that $a_1^o=x_1$, $b_1^o=y_1$. Then (see the cunstruction of the primitive object W in [8]) $W_{(a_1^o,b_1^o)}=\underline{W}_{(a_1^o,b_1^o)}\oplus F$, where F is exactly the space generated by $\widetilde{U}_{(x_1,y_1)}$, i.e. $F=\{(te_n,e_1)\}$ is a one-dimensional space (note that $[L,t,x_1]$ plays a role of [L,t]). Now it follows immediatly from the formulas (II.2) in [8] that $[L,t,x_1]^{\uparrow}=[L^{\uparrow},t,a_1]$, where $L^{\uparrow}$ is a primitive cycle of the shape $a_1\sim a-b\sim b_1-x_2\sim y_2-\dots-x_n\sim y_n$.

The case of Differentiation III is trivial because here cycles and chains in $\mathfrak{M}'$ coincide with those in $\mathfrak{M}$ (see the proof of Lemma 7).

The case of Differentiation V_3 is obvious as well. Here, all chains and cycles in $\mathfrak{M}'$, except of $a_0 \simeq b_0$, coincide with those in $\mathfrak{M}$. It is easy to see that, for a cycle L of shape $a_0 \simeq b_0$, the primitive k[t]-representation $[L,t,a_0]^{\uparrow}$ has the form $[L^{\uparrow},t]$, where $L^{\uparrow}$ is a chain $a_1,b_1 - a_2,b_2$.

In the case of Differentiation IV we don't consider cycles in $\mathfrak{M}'$ since they lie in $\mathfrak{M}$. If we have a chain L in $\mathfrak{M}'$ of a shape $\{\xi_1,\xi_2\}$-...-M, then $M \neq \{b_1,b_2\}$ and, according to the formulas (IV.2) in [8], obviously $[L,t]^{\uparrow}=[L^{\uparrow},t]$, where $L^{\uparrow}$ is the primitive chain $\{b_1,b_2\}$ $-a \sim a_1-\ldots-M$. In the dual case IV* we use use dual considerations.

In the case of Differentiation V_2 we use the formulas (V.2) from [8] and apply the considerations of the proof of Lemma 7. It isn't difficult to establish that $[L,t]^{\uparrow}=[L^{\uparrow},t]$ and, moreover, that the length of L and the dimension of the embraced space are not changed.

Finally, in the case of Differentiation I, considering a chain L in $\mathfrak{M}'$ (containing a big point), we verify without difficulties the equality $[L,t]^{\uparrow}=[L^{\uparrow},t]$ (using a construction of the integration-I and arguments of Lemma 7). Here also the length of a chain and dimension of the embraced space aren't changed.

NECESSITY OF THE CRITERION. We'll show that a poset $\mathfrak{M}$ has infinite growth if there exists a chain $A-x_1 \sim x_2-\ldots-x_{2m-1} \sim x_{2m}-B$ $(m \geqslant 1)$ with repetitions. Pick out two cases: a) A=B; b) $A \neq B$, $x_i=x_j$ for some $i \neq j$. It will follow from b) that $\mathfrak{M}$ is of infinite growth if there exist a chain and a cycle having common vertex. We'll use an induction on m.

First consider the case a). If m=1 we apply to the subset $\mathfrak{N} =$  x_j, x_i (i=1, j=2 or i=2, j=1) Differentiation IV and obtain a poset $\mathfrak{N}'$ containing a subset $\{a_1,a_2,b_1,b_2,y_1,y_2\}$ where $y \sim y_1$, $a_1,a_2 < y$ and $y_1 < b_1,b_2$. According to [4], $\mathfrak{N}'$ (and also $\mathfrak{N}$) has infinite growth (here and below, using Differentiations I,...,V, one has to verify that the corresponding series, in particular written in [4], satisfy relations used by Differentiations; in each case the relations are satisfied, but we omit the verification).

If m > 1 and $x_i \neq x_j$ for any $i \neq j$ then the poset $\mathfrak{M}_{x_1,x_2,\ldots,x_{2m-2}}(x_3,x_5,\ldots,x_{2m-1})$ has a chain of the form

$A-y_1\sim y_2-A$, and, according to Lemma 5, $\mathfrak{M}$ has infinite growth. If $x_i=x_j$ for some $i<j$ and i is odd then j is even, because, in the contrary case, $\mathfrak{M}$ has a chain $A-\dots-x_i\sim x_{i+1}-x_{j+2}\sim x_{j+3}-$ $-\dots-A$ of smaller length. Hence, if $k=\min\{i\,|\,x_i=x_j,\ i\neq j$ and $x_k=x_s,\ k\neq s\}$ then the poset $\mathfrak{M}^o_{x_1,x_3,\dots,x_k}$ has a chain $A_1-x_{k+2}\sim x_{k+3}-\dots-x_{s-3}\sim x_{s-2}-A_1$ (if $s=k+5$, the chain has the form $A_1-x_{k+2}\sim x_{k+3}-A_1$). By induction and, according to Lemma 5, $\mathfrak{M}$ has infinite growth.

Consider the case b). If $m=2$, the poset $\mathfrak{M}$ contains (up to antiisomorphism) a subset of the form $\{a_1,a_2,b_1,b_2,x_1,x_2\}$, where $a_1,a_2<x_1$ and $x_2<b_1,b_2$ (for some i,j possibly $a_i<b_j$), or $\{a_1,a_2,b_1,b_2,x_1,x_2\}$, where $a_1,a_2<x_1$; $b_1,b_2<x_2$ and $x\sim x_1$, and therefore has infinite growth [4].

If $m>2$ one can assume that $x_{2i-1}\neq x_{2j}$ (hence $x_{2i}\neq x_{2j-1}$) for $i<j$, since otherwise $\mathfrak{M}$ has a chain $A-\dots-x_{2j-1}\sim x_{2j}-x_{2i-2}\sim$ $\sim x_{2i-3}-\dots-x_2\sim x_1-A$ considered in the case a). Moreover, one can assume that, if $x_{2i-1}=x_{2j-1}$, $x_{2p-1}=x_{2q-1}$ and $i<j$, $p<q$, $i<p$, then $p<j<q$, since otherwise there exists a chain in $\mathfrak{M}$ (with repetitions) of smaller length: $A-\dots-x_{2p-1}\sim x_{2p}-x_{2q+1}\sim$ $\sim x_{2q+2}-\dots-B$.

If $x_1\neq x_i$ ($x_{2m}\neq x_j$) for any $i\neq 1$ ($j\neq 2m$), then the poset $\mathfrak{M}^o_{x_1}$ ($\mathfrak{M}^o_{x_{2m}}$) has a chain $\bar{A}-x_3\sim x_4-\dots-B$ ($A-\dots-x_{2m-3}\sim x_{2m-2}-\bar{B}$) and, according to Lemma 5, $\mathfrak{M}$ has infinite growth.

If $x_1=x_{2q-1}$, $x_{2m}=x_{2p}$ $(1<p<q<m)$ and also $p-1\leqslant 2m-q$ $(p-1\geqslant 2m-q)$, then the poset $\mathfrak{M}_{x_2}(x_3)$ ($\mathfrak{M}_{x_{2m-1}}(x_{2m-2})$) has a chain

$$A-\bar{x}_3\sim x_4-\dots-x_{2q-3}\sim x_{2q-2}-\bar{x}_3\sim x_4-\dots-x_{2p-1}\sim x_{2p}-B$$

$$(A-x_{2q-1}\sim x_{2q}-\dots-x_{2m-3}\sim \bar{x}_{2m-2}-x_{2p+1}\sim x_{2p+2}-\dots-x_{2m-3}\sim \bar{x}_{2m-2}-B)$$

of smaller length. Finally, if $x_1=x_{2m-1}$, then the elements x_i $(i\notin\{1,2,2m-1,2m\})$ are mutually different and therefore, for $m>3$ $(m=3)$, the poset $\mathfrak{M}_{x_4,\dots,x_{2m-4},x_{2m-1}}(x_5,\dots,x_{2m-3},x_{2m-2})$ ($\mathfrak{M}_{x_{2m-1}}(x_{2m-3})$) has a chain of the form $\bar{A}-y_1\sim y_2-y_1\sim y_2-B$.

We shall show that if two different chains $F=\{A-x_1\sim x_2-\dots$ $\dots-x_{2m-1}\sim x_{2m}-B\}$ and $G=\{C-y_1\sim y_2-\dots-y_{2n-1}\sim y_{2n}-D\}$ $(m,n\geqslant 0)$ have a common vertex, then $\mathfrak{M}$ has infinite growth. The proof will be carried out by induction on $m+n$.

The case $F=\{A-B\}$, $G=\{A-D\}$ follows from [11]. In a general case also one can suppose $A=C$ and, moreover, either a) $x_i \neq x_j$ for any i,j or b) $x_1=y_1$. Really, if $x_k=y_s$, where $k=2p-1$, $s=2q-1$, then among the chains $F, G, F_1=\{A-\dots-x_k \sim x_{k+1}-y_{s+2} \sim y_{s+3}-\dots-y_{2n-1} \sim y_{2n}-D\}$ and $G_1=\{C-\dots-y_s \sim y_{s+1}-x_{k+2} \sim x_{k+3}-\dots-x_{2m-1} \sim x_{2m}-B\}$ always one can choose a pair of chains with the mentioned properties, the sum of lengths of which is less or equal to the sum of lengths of F and G.

Taking into account the previous case, assume that x_i (y_j) are mutually different and, moreover, there doesn't exist an equality of the form $y_{2p}=x_{2q-1}$ (otherwise $\mathfrak{M}$ has a chain $A-y_1 \sim y_2-\dots-y_{2p-1} \sim y_{2p}-x_{2q-2} \sim x_{2q-3}-\dots-x_2 \sim x_1-A$). Then in the case a), for $m>0$ $(n>0)$, the set $\mathfrak{M}^o_{x_{2m}}$ $(\mathfrak{M}^o_{x_{2n}})$ has a chain $F_1=\{A-\dots-x_{2m-3} \sim x_{2m-2}-\bar{B}\}$ $(G_1=\{C-\dots-x_{2n-3} \sim x_{2n-2}-\bar{D}\})$ of smaller length than $F(G)$. In the case b) the poset $\mathfrak{M}^o_{x_1}$ has the chains $F_1=\{\bar{A}-x_3 \sim x_4-\dots-B\}$ and $G_1=\{\bar{C}-y_3 \sim y_4-\dots-D\}$, where $\bar{A}=\bar{C}$. In both cases $\mathfrak{M}$ has infinite growth according to Lemma 5.

Now we shall show that a poset $\mathfrak{M}$ has infinite growth if there exists a nonperiodical cycle $F=\{x_1 \sim x_2-\dots-x_{2m-1} \sim x_{2m}\}$ with repetitions. It will follow from this that $\mathfrak{M}$ has infinite growth if there exist two different (nonperiodical) cycles having common vertex. We shall use the following four standard posets of infinite growth:

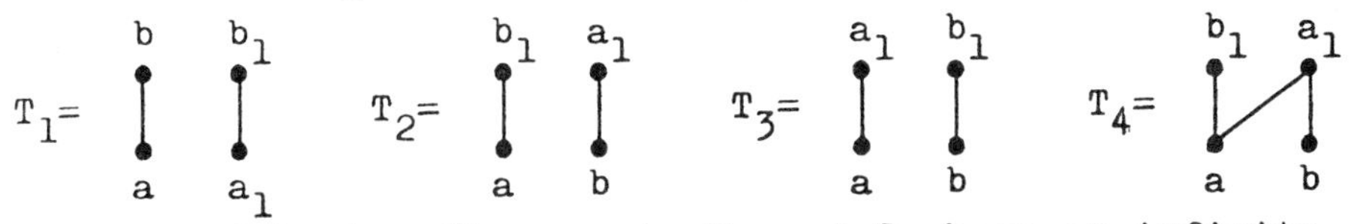

where $a \sim a_1$ and $b \sim b_1$. The posets T_1 and T_2 have an infinite growth according to [4] , T_3 according to the relation $(T_3)_a(\{b_1, b_2\}) \supset T_2$, and T_4 after the application of Differentiation II with respect to a pair (a,b) go into a poset also containing T_2.

The proof will be carried out by induction on m. For $m=3$ we have $\mathfrak{M} \supset T_1$ or $\mathfrak{M} \supset T_4$.

In the case $m>3$ assume that m is the smallest possible (and that another cycle with $m=2$ doesn't exist). Then, with a help of simple combinatorics (see, for instance, the case of a chain with repetitions), one can prove that there are no elements $x_i=x_j=x_k$ for $i<j<k$.

Denote by K_1(K_2) the set of all such points x_p that $x_p \neq x_i$ for any $i \neq p$ ($x_p = x_q$ for some $q \neq p$). An integer $|K_2|/4$ will be denoted by d_0.

If for some subset $M = \mathfrak{M}_{y_1,\dots,y_k}(Y_1,\dots,Y_k)$ there exists a nonperiodical cycle with repetitions of smaller length, then by induction (and according to Lemma 5) the poset $\mathfrak{M}$ has infinite growth. That's why we shall subsequently assume that such a poset M doesn't exist. In particular, if we have $x_i - x_{i+1}$ in the cycle, then $x_i \in K_2$ or $x_{i+1} \in K_2$ (otherwise the poset $\mathfrak{M}_{x_i}(x_{i+1})$ has a cycle of smaller length).

If $d_0 = 1$, then the cycle has length 8 and, up to a renumeration of elements, either $x_1 = x_5$, $x_2 = x_6$ or $x_1 = x_6$, $x_2 = x_5$ (for i=3,4, 7,8 $x_i \in K_1$). In the first case, $\mathfrak{M}_{x_1}(\{x_4, x_8\})$ contains a subset T_1 or T_2 (see the base of induction); in the second, the poset $\mathfrak{M}_{x_1}(\{x_7, x_8\})$ contains T_3.

In the case $d_0 > 1$, we first show that, if $x_i = x_k$, $x_j = x_s$ for some $i < k$; $i < j < s$, then $j < k < s$. Assume the contrary supposing for a definiteness that $k < j$ and $j - k < 2m + i - s$. Then, for even $k-i$ or $s-j$, it's easy to show (see, in particular, the case of a chain) that $\mathfrak{M}$ has a cycle of smaller length. If both these numbers are odd, then take k and s even, but i and j odd. Then in a new cycle, obtained from F by substitution $x_{s+1}, \dots, x_{2m}, \dots, x_{i-1}$ by the elements $x_{j-1}, x_{j-2}, \dots, x_{k+1}$, the points x_i and x_j are joined by a dash with equal points. Hence, the poset $M = \mathfrak{M}_{x_k}(x_{2k+1})$ has a cycle of smaller length.

Now we show that $x_p = x_q$ for some even p and odd q. Suppose the contrary and fix elements $x_i = x_k$ for even i,k ($i \neq k$). If x_{i+1}, $x_{k+1} \in K_1$, then the poset $\mathfrak{M}_{x_i}(\{x_{i+1}, x_{k+1}\})$ has obviously a cycle of smaller length. If x_{i+1}, $x_{k+1} \in K_2$, then $x_{i+1} = x_{k+1}$ and therefore the poset $\mathfrak{M}_{x_i}(x_{i+1})$ has a cycle of smaller length. Finally, in the case $x_{i+1} \in K_1$, $x_{k+1} \in K_2$, the points x_{i+3} and x_{k+1} are equal and, according to the minimality of m, the cycle $F \setminus \{x_{i+1}, x_{i+2}\}$ (with the dash on the place of the rupture) is periodical, then F contains the elements $x_p = x_q$, $x_{p+1} = x_{q+1}$ for even $p \neq q$. Hence, in each case we obtain a contradiction.

It follows from this that $d_0 = 2$. Really, suppose the contrary and fix points $x_p = x_q$, where $p < q$, p is even and q odd. Then the cycle F', obtained from F by placing the points $x_{p+1}, x_{p+2}, \dots, x_{q-1}$ in the contrary order, contains elements $y_i = y_k$, $y_j = y_s$

for $i<k<j<s$, but it's impossible (see the stated above).

Fix points $x_i=x_k$, $x_j=x_s$, where $i<j<k<s$ and $i-k$ is odd. The number $j-s$ is odd too (otherwise the cycle F can be replaced by a cycle F_1, obtained from F with help of placing the points $x_{i+1}, x_{i+2}, \dots, x_{k-1}$ in the contrary order). Since there exists a point $x_p \in K_2$ among the points joining in F by a dash with x_i and x_k (or x_j and x_s), then F has one of the following forms:

a) $x_1 \sim x_2 - x_3 \sim x_4 - x_2 \sim x_1 - x_4 \sim x_3$;

b) $x_1 \sim x_2 - x_3 \sim x_4 - x_5 \sim x_6 - x_4 \sim x_3 - x_6 \sim x_5$;

c) $x_1 \sim x_2 - x_3 \sim x_4 - x_5 \sim x_6 - x_7 \sim x_8 - x_4 \sim x_3 - x_6 \sim x_5$.

Clearly, one can suppose that $\mathfrak{M}$ consists of points from F only. Then according to Theorem 1 the width of $\mathfrak{M}$ is equal to 2.

In the case a) obviously $\mathfrak{M}$ is isomorphic to T_3.

In the case b) one can suppose that the following points are comparable: x_3 and x_5, x_1 and x_6, x_2 and x_4 (otherwise T_3 is contained as a subset respectively in $\mathfrak{M}$, $\mathfrak{M}_{x_2}(x_3)$, $\mathfrak{M}_{x_1}(x_5)$) and also x_1 and x_2 (otherwise $\mathfrak{M}_{x_6}(\{x_3,x_4\})$ contains a subset T_1, T_4 or T_2). Assume for a definiteness that $x_1 < x_2$. Then obviously $x_5 < x_6 < x_2$; $x_1 < x_4 < x_3$; $x_5 < x_3$; $x_1 < x_6$; $x_4 < x_2$ and, applying the sequence of Differentiations II for $(a,b)= =(x_1,x_5),(x_5,x_4),(x_6^o,x_3)$, we obtain a poset containing a subset T_4 (here x_6^o is the point generated by x_6 in the first step).

In the case c) one can assume that the following points are comparable: x_3 and x_5, x_4 and x_6 (otherwise $\mathfrak{M}$ has a cycle of the shape b)), x_1 and x_3, x_2 and x_5, x_4 and x_7, x_6 and x_8 (otherwise there exists a cycle with $d_o=1$), x_3 and x_4, x_5 and x_6 (otherwise $\mathfrak{M}$ contains a subset T_1, T_4 or T_2), x_1 and x_8, x_2 and x_7 (otherwise a subset T_3 is contained in $\mathfrak{M}_{x_2,x_7}(x_3,x_6)$ or $\mathfrak{M}_{x_1,x_8}(x_5,x_4)$).

By a symmetry assume that $x_5 < x_6$ and $x_1 < x_8$. Then $x_4 < x_3$ (otherwise $\mathfrak{M}$ contains T_3), $x_1 < x_4 < x_6$, $x_5 < x_3$, $x_5 < x_8$ and, moreover, $x_i \nless x_j$ for any $i \in \{2,3,6,7\}$ and $j \in \{1,4,5,8\}$, $(i,j) \neq \neq (2,8)$. One can also assume $x_2 \nless x_8$, since otherwise $\mathfrak{M}_{x_3}(x_6)$ contains T_4 or T_2.

Obviously, it is sufficient to consider the case $x_i > x_j$ for any $i \in \{2,3,6,7\}$, $j \in \{1,4,5,8\}$. If $x_7 < x_3$, $x_7 < x_2$, $x_6 < x_2$, then, applying a sequence of Differentiations II for (a,b)= $=(x_1,x_5),(x_5,x_4),(x_4,x_8),(x_6^0,x_3^1)$, we obtain a poset containing T_4 (here x_3^1 is the point generated by x_3 in the fird differentiation). If $x_7 > x_3$, $x_7 > x_2$, $x_6 > x_2$, then, applying a sequence of Differentiations II for $(a,b)=(x_1,x_5),(x_5,x_4),(x_4,x_8),(x_2^0,x_3^1)$, we obtain a poset containing T_2. This finishes the proof of Theorem 2.

BIBLIOGRAPHY

1. Bondarenko V.M., Zavadskij A.G., Nazarova L.A., "On representations of tame posets", Representations and quadratic forms, Kiev, 1979, 75-106.

2. Kleiner M.M., "Posets of finite type", Zap. Naucn. Sem.LOMI, 28(1972), 32-41.

3. Nazarova L.A., "Representations of posets of infinite type", Izv. AN SSSR, Ser. Mat., 39(1975), 963-991.

4. Nazarova L.A., Bondarenko V.M., Roiter A.V., "Representations of posets with involution", Preprint 86.80, Inst. Mat. AN UkSSR, Kiev, 1986, 1-26.

5. Nazarova L.A., Roiter A.V., "Categorical matrix problems and the Brauer-Thrall conjecture", Preprint 73.9, Inst. Mat. AN UkSSR, Kiev, 1973, 1-100.

6. Nazarova L.A., Roiter A.V., "Representations of posets", Zap. Naucn. Sem. LOMI, 28(1972), 5-31.

7. Nazarova L.A., Roiter A.V., "Representations and forms of weakly completed posets", Linear algebra and representation theory, Kiev, 1983, 19-54.

8. Zavadskij A.G., "An algorithm for posets with an equivalence relation", These proceedings.

9. Zavadskij A.G., "Algorithm of differentiation and classification of representation", Preprint 1606, KISI, Kiev, 1987,1-129.

10. Zavadskij A.G., Nazarova L.A., "Tame posets", Matrix problems, Kiev, 1977, 122-143.

11. Zavadskij A.G., Nazarova L.A., "On finiteness and boundness of the parameter's number", Preprint 81.27, Inst. Mat. AN UkSSR, Kiev, 1981, 21-29.

Victor Bondarenko
Institute of Mathematics
AN UkSSR
ul. Repina 3, Kiev-4,
252601, USSR

Alexander Zavadskij
Department of Mathematics
KISI
pr. Wozduchoflotskij 31,
Kiev-37, 252037, USSR

Canadian Mathematical Society
Conference Proceedings
Volume 11 (1991)

Almost Split Sequences for Bocses

W.L. BURT AND M.C.R. BUTLER

This paper contains a different treatment, with some new results, of the Auslander-Reiten theory for bocses given by Bautista and Kleiner in [BK]. The basic idea in both papers is to realise the representations of a bocs within a larger category, then use criteria like those given by Auslander and Smalø in [AS] for a subcategory to possess almost split sequences. In [BK] bocs representations are realised as induced R-modules for a suitable algebra inclusion $A \to R$, and also as induced comodules over the underlying A-coalgebra V of the bocs (a point of view previously developed in [K]). We will also use the induced R-module realisation, but induced comodules will be replaced by coinduced L-modules for another algebra inclusion $A \to L$ associated with the bocs. We call L the left algebra and R the right algebra of the bocs. One new type of result we obtain concerns the relationship of L and R; Theorem 5.1 asserts that, for the most interesting bocses covered by our theory, V is a left L and right R cotilting bimodule in the sense of

1980 *Mathematics Subject Classification* (1985 *Revision*). Primary 16G70; Secondary 16E30, 16D90, 16W30, 15A30.

Miyashita [M]. The other main new result is Theorem 3.8, where - for a large and important class of bocses - we show that the subcategories of R-mod and of L-mod which realise bocs representations are closed under extensions.

Throughout the paper the term bocs will mean an A-coalgebra V over an *algebra* A (subject to axioms stated at the end of this introduction) so differing at a formal level from standard accounts such as those of [R, D, C-B] in which A is a category. The main results apply to all 'additive Roiter bocses' [C-B], in which A is a finite-dimensional algebra over a field, hence also to the 'Drozd bocs' of a finite dimensional algebra (see [D, C-B] for details), and to bocses of finite representation type arising from many natural matrix problems. We do not, however, discuss bocses over orders, as is done in [D, BK].

§1 contains the formal definition of the left and right algebras, L and R, of a bocs $\mathfrak{A}$ = (A, V), and exhibits a natural bimodule structure ${}_LV_R$ on V compatible with ${}_AV_A$. Through §§2-5, however, our main results require that ${}_AV$ and V_A are finitely generated projective modules; this ensures that ${}_AV$, R_A and V_A, ${}_AL$ are dual pairs of modules with respect to A. It then follows from Theorem 2.5 that the representation category $\mathfrak{A}$-mod of $\mathfrak{A}$ is equivalent to each of the subcategories $\underline{I}$ of induced modules in R-mod and $\underline{C}$ of coinduced modules in L-mod; and with some additional finiteness assumptions, Theorem 2.6 states that $\underline{I}$ and $\underline{C}$ are functorially finite subcategories of R-mod and L-mod, respectively. (As is well-known, functorial finiteness of, say, $\underline{I}$ in R-mod allows one to push down almost split morphisms from R-mod to $\underline{I}$, and this property is also one of the criteria in [AS]).

In §3 we study conditions under which $\underline{I}$ and $\underline{C}$ are closed under extensions in R-mod and L-mod, respectively. The crucial condition for this, in Theorem 3.8, is that the kernel $\overline{V}$ of the counit $\epsilon : V \to A$ be 'projectivising', in the sense that the functors $\overline{V}\otimes_A -$ and $- \otimes_A \overline{V}$ map <u>all</u> A-modules to projective A-modules. This is the case for the 'free finitely generated' kernels of bocses which occur in [R, D, C-B]. As explained at the beginning of §3, the result of Theorem 3.8 is in fact far stronger than closure of $\underline{I}$ and $\underline{C}$ under extensions; the latter, though, is the specific fact needed to use [AS].

In the brief §4 we extract from §§2,3 an existence theorem for almost split sequences, Theorem 4.1: if A and V are finite-dimensional over a central subfield k, and $\overline{V}$ is projectivising, then add($\underline{I}$) and add($\underline{C}$) admit almost split sequences; if, additionally, idempotents split in $\mathfrak{A}$-mod, then $\underline{I}$, $\underline{C}$ and therefore $\mathfrak{A}$-mod also admit almost split sequences.

In §5, we again extract from §§2, 3 most of the information needed to formulate and prove the cotilting result Theorem 5.1, though finiteness of coresolutions requires an assumption that ${}_AA$ and A_A be of finite injective dimension.

§6 deals with the rather different and technically complicated problem of characterising bocses for which idempotents split in their categories of finite-dimensional representations, a key criterion in both [BK] and our Theorem 4.1. As Roiter observed in [R], idempotent splitting seems to depend on suitable triangularity and freeness properties of the bocs, and Bautista and Kleiner have described in §5 of [BK] a class of 'left triangular tensor' coalgebras with such properties. In our §6, we adopt left-right symmetric conditions and use a strategy of step-wise construction, to give a re-organised account of this complicated topic.

Thanks are offered to the S.E.R.C. (U.K.) for funding a research studentship for the first-named author, and also a visiting fellowship enabling Raymundo Bautista to visit the University of Liverpool for four weeks in 1989; to Raymundo Bautista, to U.N.A.M. (Mexico), and to the British Council for enabling the second-named author to spend six weeks at U.N.A.M. Instituto de Matematicas in 1988; to the authors of [BK] for sending us a preprint of their paper, which has strongly influenced the work presented here; and finally special thanks to Wendy Orr for typing our manuscript.

Conventions. Bocses and their representations

Throughout the paper A denotes a fixed ring, associative, with identity, subject to further qualification for specific results. The symbol $\otimes$ always means $\otimes_A$. For rings B, C, the notations ${}_BX$, Y_C, ${}_BZ_C$ indicate that X, Y, Z are respectively, a left B-module, a right C-module, and a left B, right

C-bimodule (more briefly, a B-C-bimodule).

Given ${}_BX$, ${}_BX'$, $\mathrm{Hom}_B(X, X')$ denotes the group of left B-module morphisms $X \to X'$; likewise the group of right C-module morphisms $Y_C \to Y'_C$ is denoted $\mathrm{Hom}_{C^{op}}(Y, Y')$. Groups of B-C-bimodule morphisms are denoted $\mathrm{Hom}_{B\times C^{op}}(-,-)$.

Given ${}_AZ$, W_A we often abuse notation by writing $A \otimes Z = Z$, $W \otimes A = W$.

By a <u>bocs</u> $\mathcal{Q}$ we mean a quadruple $\mathcal{Q} = (A,V,\mu,\epsilon)$, often denoted just $\mathcal{Q} = (A, V)$, in which V is an A-A-bimodule; the <u>comultiplication</u> $\mu : V \to V \otimes V$ is an A-A-morphism and is coassociative, that is, $(\mu \otimes 1)\mu = (1 \otimes \mu)\mu$; the <u>counit</u> $\epsilon : V \to A$ is an A-A-morphism, $(1 \otimes \epsilon)\mu = 1_V = (\epsilon \otimes 1)\mu$; <u>and</u> ϵ <u>is assumed throughout to be surjective</u>.

In case A is an algebra over a central subring k, we also assume tacitly that k acts centrally on V.

The categories of (finitely generated) left [right] modules over any ring B will be denoted B-Mod (B-mod) [Mod-B (mod-B)].

Now let $\mathcal{Q} = (A, V)$ be a bocs. We define $\mathcal{Q}$-mod to be the category with objects the finitely generated left A-modules $X, Y, \ldots,$ and morphism groups $\mathcal{Q}(X, Y)$ given by

$$\mathcal{Q}(X, Y) = \mathrm{Hom}_A(V \otimes X, Y),$$

(we use a variant of this definition in §6). The composition gf of $f \in \mathcal{Q}(X, Y)$ and $g \in \mathcal{Q}(Y, Z)$ is, as usual, given by the composite of the maps

$$V \otimes X \xrightarrow{\mu\otimes 1} V \otimes V \otimes X \xrightarrow{1\otimes f} V \otimes Y \xrightarrow{g} Z;$$

composition is associative, the identity of X is $\epsilon \otimes 1: V \otimes X \to X$, and in case A is a k-algebra then $\mathcal{A}$-mod is a k-category.

Finally let B be a category; recall that <u>idempotents split in B</u> if every idempotent endomorphism $e = e^2$ of every object X of B admits a factorisation $X \xrightarrow{f} Y \xrightarrow{g} X$ such that $fg = 1_Y$. In particular if B is a full subcategory of a module category then idempotents split in add(B), by which we mean the full subcategory containing all modules isomorphic to direct summands of finite direct sums of objects in B.

§1 The left and right algebras of a bocs

Let $\mathcal{A} = (A, V, \mu, \epsilon)$ be a bocs. We define its <u>left algebra</u> L to be the A-A-bimodule $\mathrm{Hom}_{A^{op}}(V, A)$ with multiplication given by the following rule:

for $e, f \in L$, $e.f$ is the composite

$$V \xrightarrow{\mu} V \otimes V \xrightarrow{f\otimes 1} A \otimes V = V \xrightarrow{e} A.$$

We omit the purely formal verification of the following facts: this multiplication is associative, with the counit ϵ as identity, and $a\epsilon = \epsilon a$ for all $a \in A$; $(a\epsilon)(b\epsilon) = (ab)\epsilon$ for all $a, b \in A$; and $\forall\ a \in A,\ e, f \in L,\ (ea)f = e(af),\ (a\epsilon)f = af$, and $e(a\epsilon) = ea$. To summarise, we obtain:

1.1 **Proposition** The bimodule ${}_AL_A$, equipped with the above multiplication $L \otimes L \to L$ and the map $a \longmapsto a\epsilon$ of A into L, is an A-algebra with identity ϵ, and the bimodule structure on L induced along the map $a \longmapsto a\epsilon$ coincides with the natural A-A-bimodule structure on L.

Similarly, the right algebra R of $\mathcal{A}$ is the A-A-bimodule $\mathrm{Hom}_A(V, A)$, with multiplication given as follows:

for s, t ∈ R, s.t is the composite

$$V \xrightarrow{\mu} V \otimes V \xrightarrow{1\otimes s} V \otimes A = V \xrightarrow{t} A.$$

1.1′ **Proposition** The bimodule ${}_A R_A$, equipped with the above multiplication $R \otimes R \to R$ and the map $a \longmapsto a\epsilon$ of A into R, is an A-algebra with identity ϵ, and the bimodule structure on R induced along the map $a \longmapsto a\epsilon$ coincides with the natural A-A-bimodule structure on R.

Remark Note that, for $a \in A$, the map $a\epsilon : V \to A$ has different values according to whether it is viewed as an element of R or of L; for $v \in V$, $(a\epsilon)(v) = a\epsilon(v)$ for $a\epsilon \in L$, whereas $(a\epsilon)(v) = \epsilon(v)a$ for $a\epsilon \in R$.

To define actions of L and R on V, we first note that there are evaluation maps

$$E_L : L \otimes V \longrightarrow A \ ; \quad e \otimes v \longmapsto e(v),$$

$$E_R : V \otimes R \longrightarrow A \ ; \quad v \otimes s \longmapsto s(v).$$

Hence there is a left action of L on V, given by

$$L \otimes V \xrightarrow{1\otimes\mu} L \otimes V \otimes V \xrightarrow{E_L\otimes 1} A \otimes V = V,$$

and a right action of R on V, given by

$$V \otimes R \xrightarrow{\mu\otimes 1} V \otimes V \otimes R \xrightarrow{1\otimes E_R} V \otimes A = V.$$

Again we omit the entirely routine verification of the following statements:

1.2 **Proposition** The actions of L and R on V just defined induce an L-R-bimodule structure ${}_LV_R$ on V, compatible with the original bimodule structure ${}_AV_A$ after restriction along the maps $a \longmapsto a\epsilon$ of A into L and A into R.

1.3 **Corollary** $\mu : V \to V \otimes V$ is an L-R-bimodule morphism.

§2 Functorially finite imbeddings of $\mathcal{A}$-mod

Throughout §2, $\mathcal{A} = (A, V)$ is a bocs such that

2.1 ${}_AV$ and V_A are finitely generated projective modules.

Therefore R_A and ${}_AL$ are finitely generated projectives, and the duality maps

$${}_AV_R \longrightarrow {}_A\mathrm{Hom}_{A^{op}}(R, A)_R \quad \text{and} \quad {}_LV_A \longrightarrow {}_L\mathrm{Hom}_A(L, A)_A$$

are bimodule isomorphisms. Also, given any left A-module X, there are natural maps

$$\beta : R \otimes X \longrightarrow \mathrm{Hom}_A(V, X) \ ; \ s \otimes x \longmapsto (v \longmapsto s(v)x),$$

$$\alpha : V \otimes X \longrightarrow \mathrm{Hom}_A(L, X) \ ; \ v \otimes x \longmapsto (e \longmapsto e(v)x),$$

and repeated use will be made of the following fact:

2.2 **Proposition** For any ${}_AX$, β is an isomorphism of R-modules and α an isomorphism of L - modules.

2.3 **Notation** Let $\underline{I}$ denote the full subcategory of R-mod with objects the induced modules - that is, modules ${}_RM$ isomorphic to $R \otimes X$ for some finitely generated ${}_AX$. Let $\underline{C}$ denote the full subcategory of L-mod with objects the coinduced modules - that is, modules ${}_LN$ isomorphic to $\mathrm{Hom}_A(L, X)$ for some finitely generated ${}_AX$.

Remark 2.1 will need to be supplemented for the proof in Theorem 2.6 that $\underline{I}$ is functorially finite in R-mod, and $\underline{C}$ is so in L-mod. We need to ensure that restriction from R to A maps R-mod into A-mod, and that $\mathrm{Hom}_L(V,-)$ maps L-mod into R-mod.

2.4 **Theorem** The functors $V \otimes_R - :$ R-mod $\to$ L-Mod and $\mathrm{Hom}_L(V,-)$: L-Mod $\to$ R-Mod restrict to mutually inverse equivalences between $\underline{I}$ and $\underline{C}$.

Proof These functors map $\underline{I}$ and $\underline{C}$ into one another, since $V\otimes_R(R \otimes X) \cong V \otimes X \cong \mathrm{Hom}_A(L, X)$ and $\mathrm{Hom}_L(V, \mathrm{Hom}_A(L, X)) \cong \mathrm{Hom}_A(V, X) \cong R \otimes X$. To prove that they induce equivalences, we consider the natural transformations

$$\nu : 1_{R\text{-Mod}} \longrightarrow \mathrm{Hom}_L(V, {}_L V \otimes_R -),$$

given on ${}_R M$ by $\nu_M(m)(v) = v \otimes_R m$ (for $m \in M$, $v \in V$), and

$$\sigma : V \otimes_R \mathrm{Hom}_L(V, -) \longrightarrow 1_{L\text{-Mod}},$$

given on ${}_L N$ by $\sigma_N(v \otimes_R h) = h(v)$ (for $v \in V$, $h \in \mathrm{Hom}_L(V, N)$).

For $M = R \otimes X \in \underline{I}$, ν_M is the composite of the chain of isomorphisms

$$R \otimes X \xrightarrow{\beta} \mathrm{Hom}_A(V, X) \longrightarrow \mathrm{Hom}_L(V, \mathrm{Hom}_A(L, X))$$
$$\xrightarrow{\alpha_*^{-1}} \mathrm{Hom}_L(V, V \otimes X) \longrightarrow \mathrm{Hom}_L(V, V \otimes_R (R \otimes X)),$$

and for $N = \mathrm{Hom}_A(L, X) \in \underline{C}$, σ_N is the composite of a second chain of isomorphisms,

$$V \otimes_R \mathrm{Hom}_L(V, \mathrm{Hom}_A(L, X)) \longrightarrow V \otimes_R \mathrm{Hom}_A(V, X)$$

$$\xrightarrow{1\otimes_R\beta} V \otimes_R (R \otimes X) \longrightarrow V \otimes X \xrightarrow{\alpha} \mathrm{Hom}_A(L, X).$$

It follows that $(V \otimes_R -)|_{\underline{I}}$ and $\mathrm{Hom}_L(V, -)|_{\underline{C}}$ are mutually inverse equivalences.

2.5 **The Imbedding Theorem** There are equivalences of categories $F_C : \mathcal{A}\text{-mod} \to \underline{C}$ and $F_I : \mathcal{A}\text{-mod} \to \underline{I}$.

Proof We construct F_C explicitly, then use the previous theorem and define $F_I = \mathrm{Hom}_L(V, -)|_{\underline{C}} \circ F_C$. On an object $X = {}_AX$ in $\mathcal{A}$-mod, we let $F_C(X) = V \otimes X$, which by Proposition 2.2 is in $\underline{C}$. On a morphism $f \in \mathcal{A}(X, Y) = \mathrm{Hom}_A(V \otimes X, Y)$, we let $F_C(f)$ be the composite of the chain of maps $V \otimes X \xrightarrow{\mu\otimes 1} V \otimes V \otimes X \xrightarrow{1\otimes f} V \otimes Y$; this is an L-morphism by 1.3. It is routine to verify that $F_C : \mathcal{A}\text{-mod} \to \underline{C}$ is a functor and is dense, so we need only show that it is full and faithful. Define

$G : \mathrm{Hom}_L(V \otimes X, V \otimes Y) \to \mathcal{A}(X, Y)$; $f' \longmapsto (\epsilon \otimes 1) \circ f'$.

Then there is a commutative diagram

$$\begin{array}{ccc}
\mathrm{Hom}_L(\mathrm{Hom}_A(L, X), \mathrm{Hom}_A(L, Y)) & \xrightarrow{\text{corestrict}} & \mathrm{Hom}_A(\mathrm{Hom}_A(L, X), Y) \\
\cong \uparrow \alpha_* & & \\
\mathrm{Hom}_L(\mathrm{Hom}_A(L, X), V \otimes Y) & & \cong \downarrow \alpha^* \\
\cong \downarrow \alpha^* & & \\
\mathrm{Hom}_L(V \otimes X, V \otimes Y) & \xrightarrow[G]{} & \mathrm{Hom}_A(V \otimes X, Y) = \mathcal{A}(X, Y)
\end{array}$$

in which all maps but G are isomorphisms, so G is also so. However $GF_C(f) = f$, from which it follows that F_C is full and faithful.

In view of 2.5, the next result exhibits $\mathcal{A}$-mod as a full functorially finite subcategory of each of R-mod and L-mod, provided some finiteness conditions on $\mathcal{A}$ hold.

2.6 Functorial Finiteness Theorem Let $\mathcal{A} = (A, V)$ be a bocs (satisfying 2.1) such that (a) $_AR$ is finitely generated, (b) $\mathrm{Hom}_L(V, -)$ maps L-mod to R-mod. Then $\underline{C}$ and $\underline{I}$ are functorially finite in R-mod and L-mod respectively.

Remark (a) and (b) can be replaced by (a) and (b') : A is left noetherian. For $\mathrm{Hom}_L(V, N) \subset \mathrm{Hom}_A(V, N) \cong R \otimes N$, which is in A-mod provided $N \in$ L-mod and (a),(b') hold; hence $\mathrm{Hom}_L(V, N)$ is also in R-mod.

Proof (i) $\underline{C}$ is covariantly finite in L-mod. Neither (a) nor (b) is needed for the well-known proof that for each $N \in$ L-mod, the covariant functor $\mathrm{Hom}_L(N, -)|_{\underline{C}}$ is a finitely generated functor on $\underline{C}$. By 2.2, $\mathrm{Hom}_A(L, N) \cong V \otimes N \in$ L-mod. So the map $N \to \mathrm{Hom}_A(L, N)$; $n \mapsto (e \mapsto en)$, induces a surjective functor morphism

$$\mathrm{Hom}_L(\mathrm{Hom}_A(L, N), -)|_{\underline{C}} \twoheadrightarrow \mathrm{Hom}_L(N, -)|_{\underline{C}}$$

from a representable functor on $\underline{C}$, as finite generation of $\mathrm{Hom}_L(N, -)|_{\underline{C}}$ requires.

(ii) $\underline{I}$ is contravariantly finite in R-mod. We only need (a). Then for any M in R-mod, $R\otimes M$ is in R-mod. Let $g : R \otimes M \to M$ be the map $s \otimes m \longmapsto sm$. Then the functor morphism induced by g,

$$\mathrm{Hom}_R(-, R\otimes M)|_{\underline{I}} \longrightarrow \mathrm{Hom}_R(-, M)|_{I},$$

is well-known to be surjective. Again using (a) to see

that $R \otimes M \in \underline{I}$, we see that the contravariant functor $\mathrm{Hom}_R(-, M)|_{\underline{I}}$ is finitely generated.

(iii) $\underline{I}$ is covariantly finite in R-mod. We need neither (a) nor (b). For any R-module M, consider the map

$$\begin{array}{ll} h : M \longrightarrow \mathrm{Hom}_A(V, V \otimes_R M); & m \longmapsto (v \longmapsto v \otimes_R m) \\ \qquad\qquad \| & \\ \qquad R \otimes V \otimes_R M & \text{by } (2.2) \end{array}$$

For $M \in$ R-mod, finite generation of ${}_AV$ implies that $R \otimes V \otimes_R M \in$ R-mod, in fact, that $R \otimes V \otimes_R M \in \underline{I}$. We show now that h actually induces a surjection of covariant functors

$$\mathrm{Hom}_R(\mathrm{Hom}_A(V, V \otimes_R M), -)|_{\underline{I}} \xrightarrow{h^*} \mathrm{Hom}_R(M, -)|_{\underline{I}},$$

which is the required result. By 2.2, modules in $\underline{I}$ are of the form $\mathrm{Hom}_A(V, X)$, where $X \in$ A-mod. Let $f \in \mathrm{Hom}_R(M, \mathrm{Hom}_A(V, X))$, then $f = \tilde{f}_* \circ h = h^*(\tilde{f}_*)$, where $\tilde{f}(v \otimes_R m) = f(m)(v)$ and $\tilde{f}_* = \mathrm{Hom}_A(1_V, \tilde{f})$. Hence h^* is surjective.

(iv) $\underline{C}$ is contravariantly finite in L-mod. We need (a) and (b). For an L-module N, consider the map

$$\mathit{1} : \tilde{N} = V \otimes \mathrm{Hom}_L(V, N) \longrightarrow N; \; v \otimes k \longmapsto k(v),$$

where $\tilde{N} \cong \mathrm{Hom}_A(L, \mathrm{Hom}_L(V, N))$ by 2.2.

The morphism $\mathit{1}_*$ of covariant functors induced by $\mathit{1}$,

$$\mathrm{Hom}_L(-, \tilde{N})|_{\underline{C}} \longrightarrow \mathrm{Hom}_L(-, N)|_{\underline{C}},$$

is surjective, for taking, by 2.2, a typical object of $\underline{C}$ to have the form $V \otimes X$, $X \in$ A-mod, any L-morphism $g : V \otimes X \to N$ lifts over $\mathit{1}_*$ to $1_V \otimes \tilde{g}$, where

$\tilde{g} : X \to Hom_L(V, N)$ is given by $\tilde{g}(x)(v) = g(v \otimes x)$. So to prove the contravariant finiteness of $\underline{C}$ in L-mod it is only necessary to check that, for $N \in$ L-mod, we have $\tilde{N} \in \underline{C}$. By (b) $Hom_L(V, N) \in$ R-mod, so by (a) ${}_A Hom_L(V, N) \in$ A-mod. Hence $\tilde{N} = V \otimes Hom_L(V, N)$ is in $\underline{C}$ as required.

This completes the proof of theorem 2.6.

We conclude this section with the observation that the bimodule ${}_L V_R$ is 'balanced'.

2.7 **Proposition** The natural ring homomorphisms

$$L \to End_{R^{op}}(V) \quad ; \quad s \longmapsto (v \longmapsto sv) \ ,$$

$$R^{op} \to End_L(V) \quad ; \quad e \longmapsto (v \longmapsto ve) \ ,$$

are isomorphisms.

Proof $L \to End_{R^{op}}(V)$ is the composite of the isomorphisms

$$L = Hom_{A^{op}}(V, A) \cong Hom_{R^{op}}(V, Hom_{A^{op}}(R, A)) \cong End_{R^{op}}(V),$$

and likewise

$R^{op} \to End_L(V)$ is the composite of the isomorphims

$$R^{op} = Hom_A(V, A) \cong Hom_L(V, Hom_A(L, A)) \cong End_L(V).$$

Remark $R^{op} \to End_L(V)$ is also realised by the map ν_R in the proof of 2.4; an analogue of 2.4 for mod-R, mod-L would include the corresponding statement about $L \to End_{R^{op}}(V)$.

§3 Extension Properties

In Theorem 3.8, we give conditions on a bocs $\mathcal{A} = (A, V)$ which ensure that the subcategories $\underline{I}$ of R-mod and $\underline{C}$ of L-mod are closed under extensions. The conditions imply 2.1, so that the functors $R \otimes -$ and

$\mathrm{Hom}_A(L, -)$ are exact on A-mod, and we will in fact obtain the much stronger result that the maps induced by these functors from Ext_A^n to $\mathrm{Ext}_R^n|_{\underline{I}\times\underline{I}}$ and to $\mathrm{Ext}_L^n|_{\underline{C}\times\underline{C}}$ are surjective for $n = 1$, and bijective for $n \geq 2$.

We need some generalities about induction and coinduction, starting with the following well-known lemma of Eckmann and Shapiro.

3.1 **Proposition** (a) Let R be an A-algebra such that R_A is projective. Given $_AZ$, $_RM$, the induction isomorphism

$$\Gamma : \mathrm{Hom}_A(Z, M) \xrightarrow{\sim} \mathrm{Hom}_R(R \otimes Z, M)$$

induces isomorphisms

$$\Gamma^{(n)} : \mathrm{Ext}_A^n(Z, M) \xrightarrow{\sim} \mathrm{Ext}_R^n(R \otimes Z, M)$$

for all $n \geq 0$.

(b) Let L be an A-algebra such that $_AL$ is projective.

Given $_AX$, $_LN$, the coinduction isomorphism

$$\Delta : \mathrm{Hom}_A(N, X) \xrightarrow{\sim} \mathrm{Hom}_L(N, \mathrm{Hom}_A(L, X))$$

induces isomorphisms

$$\Delta^{(n)} : \mathrm{Ext}_A^n(N, X) \xrightarrow{\sim} \mathrm{Ext}_L^n(N, \mathrm{Hom}_A(L, X))$$

for all $n \geq 0$.

3.2 **Notation** For A-modules $_AZ$, $_AX$, there are natural transformations

$$\gamma : {}_AX \longrightarrow {}_AR \otimes X \quad ; \; x \longmapsto 1 \otimes x,$$

$$\delta : \mathrm{Hom}_A(L, Z) \longrightarrow Z \quad ; \; h \longmapsto h(1),$$

inducing natural transformations

$$\gamma^{(n)} : \mathrm{Ext}_A^n(Z, X) \longrightarrow \mathrm{Ext}_A^n(Z, R \otimes X),$$

$$\delta^{(n)} : \mathrm{Ext}_A^n(Z, X) \longrightarrow \mathrm{Ext}_A^n(\mathrm{Hom}_A(L, Z), X)$$

for each $n \geq 0$.

We give conditions on the bocs $\mathcal{A} = (A, V)$ of which L, R are the left, right algebra (respectively) to ensure that $\gamma^{(1)}$, $\delta^{(1)}$ are surjective and $\gamma^{(n)}$, $\delta^{(n)}$ bijective for $n \geq 2$.

3.3 **Definition** The kernel $\bar{V} = {}_A\bar{V}_A$ of a bocs $\mathcal{A} = (A, V)$ is the kernel of the counit ϵ of $\mathcal{A}$. Since ϵ is surjective, the exact sequence of bimodules,

$$0 \longrightarrow \bar{V} \longrightarrow V \xrightarrow{\epsilon} A \longrightarrow 0,$$

splits as a sequence of left A-modules, and also as a sequence of right A-modules. So, for all ${}_AZ$, ${}_AX$, there are exact sequences of left A-modules

3.3 (a) $0 \longrightarrow X \xrightarrow{\epsilon^*} \mathrm{Hom}_A(V, X) \longrightarrow \mathrm{Hom}_A(\bar{V}, X) \longrightarrow 0$

3.3 (b) $0 \longrightarrow \bar{V} \otimes Z \longrightarrow V \otimes Z \xrightarrow{\epsilon\otimes 1} Z \longrightarrow 0.$

In fact, ϵ^* and $\epsilon \otimes 1$ are given by

$$\epsilon^* = \beta\gamma \ , \ \epsilon \otimes 1 = \delta\alpha$$

with β, α as defined in §2, and γ, δ as in 3.2.

3.4 **Proposition** Let $\bar{V}$ be the kernel of the bocs $\mathcal{A} = (A, V)$, and R and L the right algebra and left algebra, respectively, of $\mathcal{A}$.

(a) Assume ${}_AV$ is finitely generated projective, and X a left A-module such that ${}_A\mathrm{Hom}_A(\bar{V}, X)$ is injective. Then, for all ${}_AZ$, $\gamma^{(1)}$ is surjective and $\gamma^{(n)}$ is bijective for all $n \geq 2$.

(b) Assume V_A is finitely generated projective and Z a left A-module such that ${}_A\bar{V} \otimes Z$ is projective. Then, for all ${}_AX$, $\delta^{(1)}$ is surjective and

$\delta^{(n)}$ is bijective for all $n \geq 2$.

Proof (a) Since ${}_A V$ is finitely generated projective, Proposition 2.2 shows that β is an isomorphism, so $\gamma = \beta^{-1}\epsilon^{*}$. Hence it suffices to prove the result for the $\epsilon^{*(n)}$. The last term of 3.3 (a) is injective, so the long exact sequence obtained by applying $\mathrm{Hom}_A(Z, -)$ to 3.3 (a) shows that $\epsilon^{*(1)}$ is surjective and $\epsilon^{*(n)}$ bijective for $n \geq 2$, as required.

(b) In this case, the hypotheses imply that α is an isomorphism, so $\delta = (\epsilon \otimes 1)\alpha^{-1}$; the result follows by applying $\mathrm{Hom}_A(-, X)$ to 3.3 (b), in which the first term is, by hypothesis, projective.

We now want to combine the results of 3.1 and 3.4. On taking $M = R \otimes X$ in 3.4 (a) and $N = \mathrm{Hom}_A(L, Z)$ in 3.4(b), we obtain <u>at degree</u> 0, maps

$$\Gamma\gamma^{(0)} : \mathrm{Hom}_A(Z, X) \longrightarrow \mathrm{Hom}_R(R \otimes Z, R \otimes X)$$

$$\Delta\delta^{(0)} : \mathrm{Hom}_A(Z, X) \longrightarrow \mathrm{Hom}_L(\mathrm{Hom}_A(L, Z), \mathrm{Hom}_A(L, X))$$

which are easily verified to be given by

$$(\Gamma\gamma)(h) = 1_R \otimes h \quad , \quad (\Delta\delta)(h) = \mathrm{Hom}_A(1_L, h)$$

for $h \in \mathrm{Hom}_A(Z, X)$. Of course 3.1 gives conditions under which these composites induce maps on Ext^n -groups, and 3.4 gives further conditions under which the induced maps are surjective when $n = 1$, and bijective for $n \geq 2$. We can, however, describe these induced maps more explicitly using Yoneda's interpretation of Ext^n, $n \geq 1$, as a group of equivalence classes of 'n-fold extensions'- by which we mean exact sequences of the form

$$E_* : 0 \to U_n \to U_{n-1} \to \dots \to U_1 \to U_0 \to U \to 0.$$

3.5 **Corollary** With $\mathcal{A}$, L, R, as in Proposition 3.4

(a) assume ${}_AV$ is finitely generated projective and ${}_AX$ is a module for which ${}_A\mathrm{Hom}_A(\overline{V}, X)$ is injective. Then R_A is projective, the 'induction' maps

$$\Gamma\gamma : \mathrm{Ext}_A^n(Z, X) \longrightarrow \mathrm{Ext}_R^n(R \otimes Z, R \otimes X)$$

are defined for all ${}_AZ$ and all $n \geq 1$, are surjective for $n = 1$ and bijective for $n \geq 2$, and are induced by the map $E_* \longmapsto R \otimes E_*$ of n-fold extensions E_*.

(b) assume V_A is finitely generated projective and ${}_AZ$ is a module such that ${}_A\overline{V}\otimes Z$ is projective. Then ${}_AL$ is projective, the 'coinduction' maps

$$\Delta\delta : \mathrm{Ext}_A^n(Z, X) \longrightarrow \mathrm{Ext}_L^n(\mathrm{Hom}_A(L, Z), \mathrm{Hom}_A(L, X))$$

are defined for all ${}_AX$ and all $n \geq 1$, are surjective for $n = 1$ and bijective for $n \geq 2$, and are induced by the map $E_* \longmapsto \mathrm{Hom}_A(L, E_*)$ of n-fold extensions E_*.

Proof All but the last assertion in each part follows immediately from 3.1 and 3.4. For (a), let E_* be an n-fold extension of X by Z representing some element $\zeta \in \mathrm{Ext}_A^n(Z, X)$; we must show that $R \otimes E_*$ represents $(\Gamma\gamma)(\zeta)$. Choose a projective resolution (P_*, d_*) of ${}_AZ$, then there is a commutative diagram

$$\begin{array}{ccccccccccccc}
\dots P_{n+1} & \xrightarrow{d_n} & P_n & \xrightarrow{d_{n-1}} & P_{n-1} & \longrightarrow & \dots\dots & \longrightarrow & P_0 & \longrightarrow & Z & \longrightarrow & 0 \\
\downarrow & & \downarrow{\scriptstyle f_n} & & \downarrow & & & & \downarrow & & \| & & \\
0 & \longrightarrow & X & \longrightarrow & Y_{n-1} & \longrightarrow & \dots\dots & \longrightarrow & Y_0 & \longrightarrow & Z & \longrightarrow & 0
\end{array}$$

in which the lower row is E_* and the cocycle f_n represents ζ. Since R_A is projective, $R \otimes -$ is exact and $R \otimes P_*$ is an R-projective resolution of $R \otimes Z$, so on applying $R \otimes -$ to the above diagram, we see that the n-fold extension $R \otimes E_*$ is a Yoneda representative of the element of $\mathrm{Ext}^n_R(R \otimes Z, R \otimes X)$ determined by the cocycle $1_R \otimes f_n$. Since $1_R \otimes f_n = (\Gamma\gamma)(f_n)$, this element is $(\Gamma\gamma)(\zeta)$, as required. The proof of (b) is similar, using an injective resolution of ${}_AX$.

For the application to bocses in Theorem 3.8 below, we will symmetrise the hypotheses given separately in (a) and (b) of the preceding proposition. We begin with a definition:

Definition An A-A-bimodule U will be called a projectivising bimodule if, for all ${}_AZ$, W_A, the A-modules $U \otimes Z$ and $W \otimes U$ are projective.

3.6 **Lemma** Let ${}_AU_A$ be a bimodule. It is projectivising if and only if, for all ${}_AX$, Y_A, $\mathrm{Hom}_A(U,X)$ and $\mathrm{Hom}_{A^{op}}(U, Y)$ are injective.

Proof We just sketch the 'only if' part. If ${}_AU_A$ is projectivising ${}_AU$ and U_A are projective, so $- \otimes U$ and $U \otimes -$ are exact. For any short exact sequence ${}_AZ_*$ and module ${}_AX$, there is an isomorphism

$$\mathrm{Hom}_A(Z_*, \mathrm{Hom}_A(U, X)) \cong \mathrm{Hom}_A(U \otimes Z_*, X).$$

Since U_A is projective, $U \otimes Z_*$ is exact, and split exact since U is projectivising. Hence, for all X and Z_*, the left hand side is exact, and so for each fixed ${}_AX$, $\mathrm{Hom}_A(U, X)$ is injective. Similarly, $\mathrm{Hom}_{A^{op}}(U, Y)$ is injective for each Y_A.

3.7 **Examples** For any algebra A over a semisimple artinian subring k, $A \otimes_k A$ is projectivising, as also is any A-A-bimodule summand of any direct sum of copies of $A \otimes_k A$.

3.8 **Extension Closure Theorem**

Let $\mathcal{A} = (A, V)$ be a bocs such that ${}_AV$ and V_A are finitely generated, and ${}_A\overline{V}_A$ is a projectivising bimodule. Then for all ${}_AX$ and ${}_AZ$, the maps

$$\operatorname{Ext}^n_A(Z, X) \longrightarrow \operatorname{Ext}^n_R(R \otimes Z, R \otimes X)$$

$$\operatorname{Ext}^n_A(Z, X) \longrightarrow \operatorname{Ext}^n_L(\operatorname{Hom}_A(L, Z), \operatorname{Hom}_A(L, X))$$

induced by induction and coinduction exist, and are surjective for $n = 1$ and bijective for $n \geq 2$. Moreover, the subcategories $\underline{I}$ of R-mod and $\underline{C}$ of L-mod are closed under extensions.

Proof We have already noted that ${}_A\overline{V}$, $\overline{V}_A$ are projective, so ${}_AV$ and V_A are finitely generated projective modules. Hence all conditions of 3.5 hold, proving the assertion about the mappings of Ext -groups. We prove that $\underline{I}$ is closed under extensions.

Let $0 \to R \otimes X \to M \to R \otimes Z \to 0$ be a short exact sequence of R-modules with $X, Z \in$ A-mod. Then 3.5(a) for $n = 1$ shows that there is a short exact sequence $0 \to X \to Y \to Z \to 0$ in A-mod and an exact commutative diagram

$$\begin{array}{ccccccccc} 0 & \longrightarrow & R \otimes X & \longrightarrow & R \otimes Y & \longrightarrow & R \otimes Z & \longrightarrow & 0 \\ & & \Vert & & \downarrow & & \Vert & & \\ 0 & \longrightarrow & R \otimes X & \longrightarrow & M & \longrightarrow & R \otimes Z & \longrightarrow & 0 \end{array}$$

Hence $M \cong R \otimes Y$, so $M \in \underline{I}$. The proof for $\underline{C}$ in L-mod is similar.

§4 Almost split sequences

Let k be a field, and $\mathcal{A}$ = (A, V) a finite-dimensional bocs over k, by which we mean that A is a finite-dimensional k – algebra and $\dim_k(V) < \infty$. Then L and R are also finite-dimensional k-algebras. We now briefly give conditions for the existence of almost split sequences in $\mathcal{A}$-mod, $\underline{I}$, and $\underline{C}$, using the Auslander-Smalø criteria [AS:2.4].

4.1 **Theorem** Let $\mathcal{A}$ = (A, V) be a finite-dimensional bocs over a central subfield k, with kernel $\overline{V}$ a projectivising A-A-bimodule. Then add($\underline{I}$) and add($\underline{C}$) admit almost split sequences. If, additionally, idempotents split in $\mathcal{A}$-mod, then $\mathcal{A}$-mod, $\underline{I}$ and $\underline{C}$ admit almost split sequences.

Proof We just prove the assertions pertaining to the subcategory $\underline{I}$ of R-mod, those pertaining to $\underline{C}$ being similar. Since R is a finite-dimensional algebra, we must verify - following [AS:2.4] - that add($\underline{I}$) is a full subcategory of R-mod, closed under isomorphisms, direct summands, and extensions in R-mod, and is functorially finite in R-mod.

Since $\overline{V}$ is projectivising and V finite-dimensional, (2.1) holds. Therefore all conditions of Theorems 2.6 and 3.8 are satisfied, so that $\underline{I}$ - and hence add($\underline{I}$)- is functorially finite in R-mod and is closed under extensions. The other Auslander-Smalø requirements are, of course, true by definition of $\underline{I}$ and add($\underline{I}$). Therefore add($\underline{I}$) admits almost split sequences. If idempotents split in $\mathcal{A}$-mod, then idempotents split in $\underline{I}$, so since $\underline{I}$ is full in R-mod, $\mathcal{A}$-mod $\cong$ $\underline{I}$ = add($\underline{I}$).

4.2 **Remarks** (i) $\overline{V} \in \text{add}\,(A \otimes_k A)$ is the most frequently occurring type of projectivising kernel of a bocs.

(ii) the R-modules in add($\underline{I}$) are often called relatively projective modules, the L-modules in add($\underline{C}$) are similarly called relatively injective modules ; they are the (finitely generated) projectives and injectives for the relative homological algebras based on short exact sequences in R-mod and L-mod, respectively, which split on restriction to A.

(iii) The entirely different sort of problem of describing bocses $\mathcal{A}$ for which idempotents split in $\mathcal{A}$-mod is the subject of §6.

§5 Cotilting Properties

We assume here that $\mathcal{A} = (A, V)$ is a bocs in which A is an artin algebra and ${}_AV$, V_A are finitely generated projective.

Let $d = \max\,(\text{i.d. } {}_AA, \text{ i.d. } A_A)$ where i.d. denotes injective dimension. When $d < \infty$ it is well known that $\text{i.d. } {}_AA = d = \text{i.d. } A_A$.

Note that L, R are also artin algebras; we shall denote the usual duality associated with any of the artin algebras A, L, R by D. Our main result is as follows:

5.1 **Cotilting Theorem** Under the above hypotheses and assuming that $d < \infty$, ${}_LV$, V_R are generalised cotilting modules of i.d. at most d.

Proof We must show that ${}_LV$ satisfies

(i) $\text{i.d. } {}_LV \leq d$

(ii) $\text{Ext}^i_L(V, V) = 0 \quad (i \geq 1)$

(iii) There is an exact sequence

$$0 \longrightarrow V_d \longrightarrow \cdots \longrightarrow V_0 \longrightarrow DL \longrightarrow 0$$

with $V_i \in \operatorname{add} V$.

Consider the isomorphism of 3.1(b)

$$\operatorname{Ext}^i_A(N, X) \xrightarrow{\sim} \operatorname{Ext}^i_L(N, \operatorname{Hom}_A(L, X))$$

To prove (i) put $X = A$; since ${}_L\operatorname{Hom}_A(L, A) \cong {}_LV$, this shows that $\text{i.d. } {}_LV \le \text{i.d. } {}_AA = d$; (ii) follows by putting $N = {}_LV$, which is projective as a left A-module, and $X = A$.

Since $\text{i.d. } A_A = d$ there is a projective resolution

$$0 \longrightarrow P_d \longrightarrow \cdots \longrightarrow P_0 \longrightarrow {}_A(DA) \longrightarrow 0$$

of DA. Applying $\operatorname{Hom}_A(L, -)$ to this yields a resolution of $\operatorname{Hom}_A(L, DA) \cong DL$ in $\operatorname{add}(\operatorname{Hom}_A(L, A))$ which is $\operatorname{add}({}_LV)$. This proves (iii).

The proof of the corresponding result for V_R is similar, using the analogue of 3.1(b) for right R and right A-modules, and a projective resolution of DA as a right A-module.

Remarks (i) When $\bar{V}$ is a projectivising bimodule we have (for $i \ge 2$):

$$\operatorname{Ext}^i_A(Z, X) \xrightarrow{\sim} \operatorname{Ext}^i_L(\operatorname{Hom}_A(L, Z), \operatorname{Hom}_A(L, X))$$

so putting $X = A$ and $i = d$ shows that, if $d \ge 2$, ${}_LV$ has i.d. <u>equal</u> to d.

(ii) Since $R \cong \operatorname{End}_L(V)^{op}$ the fact that V_R is a cotilting module also follows at once from ${}_LV$ being a cotilting module.

(iii) In applications A is usually hereditary and $\bar{V}$ projectivising so ${}_LV$, V_R are either projective modules, or classical cotilting modules.

We determine here the Ext-projective and Ext-injective modules in $\underline{I}$ and $\underline{C}$. Full details will be given for $\underline{I}$ and the proof for $\underline{C}$ (which is similar) will be omitted.

Recall that $M_1(M_3)$ in $\underline{I}$ is Ext-injective (Ext-projective) if every short exact sequence $0 \longrightarrow M_1 \longrightarrow M_2 \longrightarrow M_3 \longrightarrow 0$ with each $M_i \in \underline{I}$ is split.

5.2 **Theorem** Suppose idempotents split in $\mathcal{C}$-mod and $\overline{V}$ is a projectivising bimodule; then the Ext-projectives in $\underline{I}$ are the finitely-generated projective R-modules and the Ext-injectives in $\underline{I}$ are the objects of add(DV).

Proof Since $\overline{V}$ is projectivising there is a surjection

$$\operatorname{Ext}^1_A(Z, X) \longrightarrow \operatorname{Ext}^1_R(R \otimes Z, R \otimes X)$$

Putting $X = DA$, so $R \otimes X \cong DV$, we see that all the objects of add(DV) are Ext-injective. The finitely generated projective R-modules are obviously Ext-projectives in $\underline{I}$.

If $M = R \otimes Z$ is Ext-projective consider a short exact sequence

$$0 \longrightarrow K \longrightarrow A^r \longrightarrow Z \longrightarrow 0.$$

Applying $R \otimes -$ yields a sequence in $\underline{I}$ with middle term in add(R). As $R \otimes Z$ is Ext-projective the induced sequence splits and so M is in add(R).

Likewise considering a short exact sequence

$$0 \longrightarrow X \longrightarrow (DA)^r \longrightarrow C \longrightarrow 0$$

and applying $R \otimes -$ shows that if $R \otimes X$ is Ext-injective then it is in add(DV).

5.3 **Theorem** Suppose idempotents split in $\mathcal{C}$-mod and $\overline{V}$ is a projectivising bimodule; then the Ext-projectives in $\underline{C}$ are the objects of add(V) and the Ext-injectives in $\underline{C}$ are the finitely generated injective L-modules.

Remark Since L and R are cotilted from each other the number of isomorphism classes of simple left L-modules and simple left R-modules are equal by the analogue of [M : 1.19] for cotilting modules. If this number is s then s is finite and is also equal to the number of isomorphism classes of Ext-projective or Ext-injective indecomposables in $\underline{I}$ or $\underline{C}$, provided $\mathcal{C}$ satisfies the hypotheses needed for Theorems 5.2 and 5.3.

§6 Splitting of Idempotents

In 6.4 we give sufficient conditions on bocses over fields for idempotents to split in their categories of finite-dimensional representations. The construction of such bocses starts with a so-called principal bocs and proceeds via finite sequences of elementary extensions of two basic types, which we call triangular kernel and triangular tensor algebra extensions. The bocses so obtained include all 'additive Roiter bocses' [CB: 3.5].

As mentioned before, the construction is essentially a different presentation of the material relevant to bocses over fields in [BK, §5]. As in that account, some rather extensive preliminaries are needed before the main result can be stated and proved.

Throughout §6 we keep fixed a basefield k and a commutative k-algebra S (S is an analogue of the underlying minimal category of the bocses considered by

Crawley-Boevey [C-B: 2.1, 3.6]). Each algebra considered in §6 is an S-algebra on which k acts centrally, but these algebras need not have finite k-dimension. As usual we only consider bimodules on which k acts centrally. More fundamentally we also assume that each bocs $\mathcal{A} = (A, V)$ possesses a <u>grouplike element centralised by S</u>, that is, an element ω of V such that

$$\epsilon(\omega) = 1,\ \mu(\omega) = \omega \otimes \omega,\ s\omega = \omega s \quad \text{for} \quad s \in S.$$

Hence V has direct decompositions

$${}_AV_S = {}_A\overline{V}_S + A\omega,\ {}_SV_A = {}_S\overline{V}_A + \omega A,$$

and ${}_SV \otimes V_S$ a direct decomposition

$$V \otimes V = \overline{V} \otimes \overline{V} + \overline{V} \otimes \omega A + \omega A \otimes \overline{V} + \omega A \otimes \omega;$$

(in which we have identified $\overline{V} \otimes \overline{V}$ with its canonical image in $V \otimes V$).

We also need the <u>differentials</u> [C-B: 3.3] $d_1 : A \to \overline{V}$, $d_2 : \overline{V} \to \overline{V} \otimes \overline{V}$ defined by:

$$d_1(a) = a\omega - \omega a,$$

$$d_2(v) = \mu(v) - v \otimes \omega - \omega \otimes v,$$

which one may check do indeed have images inside $\overline{V}$ and $\overline{V} \otimes \overline{V}$ [C-B : 3.4(2)]. Often we denote either differential by d; both differentials are S-S-bimodule morphisms.

We will study only representations $X = {}_AX$ of of finite k-dimension; these determine a full subcategory, to be denoted $\mathcal{A}$-k-mod, of $\mathcal{A}$-mod. For any ${}_AX$, ${}_AY$ write (X, Y) for $\mathrm{Hom}_k(X, Y)$. Then (X, Y) is an A-A-bimodule, and throughout this section we will take $\mathcal{A}(X, Y)$ to be a natural transpose of $\mathrm{Hom}_A(V \otimes X, Y)$, namely,

$$\mathcal{C}(X, Y) = \operatorname{Hom}_{A\times A^{op}}(V, (X, Y)).$$

In this form the composite of $\theta \in \mathcal{C}(X, Y)$ with $\varphi \in \mathcal{C}(Y, Z)$ is the morphism

$$V \xrightarrow{\ \mu\ } V \otimes V \xrightarrow{\ \varphi\otimes\theta\ } (Y, Z) \otimes (X, Y) \xrightarrow{\ c\ } (X, Z),$$

where c is composition of k-linear maps.

We also need a description of $\theta \in \mathcal{C}(X, Y)$ as a pair $\theta = (\theta_0, \overline{\theta})$ where $\theta_0 = \theta(\omega) \in (X, Y)$ and $\overline{\theta} = \theta|_{\overline{V}}$. Since $a\omega - \omega a = d(a)$, θ_0 and $\overline{\theta}$ are related by

$$a\theta_0 - \theta_0 a = \overline{\theta}(d(a)) \quad \text{for every} \quad a \in A.$$

Conversely any pair $(\theta_0, \overline{\theta})$ satisfying this relation determines a bocs morphism, and the correspondence set up in this way is bijective. Note that θ_0 is S-linear.

The formula for the composite of $\theta = (\theta_0, \overline{\theta}) \in \mathcal{C}(X, Y)$ and $\varphi = (\varphi_0, \overline{\varphi}) \in \mathcal{C}(Y, Z)$ is given by

$$(\varphi\theta)_0 = \varphi_0\theta_0,$$

$$\overline{\varphi\theta}(v) = \varphi_0\overline{\theta}(v) + \overline{\varphi}(v)\theta_0 + c(\overline{\varphi} \otimes \overline{\theta})\, d(v) \quad \text{for} \quad v \in \overline{V};$$

the identity morphism in $\mathcal{C}(X, X)$ is $\epsilon_X = (1_X, 0)$.

These formulae show that there is a functor, denoted ω^*,

$$\omega^* : \mathcal{C}\text{-k-mod} \to \text{k-mod}$$

given by $X \to {}_kX$, $\theta \longmapsto \theta_0$. In what follows we first exhibit a class of bocses for which ω^* reflects isomorphisms - i.e. if θ is such that $\omega^*(\theta)$ is an isomorphism then θ is an isomorphism in $\mathcal{C}$-k-mod, which means that the isomorphisms in the category

$\mathcal{A}$-k-mod are exactly the morphisms θ for which θ is invertible.

We need several concepts of one bocs being an extension of another.

6.1(i) **Definition** Let $\mathcal{A} = (A, V)$ be a bocs. We say that $\mathcal{A}$ is an extension of the bocs $\mathcal{A}' = (A', V', \mu', \epsilon')$ if

(i) A' is an S-subalgebra of A,

(ii) V' is an A'-A'-subbimodule of V containing ω (the group-like of $\mathcal{A}$), and the following diagrams commute ($\otimes'$ is the tensor product over A'):

$$\begin{array}{ccc} V' & \longrightarrow & V \\ {\scriptstyle \epsilon'}\downarrow & & \downarrow{\scriptstyle \epsilon} \\ A' & \longrightarrow & A \end{array} \quad , \quad \begin{array}{ccccc} V' & & \longrightarrow & & V \\ {\scriptstyle \mu'}\downarrow & & & & \downarrow{\scriptstyle \mu} \\ V' \otimes' V' & \longrightarrow & V \otimes' V & \longrightarrow & V \otimes V \end{array}$$

Remarks It is clear that $\overline{V}' = \overline{V} \cap V'$, and the differentials satisfy $d_1' = d_1|_{A'}$, and $d_2|_{\overline{V}'}$ is the composite

$$\overline{V}' \xrightarrow{d_2'} \overline{V}' \otimes' \overline{V}' \longrightarrow \overline{V} \otimes \overline{V}.$$

Notice that there is a 'restriction' functor $\mathcal{A}$-k-mod $\to \mathcal{A}'$-k'-mod, defined by $X \longmapsto {}_{A'}X$ and $\theta \longmapsto \theta'$ where $\theta' = \theta|_{V'} = (\theta_0', \overline{\theta}')$ and $\theta_0' = \theta_0$, $\overline{\theta}' = \overline{\theta}|_{\overline{V}'}$.

6.1 (ii) **Definition** The extension $\mathcal{A}$ of $\mathcal{A}'$ is called a triangular kernel extension if $A' = A$, and there is an S-S-subbimodule U of $\overline{V}$ such that

(a) the natural map $\overline{V}' \oplus (A \otimes_S U \otimes_S A) \to \overline{V}$ induced by the inclusions $\overline{V}' \subset \overline{V}$ and $U \subset \overline{V}$ is an isomorphism;

(b) $d(U) \subseteq \overline{V}' \otimes \overline{V}'$ (viewed as a subbimodule of $\overline{V} \otimes \overline{V}$ via (a)).

6.1 (iii) **Definition** The extension $\mathcal{A}$ of $\mathcal{A}'$ is called a <u>triangular algebra extension</u> <u>(triangular tensor algebra extension)</u> if

(a) there is an S-S-subbimodule E of A such that the morphism of A'-algebras

$$\otimes_{A'} (A' \otimes_S E \otimes_S A') \longrightarrow A$$

induced by the inclusion $E \to A$ is surjective (bijective);

(b) the natural map $A \otimes' \overline{V}' \otimes' A \to \overline{V}$ is an isomorphism;

(c) $d(E) \subseteq \overline{V}'$.

6.2 **Definition** The <u>principal bocs</u> of an algebra B, is the bocs (B, B) with comultiplication $B \cong B \otimes_B B$ and counit $1 : B \to B$. This has a grouplike (namely 1) and ω^* clearly reflects isomorphisms since (B, B)-mod is just B-mod.

6.3 **Proposition** Let $\mathcal{A}_0, \mathcal{A}_1, \ldots, \mathcal{A}_n = \mathcal{A}$ be a finite sequence of bocses such that $\mathcal{A}_0$ is principal and for $i = 0, 1, \ldots, n-1$, $\mathcal{A}_{i+1}$ is a triangular kernel or algebra extension of $\mathcal{A}_i$. Then $\omega^* : \mathcal{A}\text{-k-mod} \to \text{k-mod}$ reflects isomorphisms.

Proof Since $\omega^* : \mathcal{A}_0 - \text{k-mod} \to \text{k-mod}$ reflects isomorphisms ($\mathcal{A}_0$ is a principal bocs) it is sufficient, by induction, to prove the following lemma.

Lemma Let $\mathcal{A}$ be a triangular kernel or algebra extension of $\mathcal{A}'$, then if

$$\omega^* : \mathcal{A}'\text{-k-mod} \to \text{k-mod}$$

reflects isomorphisms so does

$$\omega^* : \mathcal{C}\text{-k-mod} \to \text{k-mod}.$$

Proof Let $\theta = (\theta_0, \overline{\theta}) \in \mathcal{C}(X, Y)$ and suppose θ_0 is invertible. We need to show that θ is an isomorphism. The restriction $\theta' = (\theta_0', \overline{\theta}') \in \mathcal{C}'(X, Y)$ is an isomorphism as $\theta_0' = \theta_0$ is an isomorphism. We will show how to extend an inverse $\varphi' = (\varphi_0', \overline{\varphi}') \in \mathcal{C}'(Y, X)$ for θ' to a left inverse $\varphi = (\varphi_0, \overline{\varphi})$ in $\mathcal{C}(Y, X)$ for θ; a similar construction gives a right inverse for θ, so θ is invertible.

We set $\varphi_0 = \varphi_0'$ so $\varphi_0\theta_0 = 1_X$, $\theta_0\varphi_0 = 1_Y$.

Case (i): Let be a triangular kernel extension of $\mathcal{C}'$, so $A' = A$, $\overline{V} = \overline{V}' \oplus (A \otimes_S U \otimes_S A)$ and $d(U) \subseteq \overline{V}' \otimes \overline{V}'$.

Define $\overline{\varphi} = \overline{\varphi}'$ on the summand $\overline{V}'$ of $\overline{V}$. For $u \in U$, since $d(u) \in \overline{V}' \otimes \overline{V}'$ we may set

$$\overline{\varphi}(u) = -\varphi_0\overline{\theta}(u)\varphi_0 - c(\overline{\varphi}' \otimes \overline{\theta}')(d(u))\,\varphi_0 \ ;$$

this gives an S-S-bimodule map $U \to (Y, X)$, and so has a canonical extension to an A-A-bimodule map of the second summand of $\overline{V}$ to (Y, X). This completes the definition of $\overline{\varphi}$ and of $\varphi = (\varphi_0, \overline{\varphi})$; φ is a bocs morphism because

$$\overline{\varphi}(d(a)) = \overline{\varphi}'(d(a)) = a\varphi_0' - \varphi_0'a = a\varphi_0 - \varphi_0 a$$

for each $a \in A$.

A short calculation shows that $\overline{\varphi\theta} : \overline{V} \to (X, X)$ is zero on $\overline{V}'$ and U, and hence on $\overline{V}$. Therefore $\varphi\theta = \epsilon_X$, as required.

Case (ii) :Let $\mathcal{C}$ be a triangular algebra extension of $\mathcal{C}'$, so A is generated by A' and the S-S-subbimodule E of A, $\overline{V} \cong A \otimes' \overline{V}' \otimes' A$, and $d(E) \subseteq \overline{V}'$. For this case we extend $\overline{\varphi}'$ in the natural way to obtain

$$\bar{\varphi} : \bar{V} \cong A \otimes' \bar{V}' \otimes A \longrightarrow (Y, X).$$

Since φ' is a left inverse of θ' we have, for $v' \in \bar{V}'$, the equation:

(1) $$0 = \varphi_0'\bar{\theta}'(v') + \bar{\varphi}(v')\theta_0' + c(\bar{\varphi}' \otimes' \bar{\theta}')\, d(v')$$

Let $e \in E$ then $d(e) \in V'$, and if we put $v' = d(e)$ in (1) and recall that $\bar{\theta}'(d(e)) = \bar{\theta}(d(e)) = e\theta_0 - \theta_0 e$, $\theta_0' = \theta_0$, $\varphi_0' = \varphi_0 = \theta_0^{-1}$, and $dd(e) = 0$ we see that the relation

(2) $$\bar{\varphi}(d(a)) = a\varphi_0 - \varphi_0 a$$

is satisfied for $a \in E$. Also (2) holds for $a \in A'$, and if it holds for a_1 and a_2 then it holds for $a_1 + a_2$ and $a_1 a_2$. Since $A' \cup E$ generates A, (2) holds for all $a \in A$. Therefore $\varphi = (\varphi_0, \bar{\varphi}) \in \mathcal{A}(Y, X)$. By definition $(\varphi\theta)_0 = 1_X$ and $\varphi\theta|_{\bar{V}'} = \overline{\varphi'\theta'} = 0$. Since $\bar{V}'$ generates $\bar{V}$ over A we conclude that $\overline{\varphi\theta} = 0$. So finally we conclude that $\varphi\theta = \epsilon_X$, as required.

We now give our main theorem concerning idempotent splitting, and will need Proposition 6.3 in its proof.

6.4 <u>**Theorem**</u> Let $\mathcal{A}_0, \mathcal{A}_1, \ldots, \mathcal{A}_n = \mathcal{A}$ be a finite sequence of bocses such that $\mathcal{A}_0$ is principal and, for $i = 0, 1, \ldots, n-1$, $\mathcal{A}_{i+1}$ is either a triangular kernel or triangular tensor algebra extension of $\mathcal{A}_i$. Then idempotents split in $\mathcal{A}$-k-mod.

<u>**Proof**</u> Following [BK] we prove that if $\sigma \in \mathcal{A}(X, X)$ is an idempotent there is an isomorphism $\alpha \in \mathcal{A}(X, Y)$, for some Y, such that $\overline{\alpha\sigma\alpha^{-1}} = 0$. It then follows that $(\alpha\sigma\alpha^{-1})_0$ is an A-module idempotent and if $Y \xrightarrow{f_0} Z \xrightarrow{g_0} Y$ is a splitting of this idempotent then

$$Y \xrightarrow{(f_0,0)} Z \xrightarrow{(g_0,0)} Y$$

is a splitting of $\overline{\alpha\sigma\alpha^{-1}}$. Hence $\sigma = (\alpha^{-1}(g_0, 0))((f_0,0)\alpha)$ is a splitting of σ. Since this property holds for all principal bocses it is enough, by induction, to prove the following lemma (in which $\mathcal{A}' = \mathcal{A}_i$ for some $i \geq 0$).

Lemma Let $\mathcal{A}$ be a triangular kernel or tensor algebra extension of $\mathcal{A}'$. Suppose that for every idempotent σ' in $\mathcal{A}'$-k-mod there is an isomorphism α' such that $\overline{\alpha'\sigma'(\alpha')^{-1}} = 0$; then for every idempotent σ in $\mathcal{A}$-k-mod there is an isomorphism α such that $\overline{\alpha\sigma\alpha^{-1}} = 0$.

Proof Given $\sigma = \sigma^2 \in \mathcal{A}(X, X)$ consider the restriction $\sigma' \in \mathcal{A}'(X, X)$.

By hypothesis there is an isomorphism $\alpha' \in \mathcal{A}'(X, Y)$, for some Y, such that $\overline{\alpha'\sigma'(\alpha')^{-1}} = 0$.

Case (i) : $\mathcal{A}$ is a triangular kernel extension of $\mathcal{A}'$, so that $A' = A$, $\overline{V} = \overline{V}' \oplus (A \otimes_S U \otimes_S A)$, $d(U) \subseteq \overline{V}' \otimes \overline{V}'$. Hence Y is already an A-module, and we define $\alpha \in \mathcal{A}(X, Y)$ by $\alpha_0 = \alpha_0'$, $\overline{\alpha}|_{\overline{V}'} = \overline{\alpha}'$ and $\overline{\alpha} = 0$ on $A \otimes_S U \otimes_S A$; it is easy to see that $\alpha = (\alpha_0, \overline{\alpha})$ is in $\mathcal{A}(X, Y)$ and since $\mathcal{A}' = \mathcal{A}_i$ for some i and $\alpha_0 = \alpha_0'$ is an isomorphism α is also an isomorphism (by 6.3).

Let $\zeta = \alpha\sigma\alpha^{-1}$. Then $\zeta = \zeta^2 \in \mathcal{A}(Y, Y)$ and $\overline{\zeta}|_{\overline{V}'} = 0$.

We must further modify ζ, and will use the fact that, since $d(A) \subseteq \overline{V}'$, if $\varphi, \theta \in \mathcal{A}(Y, Y)$ satisfy

$\bar{\varphi}|_{\bar{V}'} = 0 = \bar{\theta}|_{\bar{V}'}$, then $\overline{\varphi\theta}|_{\bar{V}'} = 0$, and hence that

$$\overline{\varphi\theta}(v) = \varphi_0\bar{\theta}(v) + \bar{\varphi}(v)\theta_0 \quad \text{for all } v \in \bar{V}.$$

Define $\beta \in \mathcal{C}(Y, Y)$ by $\beta_0 = 1_X$, $\bar{\beta}(v) = \zeta_0\bar{\zeta}(v) - \bar{\zeta}(v)\zeta_0$ for $v \in \bar{V}$. Then $2\beta - \beta^2 = \epsilon_Y$, so $\beta(v)$ is an <u>automorphism</u> of Y with $\beta^{-1} = 2\epsilon_X - \beta$. We claim that $\overline{\beta\zeta\beta^{-1}} = 0$. Using the remark above, one sees that

$$(3) \quad \overline{\beta\zeta\beta^{-1}}(v) = \bar{\zeta}(v) - \zeta_0\bar{\zeta}(v) - \bar{\zeta}(v)\zeta_0 + 2\zeta_0\bar{\zeta}(v)\zeta_0.$$

Since $\zeta^2 = \zeta$ we have $\bar{\zeta}(v) = \zeta_0\bar{\zeta}(v) + \bar{\zeta}(v)\zeta_0$, and as $\zeta_0^2 = \zeta_0$ it follows that $\zeta_0\bar{\zeta}(v)\zeta_0 = 0$. Using these two relations in (3) shows that $\overline{\beta\zeta\beta^{-1}} = 0$.

Let $\gamma = \beta\alpha$, then $\overline{\gamma\sigma\gamma^{-1}} = 0$ as required.

<u>**Case (ii)**</u> : $\mathcal{C}$ is a triangular tensor algebra extension of $\mathcal{C}'$, so that

$$A = \otimes_{A'}(A' \otimes_S E \otimes_S A'),\ \bar{V} = A\otimes' \bar{V}' \otimes' A, \text{ and } d(E) \subseteq \bar{V}.$$

We have first to extend the action of A' on Y to an action of A, and for this it suffices to define a suitable S-S-bimodule morphism of E to (Y, Y). We take this morphism to be

$$e \longmapsto \alpha_0' e(\alpha_0')^{-1} + \bar{\alpha}'(d(e))\,\alpha_0'^{-1}.$$

Next we define $\gamma \in \mathcal{C}(X, Y)$ by taking $\gamma_0 = \alpha_0'$, and $\bar{\gamma}$ to be the canonical extension of $\bar{\alpha}'$ to $\bar{V} = A \otimes' \bar{V}' \otimes' A$. We need to check the relation

$$\bar{\gamma}(d(a)) = a\gamma_0 - \gamma_0 a \quad \text{for } a \in A.$$

As in the proof of 6.3 it is sufficient to do this for $a \in A'$ (which is easy:
$\bar{\gamma}(d(a)) = \bar{\alpha}'(d(a)) = a\alpha'_0 - \alpha'_0 a = a\gamma_0 - \gamma_0 a$
if $a \in A'$), and for $a \in E$ which is a consequence of the definition of the action of E on Y.

Thus $\gamma \in \mathcal{C}(X, Y)$ is an isomorphism by 6.3 and $\overline{\gamma\sigma\gamma^{-1}} = 0$ on $\overline{V}'$ since $\overline{\gamma\sigma\gamma^{-1}}\big|_{\overline{V}'} = \overline{\alpha'\sigma'(\alpha')^{-1}} = 0$. Since $\overline{V}'$ generates $\overline{V}$ over A we conclude that $\overline{\gamma\sigma\gamma^{-1}} = 0$, as required.

REFERENCES

[AS] M. Auslander and S.O. Smalø - Almost Split Sequences in Subcategories, J. Algebra 69 (1981) 426-454.

[BK] R. Bautista and M. Kleiner - Almost Split Sequences for Relatively Projective Modules (to appear in J. Algebra).

[C-B] W.W. Crawley-Boevey - On Tame Algebras and Bocses, Proc. London Math. Soc. 56 (1988), 451-483.

[D] Yu. A. Drozd - Tame and Wild Matrix Problems, (in Russian) in 'Representations and Quadratic Forms', Inst. Mat. Akad. Nauk Ukrain. SSR, Kiev, 1979, 39-74. English translation in Amer. Math. Soc. Transl. (2) Vol. 128 (1986), 31-55.

[K] M. Kleiner - Induced Modules and Comodules and Representations of BOCS's and DGC's, in 'Representations of Algebras', Lecture Notes in Math., vol. 903, Springer-Verlag, Berlin (1981), 168-185.

[M] Y. Miyashita - Tilting Modules of Finite Projective Dimension, Math. Zeit. 193 (1986), 113-146.

[R] A.V. Roiter - Matrix Problems and Representations of BOCS's, in 'Representation Theory I', Lecture Notes in Math., vol. 831, Springer-Verlag, Berlin (1980), 288-324.

Department of Pure Mathematics,
University of Liverpool,
P.O. Box 147, Liverpool L69 3BX,
England.

Canadian Mathematical Society
Conference Proceedings
Volume 11 (1991)

. 7-DIMENSIONAL ALGEBRAS WITH MIXED BASIS-GRAPH .

THIERRY DANA-PICARD

Abstract

We prove general results in order to compute deformations of associative unitary algebras in a given dimension from lower dimensional deformations. Then we classify the 7-dimensional algebras with either loopless or mixed basis-graph.

I. Introduction.

Since the end of last century, the classification of finite dimensional associative algebras is an open question. The project of listing the isomorphism classes was carried as far as the seven dimensional non local algebras and was then abandoned early in the century. With the development of deformation theory, it became possible to organize the algebras in a deformation chart, and to introduce some structure into the classification process. In 1974, P. Gabriel [Ga] gave the deformation charts in dimension 3 and 4, and in 1979, G. Mazzola [Ma] gave a geometric classification in dimension 5 and D. Happel [Ha] gave the complete deformation chart in dimension 5. The complete classification of 6-dimensional generic algebras is given in [DP]. The aim of the present paper is to contribute to the classification of associative unitary algebras in dimension 7, by studying in detail mixed basis-graphs (by a theorem of [DP-Scl], the case of seven dimensional loopless algebras is purely combinatorial; we give the list of such generic algebras in Appendix I).

The main tool is the theory of deformations of algebras

A.M.S.Subject Classification 16A58, 16A46, 16A48.
This work is part of the author's Ph.D. Thesis at Bar-Ilan University.

using Hochschild cohomology (cf. [Ge] and [Kr]) and the appropriate invariants for the classification work are the basis-graph and the weighted basis-graph, as defined in [Sc1] by M. Schaps. In dimension equal to or greater than 6, the task of writing down a complete deformation chart is an unrewarding task, as the number of different isomorphism classes of algebras increases very fast, but it is still possible to determine good lower bounds for the number p(n) of reduced irreducible components and the total number q(n) of irreducible components of $\mathbf{Alg}_n$, including the non-reduced ones. One outcome of the classification was a proof that $\mathbf{Alg}_n$ is non-reduced for $n \geq 6$, with a non-reduced component corresponding to a connected algebra [DP-Sc2].

In section II, we recall the main definitions and theorems from the theory of deformations of associative unitary algebras (cf [Ge], [Sc1], [DP]).

In section III, we give general results which enable us to construct deformations of algebras in a given dimension from deformations in lower dimension.

In section IV, we recall results which enable us to determine the generic 7-dimensional algebras with loopless basis-graph and give a first lower bound for the number p(7) of reduced irreducible components of $\mathbf{Alg}_7$.

In section V, we give a complete list of candidates for genericity with mixed 7-dimensional basis-graph and explain for each one either why it is generic (resp. quasi-generic) or how it deforms. Then we give our main statement, namely a lower bound for the number of irreducible components of $\mathbf{Alg}_7$, either reduced or non-reduced.

For our classification work, we made much use of a computer program written by M. Schaps (described in [Sc2]), which computes, among others, $H^2(A,A)$ and the dimension of the automorphism group of A for a given algebra A. For 5-dimensional algebras, we use Happel's notations in [Ha].

II. Basis-graphs and deformations.

Denote by $\mathbf{Alg}_n$ the structure constant scheme for associative unitary algebras of dimension n over an algebraically closed field K of characteristic 0 or >>0. Since the defining equations of the scheme are defined over **Z**, there are only a finite number of primes for which the number and nature of irreducible components in $\mathbf{Alg}_n(K)$ will be different from the characteristic 0 case. Thus any characteristic which is sufficiently large relative to n will give the same classification.

The linear group GL(n,K) operates on $\mathbf{Alg}_n$ by "basis change" (cf [Kr]): the orbits of $\mathbf{Alg}_n$ under the action of GL(n) correspond to the isomorphism classes and the stabilizer of an algebra A is its automorphism group Aut A (of course automorphisms as an algebra).

DEFINITION: if an algebra or an algebraic family of algebras A lies in the closure of the orbit of a different algebra or family A', we will say that A is a **specialization** of A' or that A' is a **deformation** of A. The fact that A' is a deformation of A will be represented by A'--->A.

There is an alternative definition of deformations, that we will use as well:

DEFINITION: Let (C,t_o) be any pointed scheme, with C=Spec(R) for R a commutative affine ring over K and t_o a closed point corresponding to a maximal ideal m_o of R. A **flat deformation** B of B_o over (C,t_o) is a flat R-algebra B together with an isomorphism of B_o with $B \otimes_R R/m_o$. When the fibers over general closed points of **C** are all isomorphic to an algebra A, then A is also called a deformation of B_o.
If C is an irreducible algebraic scheme of dimension at least one, then the deformation will be called **algebraic**. If $R=K[\epsilon]$, with $\epsilon^2=0$, then the deformation is of first order. In general, if dim C=0, we call the deformation **infinitesimal**.

DEFINITION: 1) An algebra is **rigid** if its orbit under the action of Gl(n) is dense in an irreducible component of $\mathbf{Alg}_n$.
2) A family of algebras is **semi-rigid** if the union of the orbits of the algebras in this family, under the action of base changes, is dense in an irreducible component of $\mathbf{Alg}_n$.
In both of these cases, we will say that the algebra (or the family of algebras) is **generic**. An algebra or family giving the reduced part of a non reduced

component will be called **quasi-generic**.
(A family is quasi-generic if it has a non-trivial first order deformation, but no algebraic deformation).

The **basis-graph** of a finite dimensional algebra A is the diagram constructed as follows:

(i) take a number of vertices equal to the number of idempotents in a complete set of primitive orthogonal idempotents $\{e_i\}$ and label each vertex by an idempotent.

(ii) for i not equal to j, the number of arrows from e_i to e_j is equal to $\dim(e_iAe_j)$.

(iii) the number of loops from e_i to itself is equal to $\dim(e_iAe_i)-1$.

We get the **weighted basis-graph** of A by adding weightings on the arrows by the following way:

(iv) the number of arrows from e_i to e_j with weight k is $n_{ij}{}^k=\dim\ e_i(J^k/J^{k+1})e_j$, where J is the radical of A; a k-weighted arrow is an arrow with k barbs. A weight-1 loop having only trivial products with other basis-elements is called **a trivial loop** and is denoted by .

(v) matrix units are weighted by ∞ and marked by a solid triangular barb.

<u>REMARK:</u> the quiver of an algebra includes arrows only for elements of $A/(\text{Rad } A)^2$. Hence, unless $(\text{Rad } A)^2= 0$, the basis-graph and the quiver of an algebra are different.

In [Scl], M. Schaps defined the basis-graph as an extension of the quiver. In this work, in order that the composition of arrows should appear in a natural order corresponding to the multiplication in the algebra, we dualize the definition. As far as the classification is concerned, this dualization is irrelevant, since an algebra is generic if, and only if, its dual is generic.
On these weightings, we define a partial ordering: suppose Φ and Φ' are two weightings on the same basis-graph Q. Then $\Phi \leq \Phi'$ if for each pair i,j and for each natural number k, the number of arrows from i to j of weight less than or equal to k in Φ is greater than or equal to the corresponding number in Φ'. A weighting Φ is called **maximal** if it has the following property: given another weighting Φ' on the same basis-graph, either Φ and Φ' cannot be compared or $\Phi'\leq \Phi$.

The basis-graphs, both weighted and unweighted, are the appropriate invariants for the classification work. In fact, M. Schaps proved that if A' is a deformation of A, then the basis-graph of A either equals the basis-graph of A', or is obtained by coallescing vertices, replacing each vanishing vertex by a loop. This preserves the dimensions in the Peirce decomposition of A (cf.[Sc1]). These graphs have the suitable semi-continuity properties as requested by M. Gerstenhaber in [Ge]; in particular, the partial ordering defined on the weightings of a given basis-graph is compatible with specialization. For these reasons we will consider in the following only maximal weightings. (cf. [DP],[DP-Sc1]).

DEFINITION: an arrow x is called **reducible** if there are two arrows not including x and forming a path from the initial point of x to its end point. Other arrows will be called **irreducible.** An arrow through which a reducible arrow can factor is called **reducing**.

Note that only reducible arrows can have more than a single barb.

LEMMA [DP-Sc1]: a set of n vertices in a basis-graph algebra A_0 can be deformed to a matrix block if and only if they are connected, they are completely symmetric with regard to all permutations, and the number of loops at each vertex is one less than the number of arrows from each point to another. In particular, a matrix block configuration must contain a two arrow cycle.

In order to find the candidates for genericity and to check their actual genericity, we act as follows: we determine the **type-generic** algebras or families, i.e. the "most generic" ones, and then, generally by a computer calculation we determine $H^2(A,A)$. If it is zero for an algebra A in one component, then the component contains a dense orbit of a single rigid algebra (cf [Ge]). For a family of algebras, in order to prove semi-genericity or to show that a first order deformation is associative, it suffices to look at special values of the parameters; if for an algebra A in the family the dimension of $H^2(A,A)$ is equal to the number of parameters in the family, then the family is semi-rigid (cf [DP-Sc2]).

The problem of proving that a type-generic component is not generic is more difficult. If W contains an open dense orbit, and for some algebra A in this orbit, $H^2(A,A)$ is not zero, then the algebra is not rigid. It may, however, represent a generic closed point in a non-reduced component of **Alg$_n$**. When W does not have a single orbit, but must be given by a family, the situation is yet more difficult, for in order to demonstrate that this family is not semi-rigid, it is necessary to construct an infinitesimal deformation which is valid for all values of the parameters in an open dense set. This situation already appeared in dimension 6 with one family with mixed basis-graph and permitted us to show that **Alg$_n$** has non reduced components for all $n \geq 6$ (cf [DP], [DP-Sc2]), but since the family has a single parameter the deformation was not difficult to construct. This problem is treated in general in the works cited above.

III. Construction of deformations in higher dimensions.

In low dimension, it is already possible to write down lists of algebras and to check them for genericity. By increasing the dimension, it is necessary to get some general results in order to eliminate most cases as quickly as possible.

It is obvious that adding a trivial loop on one idempotent of the basis-graph of the algebra A (with radical J) determines a new algebra $A[z]/(z^2,zJ,Jz)$, which is trivially deformable.

If A'_1 and A'_2 are respective deformations of A_1 and A_2, then $A'_1 x A'_2$ is clearly a deformation of $A_1 x A_2$; therefore we will be mainly interested in "connected" algebras.

DEFINITION: let A_1 and A_2 be two algebras, and let $e_1 \epsilon A_1$ and $e_2 \epsilon A_2$ be primitive basic idempotents. Let N_1, N_2 be ideals of A_1, A_2 respectively, such that $A_j/N_j \xrightarrow{\sim} K$ has the residue of e_j as identity. The **fiber product** $A_1 x A_2$ of A_1 with A_2 along (e_1, e_2) is the set of all elements in $A_1 x A_2$ such that both components have the same residue modulo $N_1 x N_2$.

LEMMA 1: Let A be an algebra with primitive basic idempotent e, and B the algebra which is the fiber product of the arrow algebra T generated by $e_1 . . e_o$ with A along (e_o, e). The deformations of B are completely determined by those of A, i.e. if $B^\sim$ is a deformation of B, then $B^\sim$ is the fiber product of T with $A^\sim$ along (e_o, $e^\sim$), where $A^\sim$ is a deformation of A and $e^\sim$ is a basic primitive idempotent of $A^\sim$.

REMARK: by duality, the corresponding result is evidently true for the fiber product TxA along (e_1,e). This lemma is a major component in the proof that $\mathbf{Alg_n}$ is non reduced and the proof will appear in an article on that topic [DP-Sc2].

LEMMA 2: Let A be a basic algebra with a complete orthogonal idempotent set $e_1, \ldots, e_r$, $A^\sim$ be a deformation of A, in which the idempotents lift to $e_1^\sim, \ldots, e_r^\sim$, and let $x_1, \ldots, x_n$ be a basis for $A^\sim$ which respects the idempotent set $e_1^\sim, \ldots, e_r^\sim$. Let $\{f_{ij}, 1 \le i \le r, 1 \le j \le r\}$ be a complete idempotent set for $A^\sim$ such that $\Sigma_j f_{ij}$ is a lifting of e_i. If f_{ij} is basic in $A^\sim$ and if x_q is a basis element contained in $e_i^\sim A^\sim e_i^\sim$, then the coefficient of f_{ij} in x_q lies in $\mathfrak{m}_o$ (where $\mathfrak{m}_o$ is the distinguished maximal ideal of the parameter ring R).

Proof: Let $\phi: R \longrightarrow K$ be the reduction modulo $\mathfrak{m}_o$. If $z \in A^\sim$, then $z = \Sigma b_i x_i$ and its evaluation at the distinguished point is $z(t_o) = \Sigma \phi(b_i) x_i$. Thus if $z(t_o) = 0$, then $\phi(b_i) = 0$ for all i, and thus $b_i \in \mathfrak{m}_o$.

Let x_i be a basis element of $e_i^\sim A^\sim e_i^\sim$, different from $e_i^\sim$. Then $x_i(t_o)$ is a basis element of $e_i A e_i$, different from e_i, and that lies in Rad $e_i A e_i$. Therefore $x_i(t_o)$ is nilpotent and there is an integer n such that $x_i(t_o)^n = 0$. Since f_{ij} is basic, no product of other basis elements produces it. Therefore, if $x_i = c_{ij} f_{ij} + \ldots$, then $x_i^n = c_{ij}^n f_{ij} + \ldots$. This implies $c_{ij}^n \in \mathfrak{m}_o$, so $c_{ij} \in \sqrt{\mathfrak{m}_o}$; as $\mathfrak{m}_o$ is maximal, $\sqrt{\mathfrak{m}_o} = \mathfrak{m}_o$ and $c_{ij} \in \mathfrak{m}_o$.

All over this work, Kr_p denotes the Kronecker algebra of dimension p+2.

PROPOSITION 1: Let $A^{\sim}$ be a deformation of the algebra A, and let f_{1j} be a primitive idempotent which is defined at the general fiber of $A^{\sim}$. Let $B^{\sim}$ be an algebra which contains $A^{\sim}$ as a subalgebra, in such a way that a basis for $B^{\sim}$ is obtained by adding to a basis of $A^{\sim}$ containing f_{1j} exactly one idempotent f for which dim $f_{1j}B^{\sim}f=p$, dim $fB^{\sim}f=1$ and every other Peirce component either starting or ending at f equal to 0. Then $B^{\sim}$ is a deformation of the algebra B, the fiber product of Kr_p with A along (e,e_1) where e is the right idempotent of Kr_p.

REMARK: reversing the arrows of Kr_p, we get the dual result.

$$\cdot \xrightarrow{\ p\ } \cdot A \qquad (\cdot \xleftarrow{\ p\ } \cdot A)$$

Proof: let $z_0, \ldots, z_p, z_{p+1}, \ldots$ be the natural basis of $B^{\sim}$, with z_0 as f, $z_1, \ldots, z_p$ as arrows and z_{p+1} as f_1. Thus we have $z_i.z_s=0$ for every $s>p+1$ and $0\le i\le p$, and $z_i.z_{p+1}=z_i$ for $0\le i\le p$. By transport of structure, we get the deformation basis $x_0, \ldots, x_p, x_{p+1}, \ldots$ where $x_i=z_i$ for every $i\le p+1$ and $x_{p+1}, x_{p+2}, \ldots$ is the deformation basis defined for the given deformation of A to $A^{\sim}$. If we have $x_r=\Sigma_{j\ge p+1}a_jz_j$, then $x_i.x_r=a_{p+1}z_i$, for every $1\le i\le p$. By the previous lemma, if x_r reduces to a radical element at the distinguished point, we know that $a_{p+1}\in m_0$; therefore, at the special fiber we get $y_i.y_s=0$ for every y_s in Rad A. Thus we get the fiber product of Kr_p with A along (e,e_1), as required.

PROPOSITION 2: let A be a local algebra having a local deformation $A^{\sim}$. Let J be the radical of A, and let I be an ideal of codimension ≤ 1 in J. Let B be an algebra with two idempotents e and f, two arrows z and w from e to f and fBf=A. Suppose that w.J=0, that z.J is contained in <w> and that I is the annihilator of z in J. Then B has a non trivial deformation $B^{\sim}$ with two idempotents $e^{\sim}$, $f^{\sim}$, two arrows $z^{\sim}$ and $w^{\sim}$ and $f^{\sim}B^{\sim}f^{\sim}=A^{\sim}$.

$$\cdot \rightrightarrows \cdot A \qquad (\cdot \leftleftarrows \cdot A)$$

Proof: We first consider the case $\mathrm{codim}_J I=1$. We claim that $I>J^2$. If not, then there exists an element $x^{\wedge}\in J^2$, but $x^{\wedge}\in I$. We take $x_1, \ldots, x_r \in J$ generating J as an ideal,

and linearly independent modulo J^2. Replace each x_i by $x'_i=x_i-c_ix^\wedge$, where $zx_i=c_iw$. Then $zx'_i=0$ for all i; since the x'_i generate J, z.J=0. Contradiction.
Now let 1, y_1, y_2, ... be a basis for A, such that $zy_1=w$ and $zy_i=0$ for $i\neq 1$, and let x_1 be the element of the deformation basis of $A^\sim$ corresponding to y_1.
Since $y_1\notin I$, we know that $y_1\notin J^2$, and thus $x_1\notin J^{\sim 2}$. Choose a flat ideal L equal to $J^{\sim 2}$ at the generic fiber. Such an ideal exists because the parameter space is a curve, as in Prop. 2 of [Sc]. By the same proposition, we can choose a lifting $x_1^\sim$, $x_2^\sim$, ..., $x_q^\sim$ of y_1, y_2, ..., y_q such that $L=\langle x_{r+1}^\sim, x_{r+2}^\sim, \ldots, x_q^\sim\rangle$. Let $I^\sim$ be the ideal generated by $x_2^\sim$, ..., $x_q^\sim$ and define the multiplication in $B^\sim$ by $z^\sim x_1^\sim=w^\sim$, $z^\sim x_i^\sim=0$ for $i\neq 1$.
In the case $\mathrm{codim}_J I=0$, then choose a flat ideal $I^\sim$ of codimension 1 containing $J^{\sim 2}$; let $x_1^\sim$ be an element of $J^\sim$, not in $I^\sim$, and define $z^\sim x_1^\sim=tw^\sim$, where $t\in m_o$, the ideal at the distinguished point.

REMARK: The deformations given in Prop. 2 are not idempotent-splitting; in the case $\mathrm{codim}_J I=1$, they require that A has a non trivial local deformation. In lemma 3, we will show that we get a non trivial deformation, this time idempotent-splitting, in the case where A has a trivial loop y_1.

As a consequence of Prop. 2 and the subsequent remark, given an algebra B as in Prop. 2, we need only consider it a candidate for genericity if A has no local deformation and no trivial loop.

DEFINITION: an idempotent e in an algebra A is called **loopless** if the dimension of eAe is 1, i.e. there is no loop on the basis-graph at that point.

PROPOSITION 3: let A_1 and A_2 be connected algebras consisting of more than one point, $e_1\in A_1$ and $e_2\in A_2$ primitive basic loopless idempotents. Then the deformations of the fiber product B of A_1 and A_2 along (e_1,e_2) correspond to pairs of deformations $(A_1^\sim,e_1^\sim)$ and $(A_2^\sim,e_2^\sim)$ of (A_1,e_1) and (A_2,e_2) such that $e_1^\sim$ and $e_2^\sim$ are still basic.

Proof: on the one hand, given a deformation $B^{\sim}$ of B, we can choose an idempotent section $e^{\sim}$ deforming the joint idempotent e of B. The idempotent e cannot deform to a non-basic idempotent, since $e^{\sim}$ is connected to other idempotents in A_1 and A_2, while the other idempotents in A_1 and A_2 are not connected. The basis-graph of a matrix block must be symmetric with respect to interchanging idempotents in the block. Since dimensions of Peirce components are preserved, $\dim e^{\sim}B^{\sim}e^{\sim}=1$, so $e^{\sim}$ is primitive and loopless.

Let $f_1^{\sim}$ be the sum of the deformations of all idempotents of A_1, and let $N^{\sim}=e^{\sim}B^{\sim}(1-e^{\sim})+(1-e^{\sim})B^{\sim}e^{\sim}+(1-e^{\sim})B^{\sim}(1-e^{\sim})$, which is an ideal because $e^{\sim}$ is basic and primitive. Then $f_1^{\sim}B^{\sim}f_1^{\sim}$ is a deformation of A_1 such that $N_1=e_1A(1-e_1)+(1-e_1)Ae_1+(1-e_1)A(1-e_1)$ has a lifting $N_1^{\sim}=f_1^{\sim}Nf_1^{\sim}$ with $\dim f_1^{\sim}(B^{\sim}/N_1^{\sim})f_1^{\sim}=\dim A_1/N_1=1$. As the residue of e_1 modulo N_1 is the identity of K, so is the residue of $e^{\sim}$ modulo $N_1^{\sim}$.

Let $A_1^{\sim}=f_1^{\sim}B^{\sim}f_1^{\sim}$; then $B^{\sim}$ is the fiber product of $A_1^{\sim}$ with $A_2^{\sim}$ along their "common" idempotent.

On the other hand, suppose we are given a pair $(A_1^{\sim}, e_1^{\sim})$ and $(A_2^{\sim}, e_2^{\sim})$ of deformations of (A_1, e_1) and (A_2,e_2) with $e_1^{\sim}$ and $e_2^{\sim}$ basic. Then $\dim e_1^{\sim}A_1^{\sim}e_1^{\sim}$ is equal to $\dim e_1A_1e_1$ and the fiber product of $A_1^{\sim}$ with $A_2^{\sim}$ along $(e_1^{\sim}, e_2^{\sim})$ is a deformation of the fiber product of A_1 with A_2 along (e_1, e_2).

PROPOSITION 4 (one-way bridge): let A be an algebra whose basis-graph can be represented by $A_1. \xrightarrow{p} .A_2$, where the points represent loopless idempotents and $\xrightarrow{p}$ means that there are p arrows in the same direction. Assume that each of the p arrows is of weight 1. Then every deformation of A has a basis-graph of the form $A_1^{\sim}.\xrightarrow{p}.A_2^{\sim}$, where $A_1^{\sim}$ is a deformation of A_1.

Proof: let e' and e" be the left and the right idempotent of the basis-graph of Kr_p, respectively and let A_1, A_2 be two algebras with respective basic primitive idempotents e_1, e_2. The fiber product of Kr_p with A_2 along (e", e_2) will be denoted by B". As Kr_p is rigid, by Prop. 3, any deformation of B" is the fiber product of Kr_p with a deformation $A_2^{\sim}$ of A_2 along $(e'', e_2^{\sim})$, where $e_2^{\sim}$ is a lifting of e_2 which is still basic. We apply now Prop. 3 to the fiber product of A_1 with B'' along (e_1, e') to get our result.

LEMMA 3: Let A_1, A_2 be two connected algebras and $e_1 \epsilon A_1$, $e_2 \epsilon A_2$ two basic primitive idempotents, possibly looped. If A_1 and/or A_2 have basic deformations, then the fiber product of A_1 and A_2 along (e_1, e_2) is deformable.

Proof: If the idempotents are loopless, we are done, by proposition 3.
Suppose e_1 is looped and A_1 has a deformation $A_1^{\sim}$ over (C, t_o), where C=Spec R and t_o is the closed point corresponding to the maximal ideal m_o. Let f be a lifting of e_1, with $f=\Sigma f_i$, where every f_i is a primitive basic idempotent.

Let f_1 be one of these idempotents and A' the fiber product of $A_1^{\sim}$ with A_2 along (f_1, e_2). A basis of A' is obtained by the union of a basis for $A_1^{\sim}$ and a basis for A_2, identifying e_1 and e_2; the product is defined by $xy=yx=0$ for every $x \epsilon A_1^{\sim}$ and $y \epsilon A_2$. For every $x \epsilon f A_1^{\sim} f$ and every $y \epsilon A_2$ different of e_2, we have then $xy=yx=0$. At the special fiber, x reduces to a loop on $e_1 \equiv e_2$ with trivial product with every basis element of A_2, except e_2. In order to involve also a deformation of A_2, we make the previous construction twice.

REMARK: in the case of a fiber product A of two algebras $A_1 x A_2$ along (e_1, e_2) where e_1 and e_2 are looped, A can deform even when A_1 and A_2, separately, are generic. For example, if $A_1=A_2=F^4=K\langle x,y\rangle/(x^2,y^2,yx-sxy)$, which is the generic 4-dimensional local family, the fiber product $A_1 x A_2$ is a family in Scorza(4,2) which is a non generic 7-dimensional family (cf [DP-Sc2]).

COROLLARY: Let Q be a basis-graph, let f_i be an idempotent. Suppose that f_i is connected to all other idempotents by at most one arrow. Then the local ring at f_i for any algebra with basis-graph Q deforms independently of the remainder of the algebra. In particular, if there are exactly one or two loops at

f_1, then every algebra with that basis-graph has idempotent-splitting deformations.

Proof: The idempotent f_1 cannot be part of a matrix block, for if it were it would be connected to some other idempotent by one arrow in each direction. The loops at f_1 are neither reducible nor reducing with respect to the rest of the algebra. Therefore the algebra has the form of a fiber product at f_1 of two algebras (A_1,e_1) local and (A_2,e_2) with a loopless basic idempotent e_2. All deformations of A_1 leaving at least one basic idempotent will give deformations of this fiber product, as well all deformations of A_2. (Here, unlike Lemma 2 above, we need not worry that e_2 will stop being basic). Any deformation of (A_2,e_2) leaves the basis-graph unchanged at e_2, since e_2 is loopless. Therefore, in any deformation, the local ring at f_1 will remain independent of the rest of the algebra, showing that it is a fiber product of two independent deformations.

This allows us to eliminate immediatly all algebras in which there is a local ring at one of the idempotents which is a truncated polynomial ring, whose multiplication is independent of the remaining algebras. It also eliminates every algebra with a trivial loop. In general it eliminates all basis-graphs with one of the following configurations (a line means an arrow in either direction):

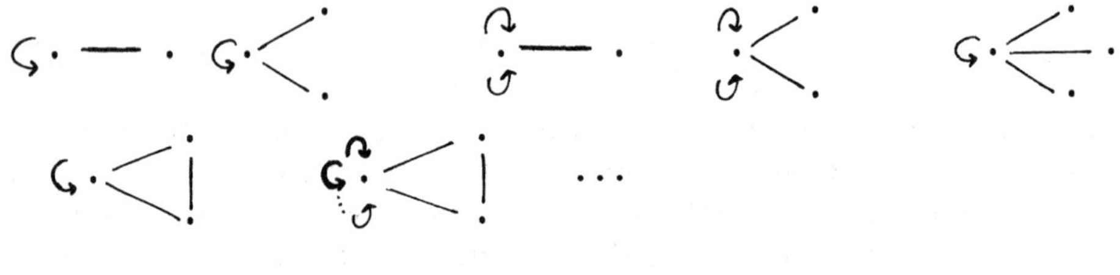

DEFINITION: a weighted basis-graph is called **locally maximal** if there are algebras with this weighted basis-graph which have no deformation to algebras with the same underlying basis-graph.

This is the natural generalization to the mixed case of the concept of maximal weighting used to classify generic algebras in the loopless case (cf [DP-Scl]; v.i. section IV). The generic algebras with mixed weighted basis-graph have to be looked for among those with locally maximal weighting.

Since Prop. 4 eliminates most connected algebras in dimension 7 with three or more idempotents, the major classification effort is concentrated on connected algebras with two idempotents. To enumerate the cases we use the following

DEFINITION: a connected basis-graph with two idempotents e_0, e_1 has **2-type (a,b,c,d)** when $a=\dim e_0Ae_0$, $b=\dim e_0Ae_1$, $c=\dim e_1Ae_0$ and $d=\dim e_1Ae_1$.

LEMMA 4 [DP]: let A be an algebra with a basis-graph containing exactly two idempotents and of 2-type (1,0,2c,2), where c is a non zero integer. Then A has an idempotent-splitting deformation.

Proof: We will give the actual interleaving. Let x be the loop on e_1.

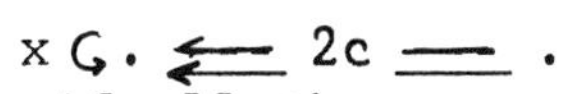

(i) By Prop. 1, if all the arrows are in Rad A but not in $(\text{Rad } A)^2$, we can pull off an idempotent.

(ii) Denote by B the algebra with the given basis-graph but weighting $\emptyset_B$, in which half the arrows are of weight 2, organized in pairs z_k, z_kx for $1 \le k \le c$; let now $v=x$, $y_i=z_i$ for every $1 \le i \le c-1$, $y_c=tz_c$ and $y'_c=z_cx$. We get a specialization of B to an algebra A with only c-1 arrows of weight 2. By induction, we can get a specialization to any algebra A with the same basis-graph and less than c arrows of weight 2. Since deformation is transitive, it suffices to show that B has an idempotent-splitting deformation. B has a deformation to an algebra with the following basis-graph, all the arrows being out of e_0.

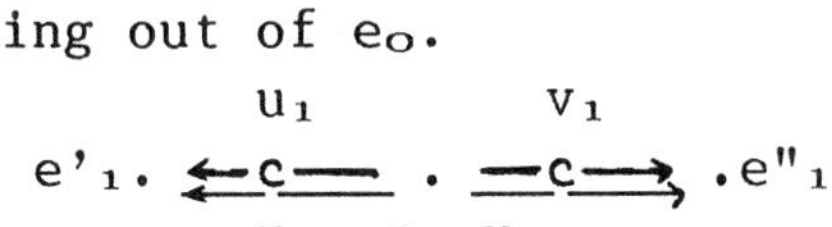

An appropriate interleaving is given by $e_1=e'_1+e''_1$, $x=(t/2)(e'_1-e''_1)$ and $z_k=u_k+v_k$ for $1 \le k \le c$.

LEMMA 5: If A is a local algebra and B is a mixed algebra with two idempotents e_0, e_1 such that e_0Be_0 is isomorphic to A and $\dim(e_iBe_j)=1$ for i,j not both zero, then the product of the two arrows in B must lie in the socle of A.

$$\bullet \rightleftarrows \bullet A$$

Proof: let x,y be the arrows with $xy \in B_0 = e_0Be_0$. If there is a z in Rad B_0 such that $zxy=0$ or $xyz=0$, then $zx=0$ or $yz=0$, in contradiction to $\dim(e_iBe_j)=1$.

LEMMA 6: Let A be a local algebra which socle is not contained in $(\text{Rad } A)^2$ and let B be an algebra having a basis containing two idempotents e_1, e_2 such that the local ring at e_1 is isomorphic to A, $\dim e_2Be_1=2$, $\dim e_1Be_2=0$ and $\dim e_2Be_2=1$. Then B has an idempotent-splitting deformation.

$$\bullet \rightrightarrows \bullet A$$

Proof: let $x \in \text{Soc } A \backslash (\text{Rad } A)^2$; we distinguish two cases:
1) if $\dim e_2Bx=0$, then $x \in \text{Soc } B$, and therefore, as the basis-graph contains a trivial loop, we can push out an idempotent.

2) if $\dim e_2Bx=1$, let $y \in e_1Be_1 \backslash (\text{Rad } A)^2$ such that $\dim e_2By=1$. By an automorphism of the form $y \longmapsto y'=y-\alpha x$, we get $\dim e_2By'=0$ and the algebra B is the fiber product of the algebra with weighted basis-graph $\circlearrowleft\bullet \rightrightarrows \bullet$ with a local algebra of dimension equal to $(\dim A - 1)$. As the first one has an idempotent-splitting deformation, so does B.

LEMMA 7: Let A be a local algebra having a local deformation $A^{\sim}$ such that the dimension of the socle is fixed. We define an algebra B with mixed basis-graph containing two idempotents e_0, e_1 such that e_0Be_0 is isomorphic to A, $\dim e_iBe_j=1$ for i,j not both zero and $\dim e_1Be_1=1$. Then B has a deformation.

$$\bullet \rightleftarrows \bullet A$$

Proof: Denote by C=Spec R the parametrizing space of the deformation of A to $A^{\sim}$. Suppose now that we have a basis $x_1, \ldots, x_p, x_{p+1}, \ldots, x_n$ of A such that $x_{p+1}, \ldots, x_n$ is a basis of the socle of A. These basis elements have liftings $x_1^{\sim}, \ldots, x_p^{\sim}, x_{p+1}^{\sim}, \ldots, x_n^{\sim}$ to $A^{\sim}$. We add a primitive idempotent e_1 and two arrows z_1 and z_2 such that $z_1z_2\in(\mathrm{Soc}\ A)$ and $z_2z_1=0$. We have

$$z_1z_2 = \Sigma_{p+1\le i\le n}\ a_ix_i$$

where the coefficients are not all equal to zero.

Take liftings $a_i^{\sim}$'s over R of the a_i's. As the coefficients a_i's are not all zero at the special fiber, so are the $a_i^{\sim}$'s outside this special fiber. We take now a primitive idempotent $e_1^{\sim}$ and two arrows w_1 and w_2 such that

$$w_1w_2 = \Sigma_{p+1\le i\le n}\ a_i^{\sim}x_i^{\sim}$$

and that defines the desired deformation.

As w_1w_2 is an element of the socle, every product involving this element is equal to 0 and associativity relations are trivially verified.

PROPOSITION 5: Let A' be an algebra with mixed basis-graph containing exactly two idempotents e_o, e_1 such that $\dim\ e_1A'e_1=1$. Suppose A' contains an arrow x of $e_1A'e_o$ such that $x\in(\mathrm{Rad}\ A')^2$ and x lies in the socle, so that A'(x)A' is an ideal of dimension 1. Let A=A'/(x). Then if A has a basic deformation, so does A'.

Proof:

1) A deforms to $A^{\sim}$ without creation of an idempotent.

Let $z_o, \ldots, z_p, z_{p+1}, \ldots z_n$ be the natural basis of $A^{\sim}$, where $z_o, \ldots, z_p$ are the basis idempotents. By transport of structure, we get the deformation basis $z_o, \ldots, z_p, z'_{p+1}, \ldots z'_n$ with respect to which the structure constants of $A^{\sim}$ reduce to those of A at the special fiber. Add one arrow $z_{n+1}\in z_oA^{\sim}z_1$ to the natural basis of $A^{\sim}$ such that $z_{n+1}z_i=z_iz_{n+1}=0$ for every $i\in\{p+1, \ldots, n\}$. We have $z'_r=\Sigma a_{rj}z_j$ for every r, where $a_{rj}\in m_o$ if z_j is an idempotent and the residue of z'_r is in the radical of A. Then $z'_rz'_{n+1}=a_{ro}z_{n+1}$. As $a_{ro}\in m_o$, at the special fiber we get $z'_rz'_{n+1}=0$.

2) A deforms to $A^{\sim}$ with creation of new idempotents.

Let $e_1, \ldots, e_r$ be a complete set of orthogonal idempotents for A with liftings $e_1^{\sim}, \ldots, e_r^{\sim}$ to $A^{\sim}$, and $x_1, \ldots, x_n$, a basis of $A^{\sim}$ which respects $e_1^{\sim}, \ldots, e_r^{\sim}$. We denote by $\{f_{ij},\ 1\le i,j\le r\}$ a complete set of basic idempotents for $A^{\sim}$, such that $\Sigma_j f_{ij}=e_i^{\sim}$.

We add a new arrow x_{n+1} to the basis of $A^{\sim}$, with source f_{0j} and terminus f_{1j}. Then for every r, the coefficient of f_{1j} in the product $x_r x_{n+1}$ is an element of the special ideal $\mathfrak{m}_0$, for every i different from 1, and at the special fiber we get $x_r x_{n+1}=0$. The same argument works for products $x_{n+1}x_r$.

An example of the importance of the assumption that A has a basic deformation is 6-dimensional (M8)' in [DP]; the 5-dimensional algebra with weighted basis-graph

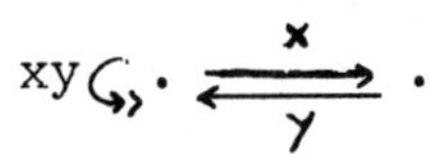

deforms to $K \times M_2(K)$, but the 6-dimensional algebra (M8)', whose weighted basis-graph is obtained from the previous one by simply adding an arrow with only trivial products, doesn't deform at all!

As a simple consequence of lemma 4 and Prop. 5, we get then the following result:

LEMMA 8: an algebra with 2-idempotent basis-graph of 2-type (1,2c+1,0,0) has an idempotent-splitting deformation.

IV. Generic 7-dimensional algebras with loopless basis-graph.

We recall without proof results from [DP-Sc1] which enable us to write down the complete list of 7-dimensional rigid (thus generic) algebras with loopless basis-graph.

Let Q be a basis-graph; the **basis-graph algebra A_Q of Q** is the algebra with basis-graph Q such that every arrow is of weight 1. A **complete lexicographically ordered subgraph on n vertices** is a subgraph containing n vertices, numbered 1, 2, ...,n such that for each pair i<j there exists in the graph exactly one arrow from the vertex i to the vertex j, and no arrow in the other cases.

THEOREM [DP-Sc1]: If Q contains no matrix block configuration and no complete lexicographically ordered subgraph on 4 vertices, then there is a one-to-one correspondence between irreducible components containing A_Q and the maximal weightings of Q.

As a complete lexicographically ordered subgraph on 4

vertices corresponds to a 10-dimensional subspace, the second condition is trivially fulfilled for loopless basis-graphs in dimension 7. In that dimension, a matrix configuration block (contained in a connected basis-graph) can be only formed by two idempotents connected by two reversed arrows; it appears in two cases (one the dual of the other), which will be studied explicitly hereafter. Writing down the rest of the list of 7-dimensional rigid algebras with loopless basis-graph is pure combinatorial work.

As already mentioned, the only special case we have to study is the following deformation:

given by: $x=tE_{01}$, $z=tE_{10}$ and $y=(1/t)v$. We get that $xz \dashrightarrow 0$, $zx \dashrightarrow 0$ and $xy \dashrightarrow u$ when $t \dashrightarrow 0$.

PROPOSITION 6: $p(7) \geq 55$, i.e. **Alg$_7$** contains at least 55 reduced irreducible components, dominated by the algebras whose basis-graphs are listed in Appendix I.

Appendix Ia contains the list of loopless basis-graphs organized according to increasing automorphism dimension. In order to make it easier to read, we organize the same list according to decreasing number of idempotents in Appendix Ib.

V. Generic 7-dimensional algebras with mixed basis-graph.

We can classify the 7-dimensional algebras and families of algebras with mixed basis-graph in 3 categories:

1) A = KxA', where A' is 6-dimensional and generic, with mixed basis-graph. There are 2 generic algebras and 2 quasi-generic families of this kind, with respective basis-graphs

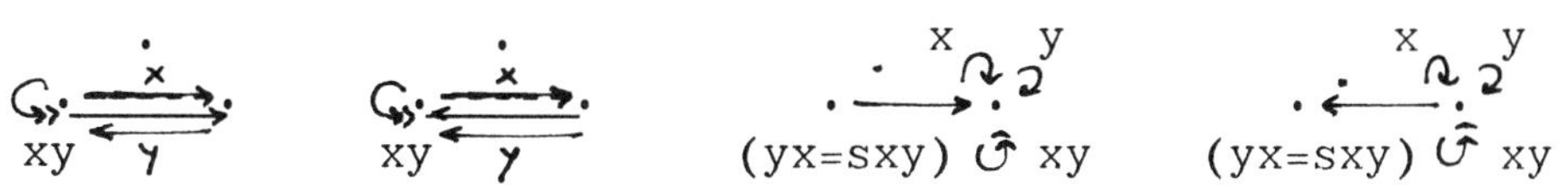

2) the second category contains only one family A = A'xA",

where A' is the arrow-algebra with basis-graph $\cdot \longrightarrow \cdot$, i.e. the algebra $T_2(K)$ of triangular matrices of order 2, and A" is the semi-rigid family of dimension 4 (i.e. $A''=K\langle x,y\rangle/(x^2,y^2,yx-sxy)$).

3) the third category is composed of the "connected" algebras. Let us list the connected mixed basis-graphs, together with any possible candidates to be generic algebras. When there are no compositions giving arrows of weight 3 or more, they are given by maximal weightings.

<u>Determination of candidates for genericity or quasi-genericity:</u>

We now give a case by case treatment of 7-dimensional mixed basis-graphs. Our fundamental invariant is the unweighted basis-graph (although representation theorists are accustomed to classify finite dimensional algebras by their quivers, here the quiver is not the appropriate invariant). In [DP], the entire list of 6-dimensional mixed weighted basis-graphs is given, but here in dimension 7, the corresponding list is too long. For each basis-graph, there is a number of possible weightings; some can be eliminated immediately by various considerations, as the presence of a trivial loop, etc... The algebras listed as candidates for generic algebras have no trivial loops and are locally maximally weighted (in order for the reader to get an example of what happens for a non locally maximal weighting, we leave two such weightings, namely basis-graph (3) with weighting (3)" and basis-graph (4) with weighting (4)").

For some of the weighted basis-graphs, there can be various parametrizations. All possible parametrizations cannot be listed without expanding too much this paper's length. We can reduce this number by the following "normalization" technique, explained on an example:

Let us see a case of an algebra with basis-graph (number (2) in the subsequent list)

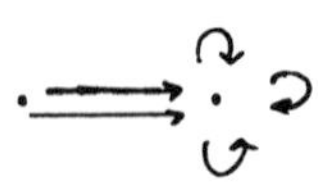

where the local ring at the idempotent on the right contains a trivial loop: take for this local ring the algebra $A=K[x,y]/(x^3,xy,y^2)$.

α) The algebra with weighted basis-graph

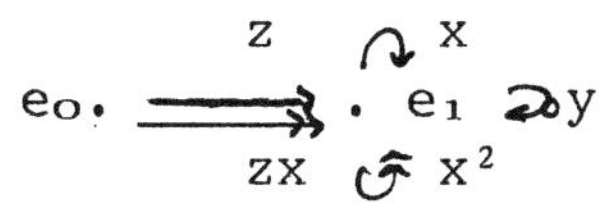

trivially deforms, as we can split off an idempotent.

β) Suppose we have the algebra with weighted basis-graph

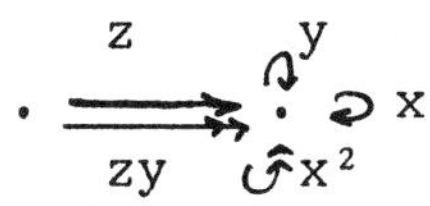

If this algebra is the fiber product of the 2 idempotent algebra which radical is generated by z and y, by $K[x]/(x^2)$, we are done. Suppose now that we have $z(ay-bx)=0$; if $b=0$, then $zy=0$; if $a=0$, then $zx=0$ and we get one of the first cases above. If $azy=bzx=w$, by the basis change $y'=y$, $x'=x-(b/a)y$, we return to the previous case.

In addition, in order to eliminate superfluous parameters from the defining equations of algebras with mixed basis-graph, we will freely use the trivial automorphisms multiplying basis elements by constants. For example, consider the following basis-graph:

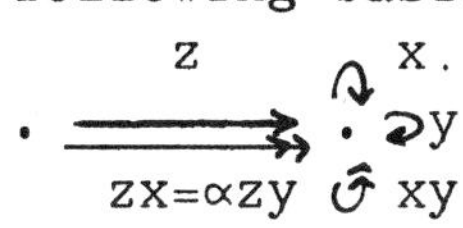

With the automorphism $y \longmapsto (1/\alpha)y$, we reduce to the case where $zx=zy$. (Note that this kind of reduction is especially important in cases similar to the case given here, with $yx=s\,xy$ and $x^2=y^2=0$, as the generators x and y are particular generators).

We indicate by a (*) the algebras and families which will be shown to be generic and by (**) the families which will be shown to be quasi-generic.

2 idempotents.

2-type basis-graph candidates for genericity

(1) (0,1,0,4) || (4,0,1,0)

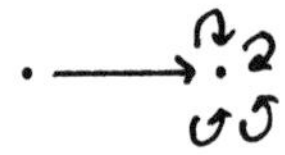

(2) (0,2,0,3) || (3,0,2,0)

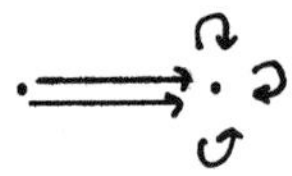

$(2)_1$

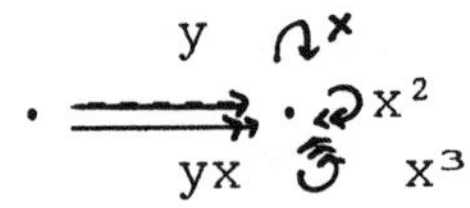

$(2)_2$

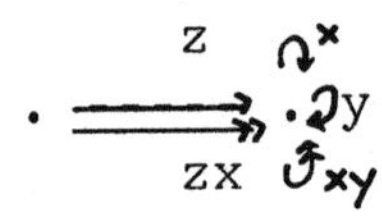

(**)

(yx=sxy;zy=zx)

By lemma 6, Prop. 1, Prop. 2 and Prop. 4, we do not need to consider other cases with 2-type (0,2,0,3).

(3) (0,3,0,2) || (2,0,3,0)

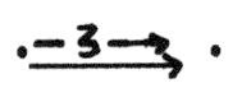

(3)'

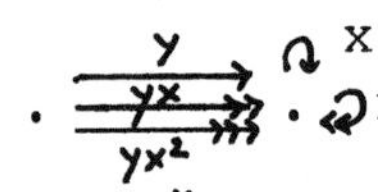

(3)"

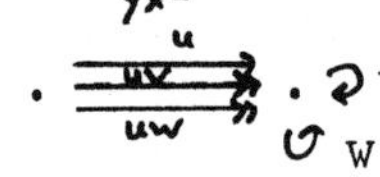

(4) (0,4,0,1) || (1,0,4,0)

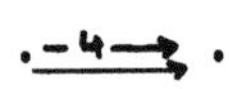

(4)'

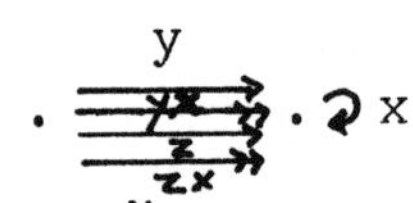

(4)"

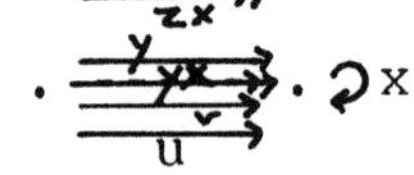

By Prop. 2 and the use of some trivial automorphisms, we eliminate the other possible cases with 2-type (0,3,0,2) or (0,4,0,1).

The 2-types (0,0,1,4)=(4,1,0,0), (0,0,2,3,)=(3,2,0,0), (0,0,3,2)=(2,3,0,0) and (0,0,4,1)=(1,4,0,0) are the respective opposites of the precedent ones.

(5) (0,1,1,3) || (3,1,1,0)

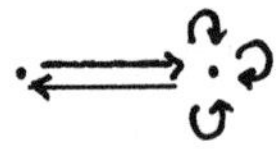

$(5)_1$

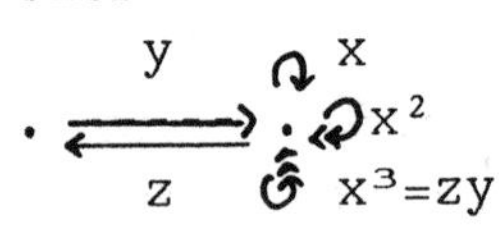

$(5)_2$

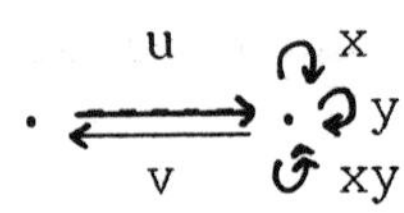

(**)

(yx=sxy;vu=xy)

By lemmas 5 and 7, we know that these cases are the only ones to consider here.

2-type	basis-graph	candidates for genericity
(6) (0,2,1,2) = (2,1,2,0)	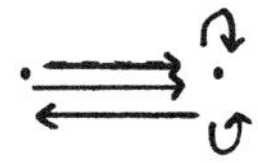	(6)' 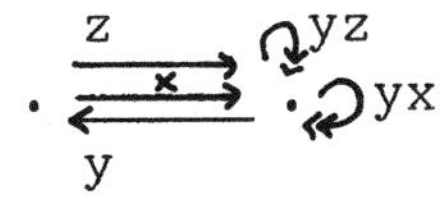
		(6)" 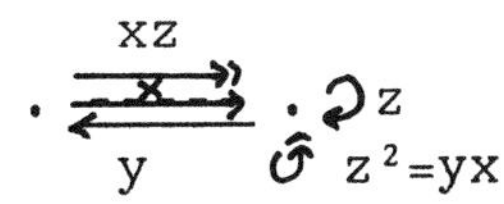

The 2-type (0,1,2,2)=(2,2,1,0) is the opposite of (6).

2-type	basis-graph	candidates for genericity
(7) (0,3,1,1) = (1,1,3,0)	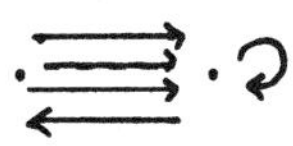	(7)' 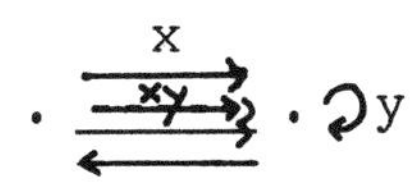
		(7)" 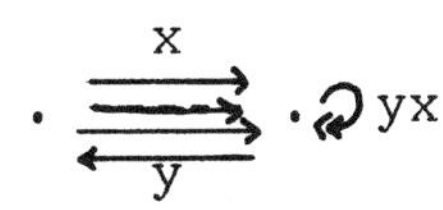 (*)

Using trivial automorphisms as indicated above, we do not need to consider other cases with this 2-type. The 2-type (0,1,3,1)=(1,3,1,0) is the opposite of (7).

2-type	basis-graph	candidates for genericity
(8) (1,1,2,1) = (1,2,1,1)	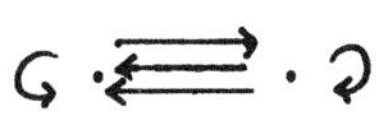	$(8)_1$ xz=xy 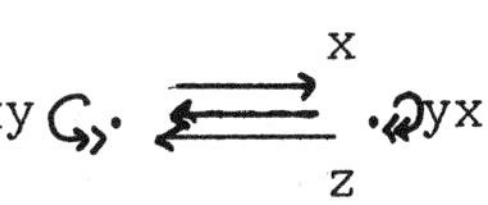
		$(8)_2$ xy 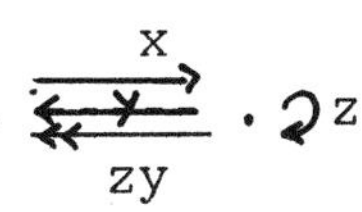
		$(8)_3$ xy 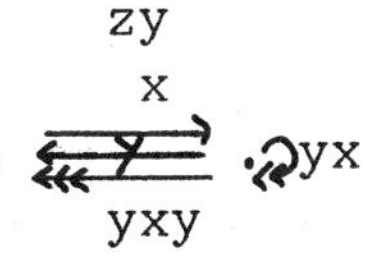
		$(8)_4$ xy 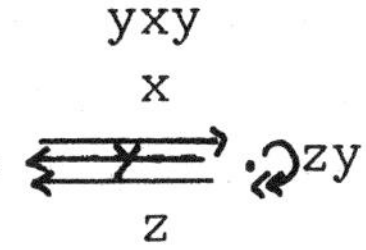
		$(8)_5$ xy 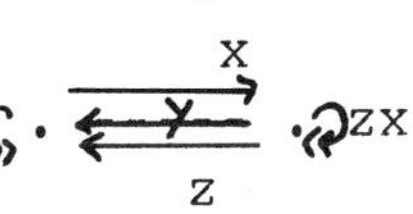 (*)

Trivial automorphisms reduce every other case to one of the listed basis-graphs $(8)_1$ to $(8)_4$. This case is equal to its own opposite.

2-type	basis-graph	candidates for genericity
(9) (1,1,1,2) ‖ (2,1,1,1)		(9)' 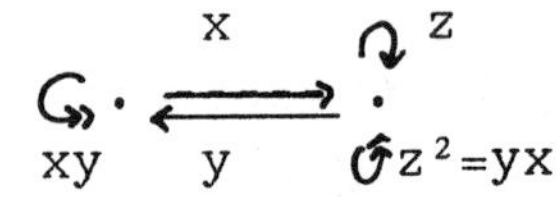
		(9)" 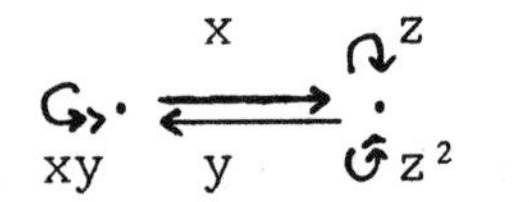

By lemma 5, there is no algebra with such a basis-graph and yx=z. As already mentioned, we do not list weightings with a trivial loop. This case is equal to its own opposite.

(10) (1,2,0,2) ‖ (2,0,2,1)		(10)' 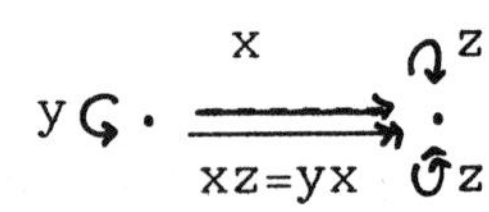
		(10)" 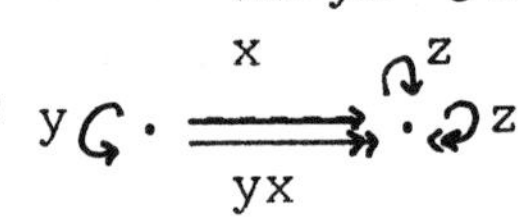

Any other case would contain a trivial loop and thus define a trivially deformable algebra. The 2-type (1,0,2,2)=(2,2,0,1) is the opposite of (10).

(11) (2,1,0,2) ‖ (2,0,1,2)	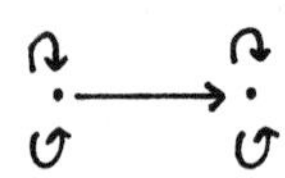	(11)' 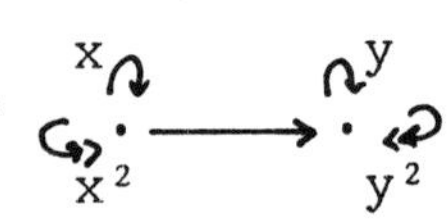

This 2-type is its own opposite. Any other case would contain a trivial loop and thus define a trivially deformable algebra.

(12) (3,1,0,1) ‖ (1,0,1,3)	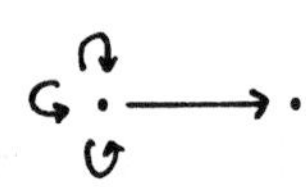	

The 2-type (3,0,1,1)=(1,1,0,3) is the opposite of (12).

(13) (1,3,0,1) ‖ (1,0,3,1)		(13)' 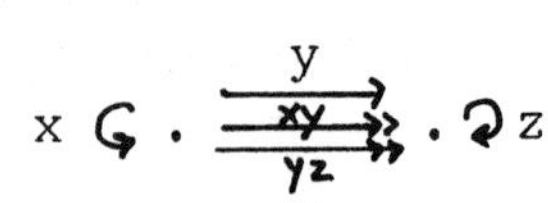
		(13)" 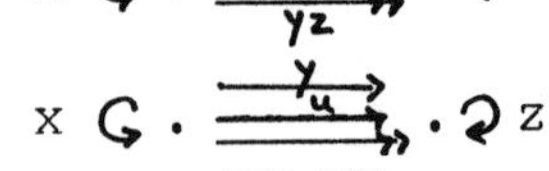
		(13)"' 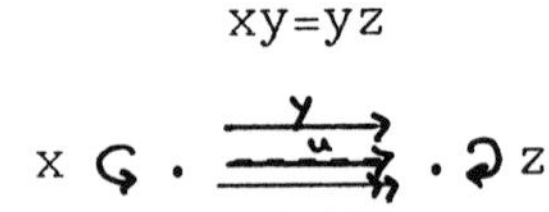

2-type	basis-graph	candidates for genericity
(14) (0,2,2,1) || (1,2,2,0)	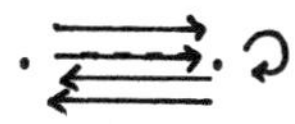	(14)$_1$ 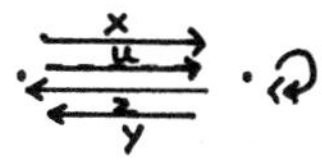 yx=zu (*)
		(14)$_2$ 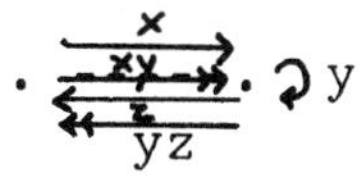

Each one of these 2-types is its own opposite. Note that for (13) and (14), every other case contains a trivial loop, defining thus a non-generic algebra.

3 idempotents.

	basis-graph	candidates for genericity
(15)	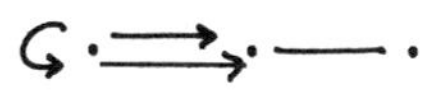	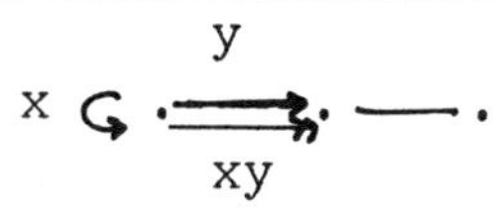
(16)	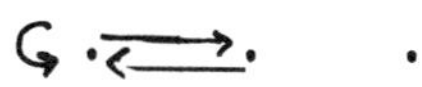	 (*)
(17)	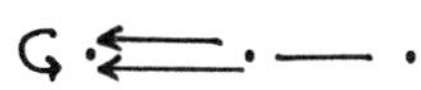	
(18)	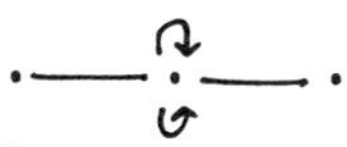	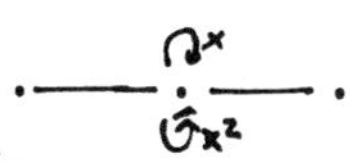
(19)		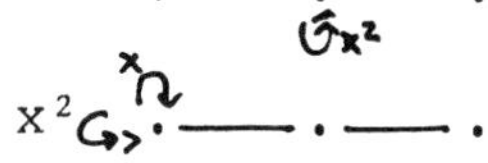
(20)	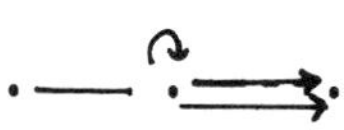	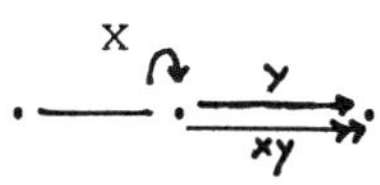
(21)	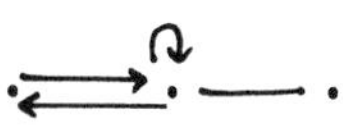	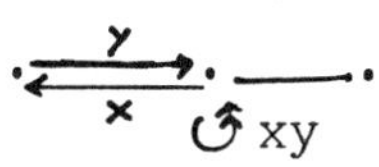

A "line" indicates that there can be an arrow in any direction. We did not indicate the respective opposites of these basis-graphs.

All the algebras whose basis-graph involves a triangle are eliminated by the corollary to lemma 3.

Case by case treatment of genericity and quasi-genericity:

Let us now study the genericity or the non-genericity of the algebras which weighted basis-graphs are listed in the previous pages.

2 idempotents.

By lemma 1, there is no candidate for being a generic algebra with basis-graph (1), because there is no 5-dimensional generic local algebra.

For $(2)_1$, we have the following deformation:

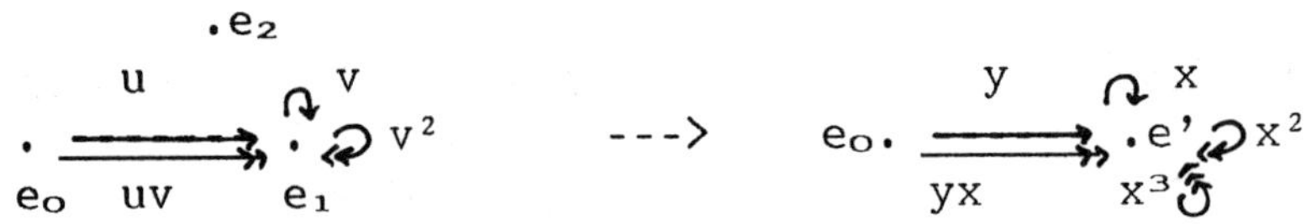

given by $e'=e_1+e_2$, $x=t(e_1-e_2)+v$, $y=u$. (We get $yx=uv+tu$, $x^2=t^2e'+2tv$ and $x^3=3t^2v+t^3(e_1-e_2)$).

$(2)_2$ is not generic; a computer calculation indicates that for an algebra A with weighted basis-graph $(2)_2$, the cohomology space $H^2(A,A)$ contains first order deformations, the multiplication table of one of them being the following:

	x_2	x_3	x_4	x_5	x_6
x_2		$x_4+\epsilon x_2$			
x_3	rx_4	$\epsilon\sigma x_3$	$\epsilon\sigma x_4$		
x_4		$\epsilon\theta x_4$			
x_5					
x_6	x_6	ϵx_6			

where $\sigma=2r/(r-1)$ and $\theta=(r+1)/(r-1)$.

It can easily be checked that this multiplication is associative if $\epsilon^2=0$ (By [DP] Prop. 8 (ii) p. 32, it suffices to check that this first order deformation appears among the first order deformations in $Z^2(A,A)$ in the computer output for 3 different values of r).

The passage from $Z^2(A,A)$ to $H^2(A,A)$ is made on the computer by taking the elements of $H^2(A,A)$ orthogonal, in the ordinary inner product in $\mathbb{K}^{n^3}$, to the elements of $B^2(A,A)$. Thus we could show that the given element of $Z^2(A,A)$ was non trivial by finding an element of $H^2(A,A)$ to which it was not orthogonal. Consider an algebra A with basis-graph $(2)_2$; the inner product of the deformation part of the structure constant tensor $(b_{ij}{}^{k})$ for an element of $H^2(A,A)$such that $b_{22}{}^{3}=0$ with the deformation part of the tensor defined in the table above is non zero for general values of the parameters; hence the infinitesimal deformation given by the tensor is not trivial, and the given algebra is not generic. This family is quasi-generic, as shown in [DP], theorem p.119.

As already mentioned, basis-graph (3) provides an example of a non locally maximal weighting without trivial loops. The algebra with weighted basis-graph (3)" deforms to (3)'; the deformation is given by $u=(1/t)y$, $v=tx$ and $w=tx^2$.

(3)' has the following deformation:

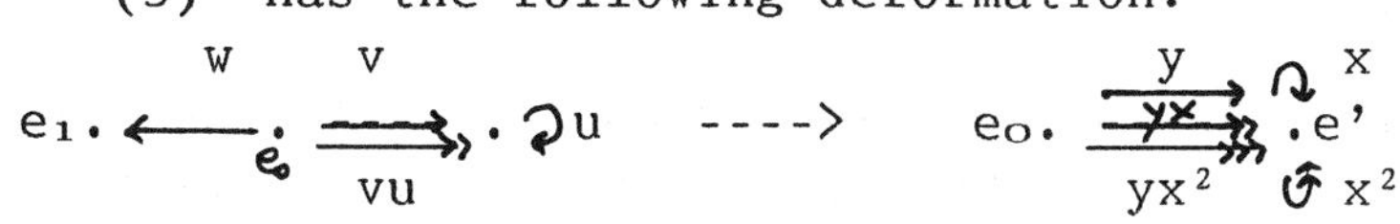

given by $e'=e_1+e_2$, $x=t(e_1-e_2)+u$ and $y=v+w$. We get then $x^2=t^2(e_1+e_2)-2tu$, $x^3=3t^2u+t^3(e_1-e_2)$, $yx=vu-tv$ and $yx^2=-2tvu+t^2v$.

(4)' is of 2-type (1,4,0,0); it has the following deformation:

given by $e'=e_1+e_2$, $x=t(e_1-e_2)$, $y=(1/t)(u_1+u_2)$ and $z=(1/t)(v_1+v_2)$. In fact it is a special case of 2-type (1,2c,0,0) with two idempotents, for which we have lemma 4.

(4)" has the following deformation:

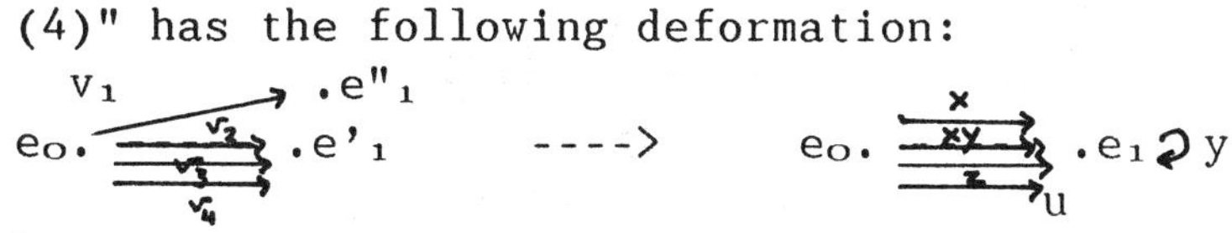

with $e_1=e'_1+e''_1$, $x=v_1+v_2$, $y=t(e'_1-e''_1)$, $z=v_3$ and $u=v_4$. Actually, the existence of an idempotent-splitting deformation for this algebra is a consequence either of lemma 4 or of Prop. 2.

$(5)_1$ defines an algebra which has the following deformation:

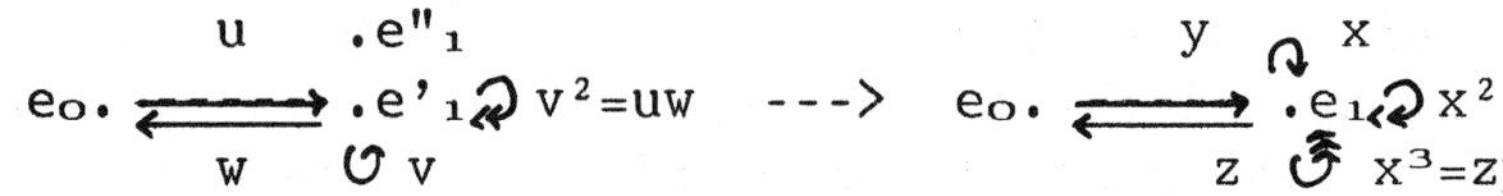

given by $e_1=e'_1+e''_1$, $x=(1/t)v+(t^2/2)(e'_1-e''_1)$, $y=u$ and $z=w$.

$(5)_2$ is an example of a non generic algebra which has a parameter dependent first order deformation, given by

	x_2	x_3	x_4	x_5	x_6
x_2			ϵx_2		
x_3	x_6				
x_4			$\epsilon\sigma x_2$	$\epsilon\sigma x_2+x_6$	
x_5			$\epsilon\theta x_2+rx_6$		
x_6			ϵx_6		

where $\sigma=2/(1-r)$ and $\theta=(1+r)/(1-r)\leq$ and $\epsilon^2=0$. It can easily be checked that this multiplication table defines an associative operation. Let us now consider the structure constant tensor given by this multiplication table. For an algebra A with weighted basis-graph $(5)_2$, there is a basis element of $H^2(A,A)$ such that the inner product of the deformation part of the structure constant tensor for A with deformation part of the structure constant tensor defined in the previous multiplication table is non zero for general values of the parameters; hence the infinitesimal deformation given by the tensor is not trivial, and the given algebra is not generic. By lemma 1, this family is quasi-generic ([DP] theorem p. 119).

(6)' has the following deformation:

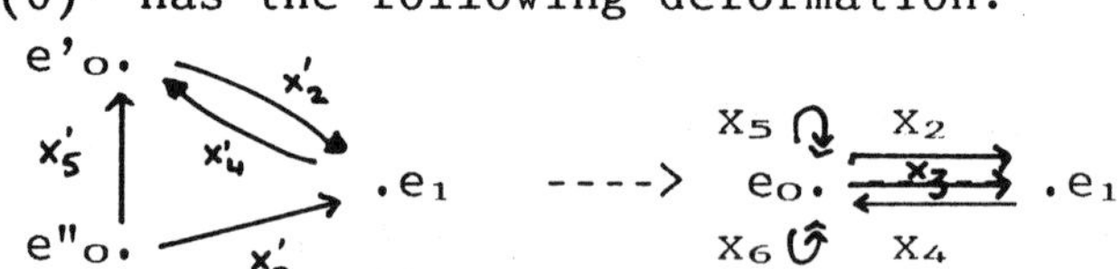

$(x_5=x_3x_4;x_6=x_2x_4)$

where $e_0=e'_0+e_0$, $x_2=tx'_2$, $x_3=x'_3$, $x_4=tx'_4$, $x_5=tx'_5$ and $x_6=t^2e'_0$. Note the presence of a matrix block in the deformed algebra.

(6)" has the following deformation:

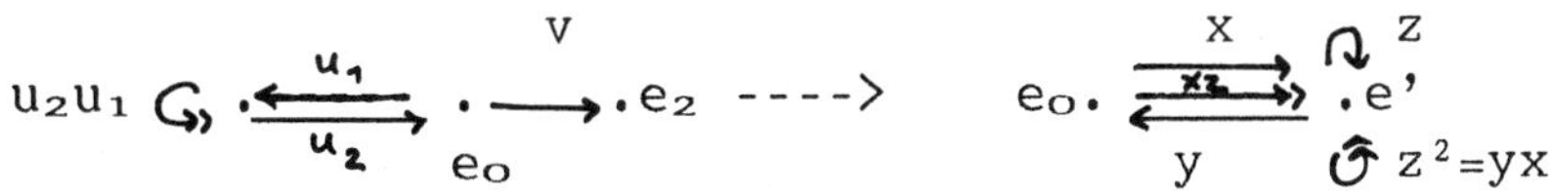

where $e'=e_1+e_2$, $x=u+(1/t)v, y=u_2$ and $z=t(e_1-e_2)+(1/2)u_2u_1$.

(7)' has the following deformation

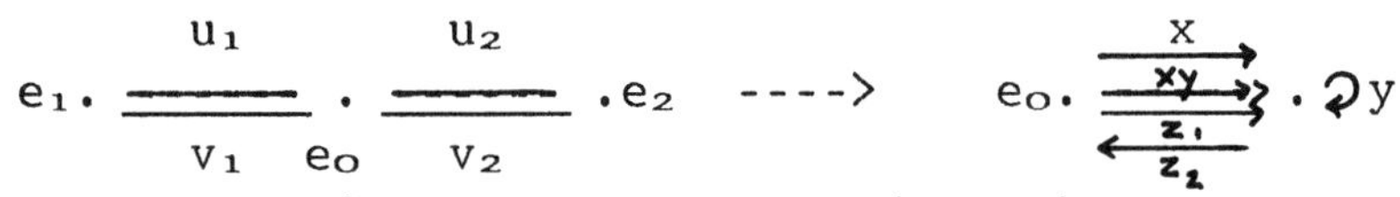

where $e'=e_1+e_2$, $y=t(e_1-e_2)$, $x=v_1+v_2$, $z_1=u_2, z_2=u_1$ (we have then $xy=t(v_1-v_2)$).

(7)" determines a rigid algebra (a computer calculation indicates that every algebra A with a such weighted basis-graph has its second cohomology group equal to 0), with dim Aut A=12.

$(8)_1$ determines a rigid algebra, for the same reason. This algebra is its own opposite, after replacing z by y-z.

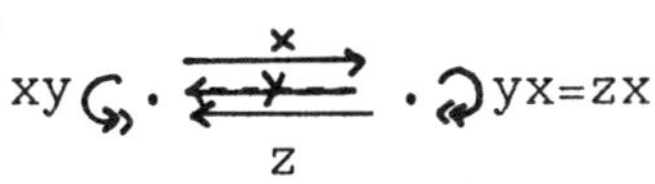

(dim Aut A=7)

$(8)_2$ has the following deformation:

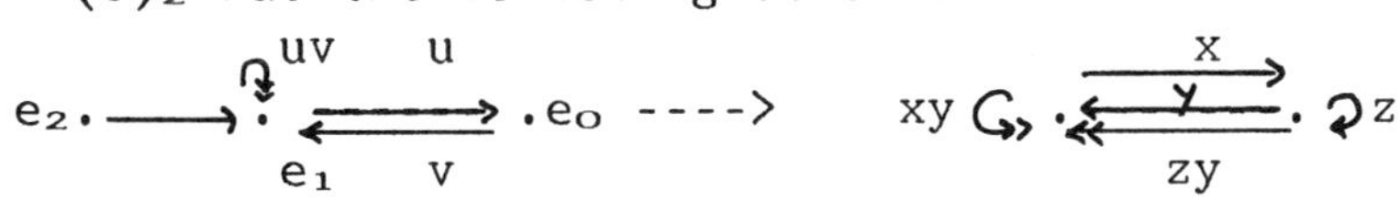

where $e'=e_0+e_2, z=(1/t)(e_0-e_2), x=u$ and $y=v+w$.

$(8)_3$ has the following deformation to $T_2(K)xM_2(K)$, where $T_2(K)$ is the algebra of triangular 2x2 matrices:

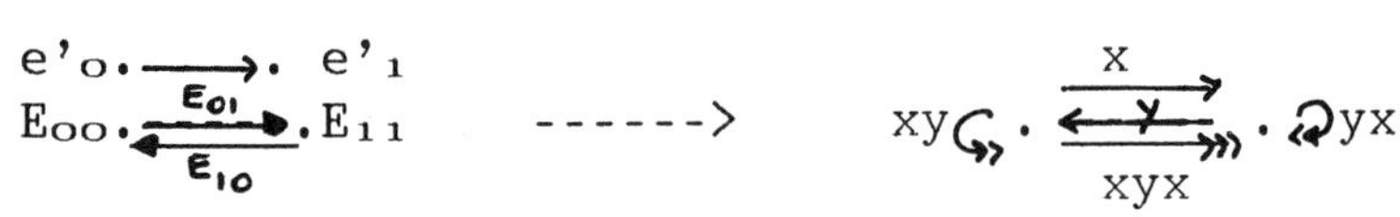

The deformation is given by: $e_0=e'_0+E_{00}$, $e_1=e'_1+E_{11}$, $x=tE_{01}+u$, $y=tE_{10}$.

$(8)_4$ deforms to $(8)_3$, by the following formulae: $u=x$, $v=y$, $z=(1/t)yxy$.

$(8)_5$ defines a rigid algebra, with dim Aut=11 (a computer calculation indicates that Hochschild cohomology vanishes in degree 2 for every algebra with this weighted basis-graph). Note that the rigidity of this algebra can be obtained as a consequence of Cor 3.11 of [Ci], determining the parallel quiver of the algebra and checking the connected components of it; I thank here C. Cibils for giving me a detailed proof.

(9)' has the following deformation:

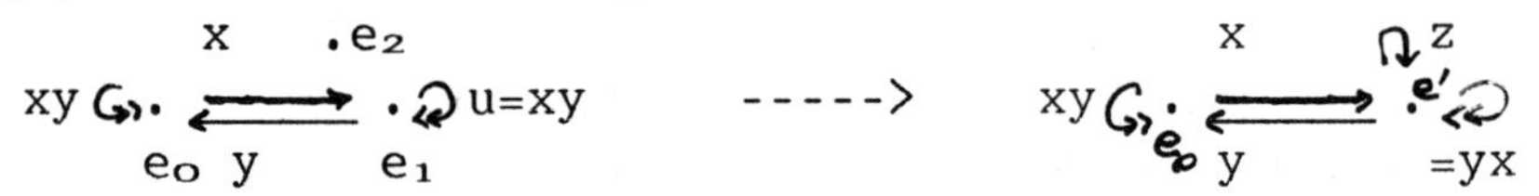

where $e'=e_1+e_2$ and $z=(t/2)(e_1-e_2)+(1/t)u$.

(9)" is the fiber product of $K[z]/(z^3)$ with the algebra with weighted basis-graph

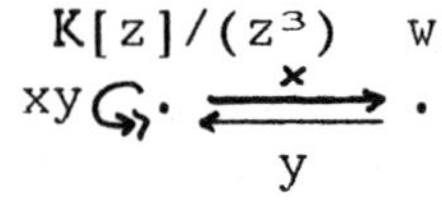

along the non-looped idempotent of the second one, hence has an idempotent-splitting deformation. For similar reasons, (10)" defines a non generic algebra.

(10)' has the following deformation:

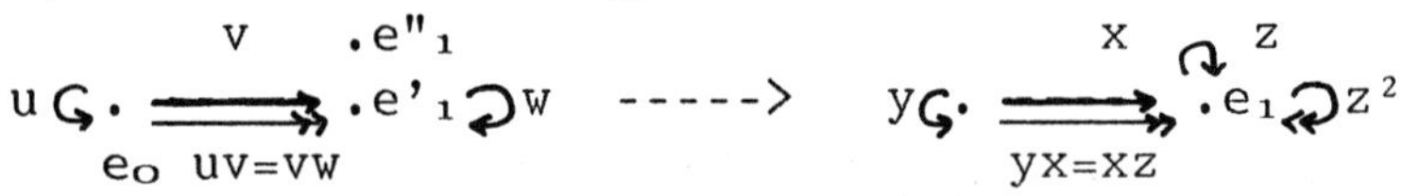

where $e_1= e'_1+e"_1$, $x=v+(1/2t)uv$, $y=u+w$ and $z=t(e'_1-e"_1)+(1/2)w$.

Applying twice lemma 1, we show that an algebra with weighted basis-graph (11)' trivially deforms.

Because of the presence of a trivial loop on the right idempotent, every algebra with basis-graph (12) deforms (there can be a great number of deformations, depending on the local ring at the left idempotent).

(13)' has the following deformation:

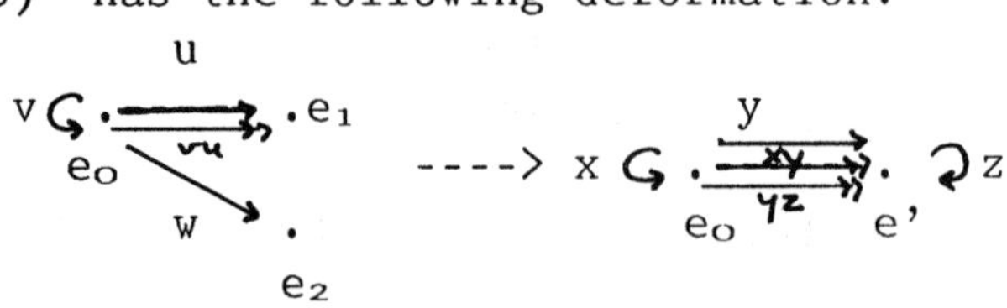

given by $e'=e_1+e_2$, $x=v$, $y=u+w$, $z=t(e_1-e_2)$; we get

$xy=vu$, $yz=t(u-w)$.

(13)" has the following deformation:

given by $e_o=e'_o+e''_o$, $e_1=e'_1+e''_1$, $x=t(e'_o-e''_o)$, $y=v+w_1$, $z=t(e'_1-e''_1)$, $u=w_2$. Actually, the existence of an idempotent-splitting deformation for this algebra is a consequence of Prop. 5.

(13)"' has the following deformation:

with $e_1=e'_1+e''_1$, $z=t(e'_1-e''_1)$, $u=w_1+(1/t)vw_2$, $x=v$ and $y=w_2$. We get then $xu=0$, $xy=uz$, and when $t\to 0$, $uz\to 0$ and $yz\to 0$.

$(14)_1$ is rigid, with dim Aut A=12 (a computer calculation indicates that every algebra A with a such weighted basis-graph has its second cohomology group equal to 0).

$(14)_2$ has the following deformation:

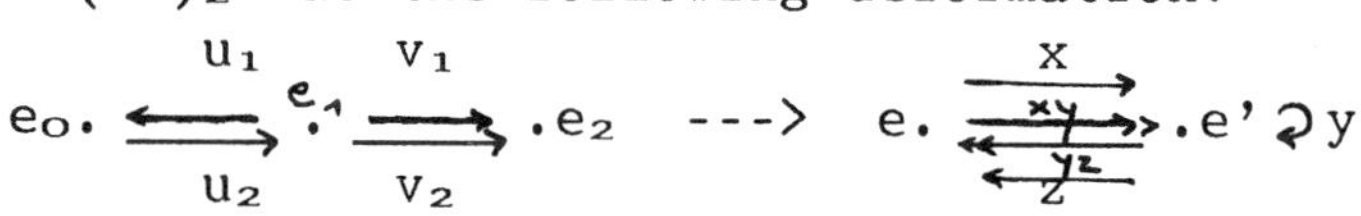

with $e'=e_o+e_2$, $e=e_1$, $x=u_1+v_1$, $y=(t/2)(e_o-e_2)$ and $z=u_2+v_2$.

3 idempotents.

Except the weighted basis-graph marked by (*), namely number (16), all these weighted basis-graphs determine non generic algebras. There are two algebras A with weighted basis-graph (16), with the right arrow in either direction; these algebras are rigid, as $H^2(A,A)=0$, with dim Aut A=6.

The basis-graphs (15), (17) and (20) determine deformable algebras, as a consequence of Lemma 3. The basis-graphs (18) and (19) determine deformable algebras, as a consequence of the Corollary of Lemma 3.

For (21), we have the following deformation:

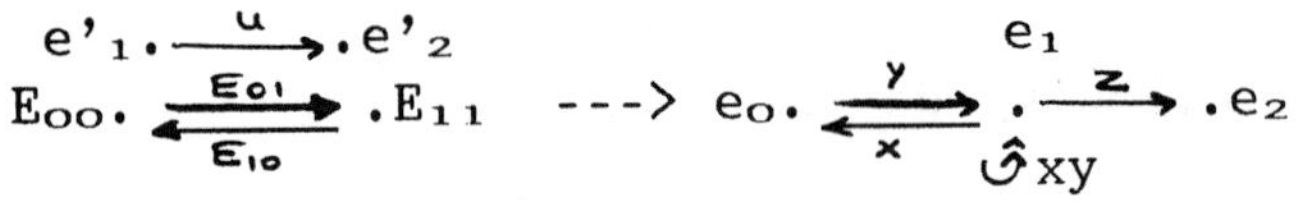

with $e_0=E_{00}$, $e_1=e'_1+E_{11}$, $e_2=e'_2$, $x=tE_{10}$, $y=tE_{01}$ and $z=u$. We get $xy=t^2E_{11}=t^2(e_1-e'_1)$, and $yx\to 0$ when $t\to 0$.

As before, we denote by p(n) the number of reduced irreducible components of $\mathbf{Alg_n}$. The total number of irreducible components of $\mathbf{Alg_n}$, including the non reduced ones, will be denoted by q(n). Using Proposition 5 and the results of the present section, we get our main statement:

<u>THEOREM:</u> $p(7) \geq 65$ and $q(7) \geq 70$ i.e. $\mathbf{Alg_7}$ contains at least 65 reduced irreducible components, each of them dominated by one of the algebras whose basis-graphs are listed in Appendices I and II, and at least 5 non reduced irreducible components, dominated by the families listed in Appendix III.

<u>REMARK:</u> in order to get better lower bounds for p(7) and q(7), we have to study the local 7-dimensional algebras and families. This requires heavy machinery and is treated elsewhere (see [DP-Sc2]).

.Appendix I: rigid loopless algebras of dimension 7 .
Ia: classified by increasing automorphism dimension.

dim Aut A

0 

2 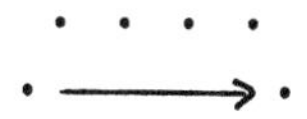

3 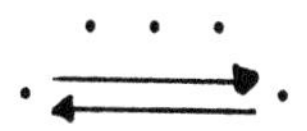

4 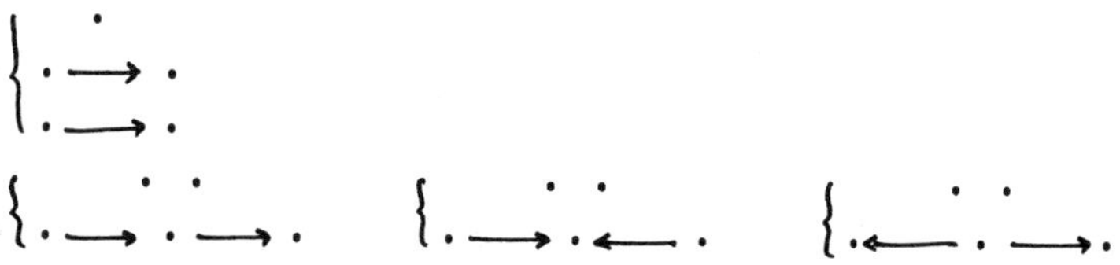

5 

6 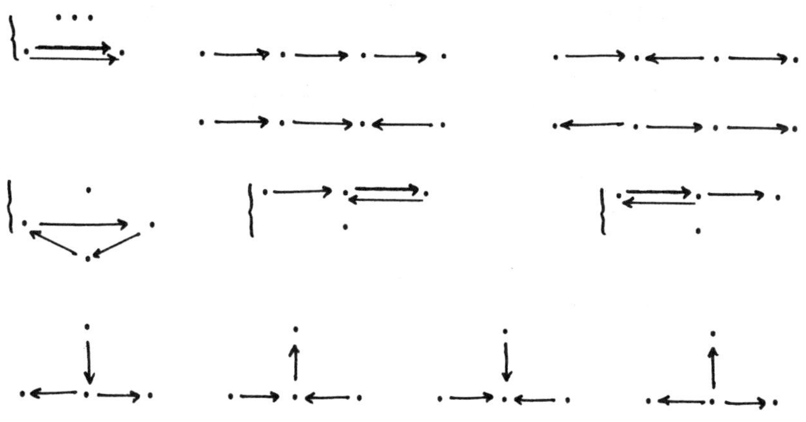

7 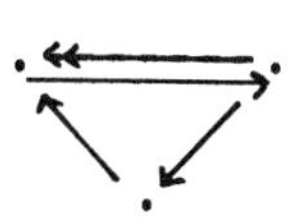

<u>dim Aut A</u>

8

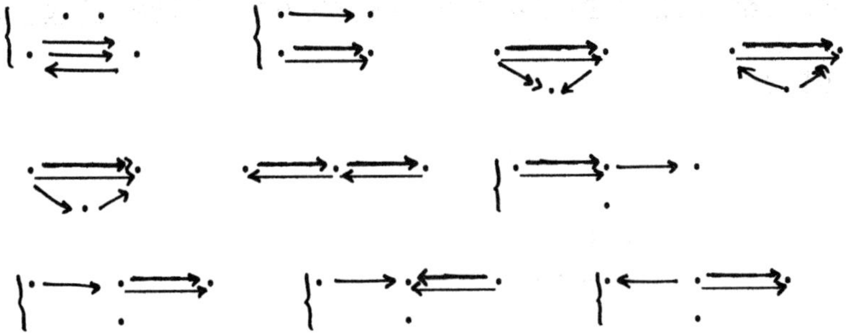

10

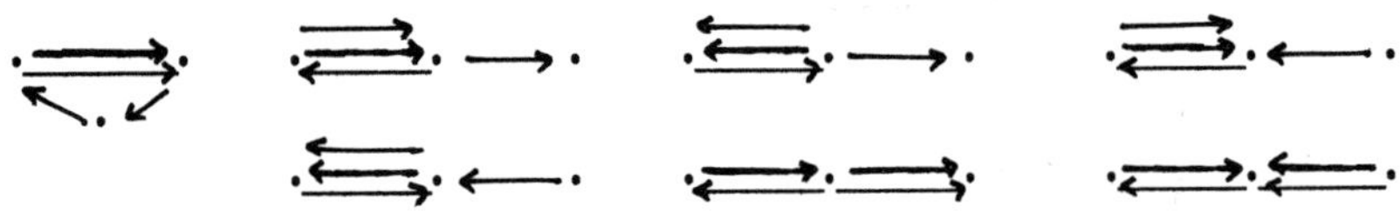

12

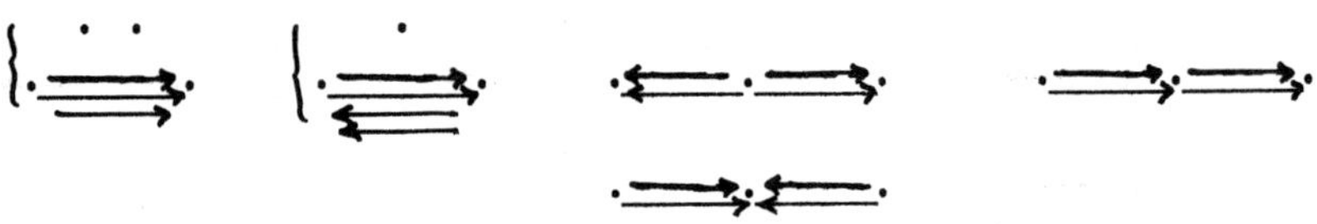

14

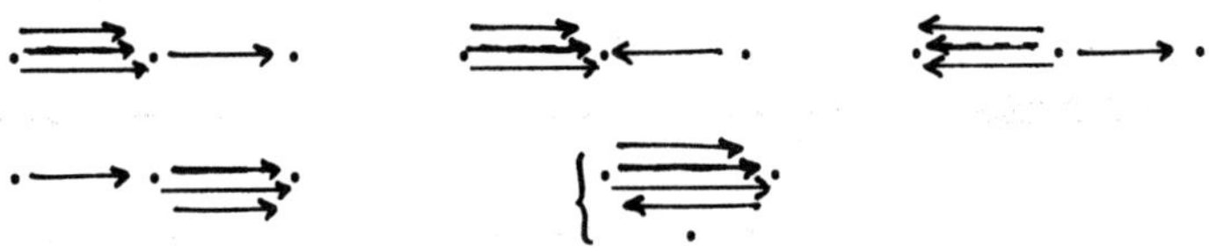

18

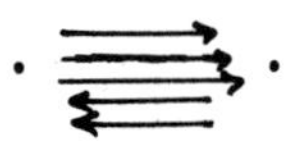

20

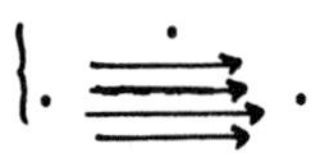

22

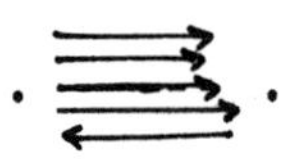

30

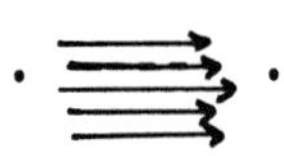

(Kronecker algebra)

Ib: classified by decreasing number of idempotents

7 idempotents

6 idempotents

5 idempotents

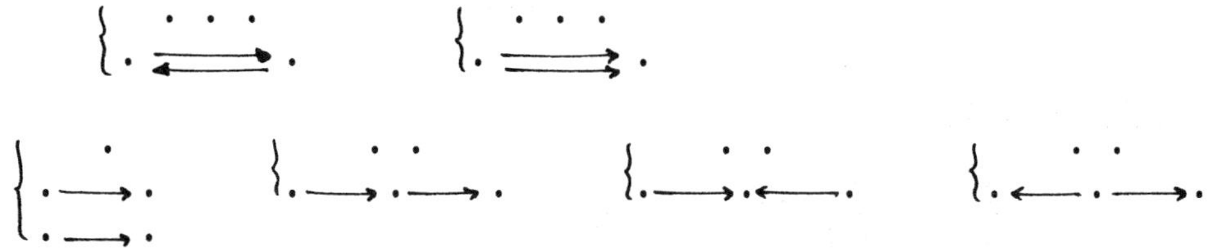

4 idempotents

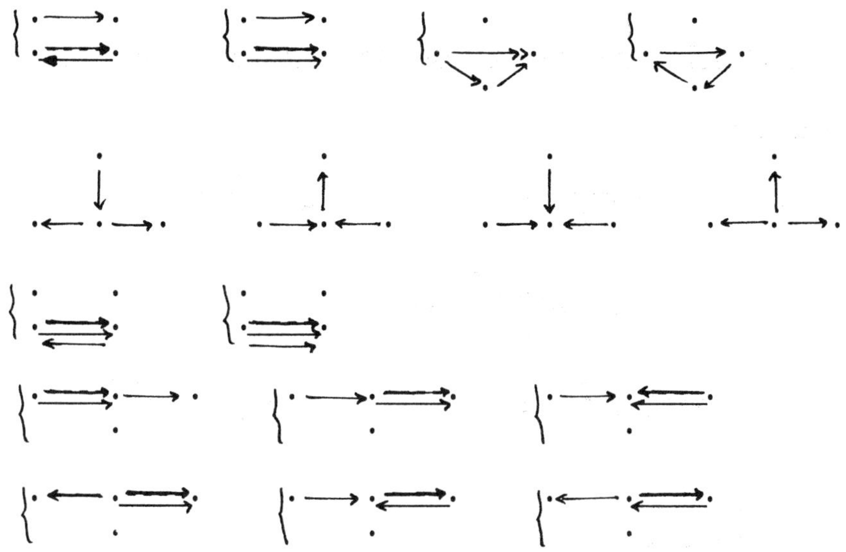

3 idempotents

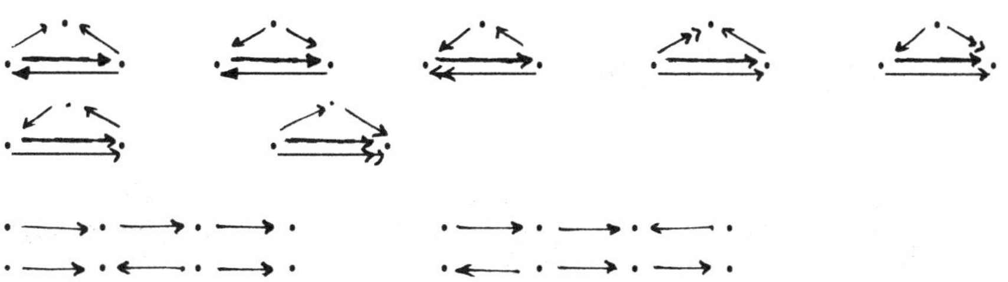

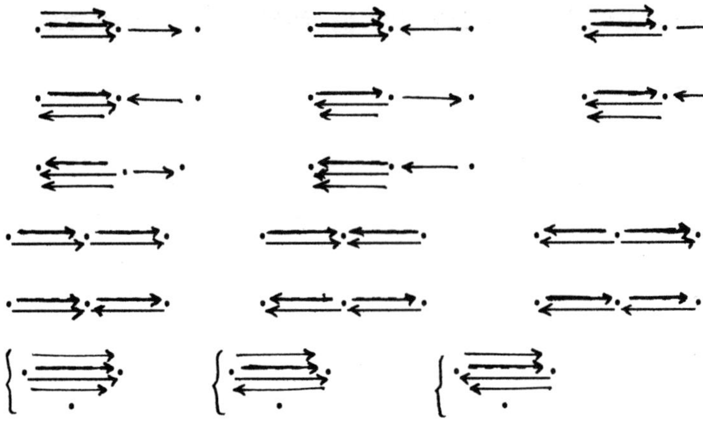

2 idempotents

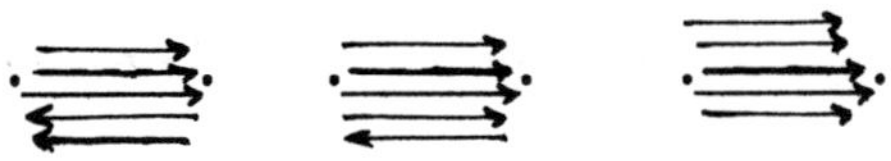

(Kronecker algebra)

.Appendix II: generic mixed basis-graphs of dimension 7.

dim Aut A

6

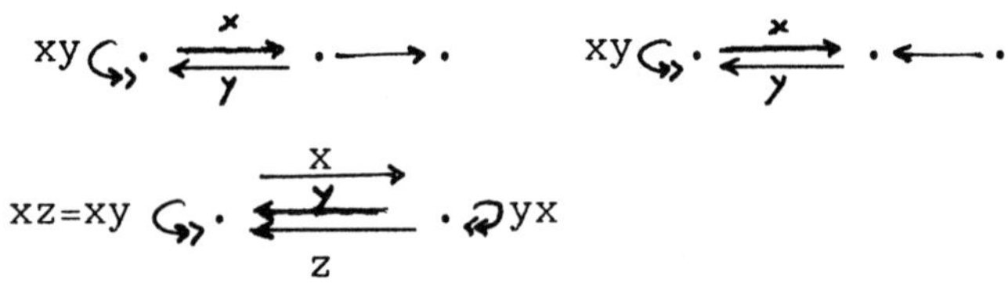

7

8

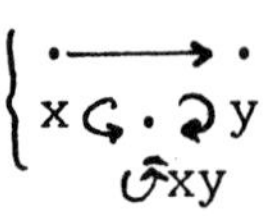

(yx= s xy)

11

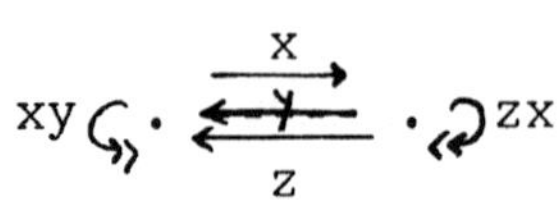

12

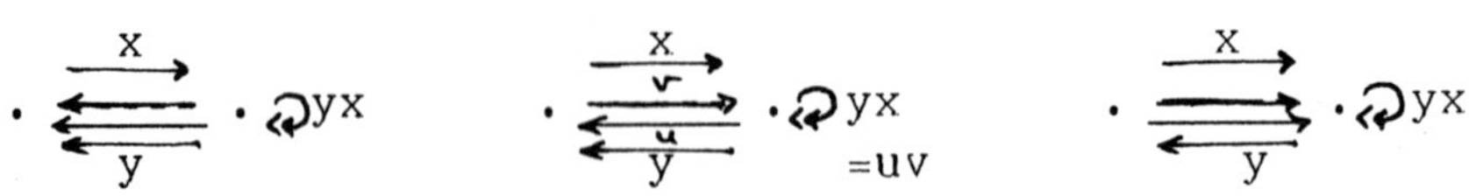

.Appendix III: known quasi-generic mixed basis-graphs of dimension 7.

dim Aut A

5

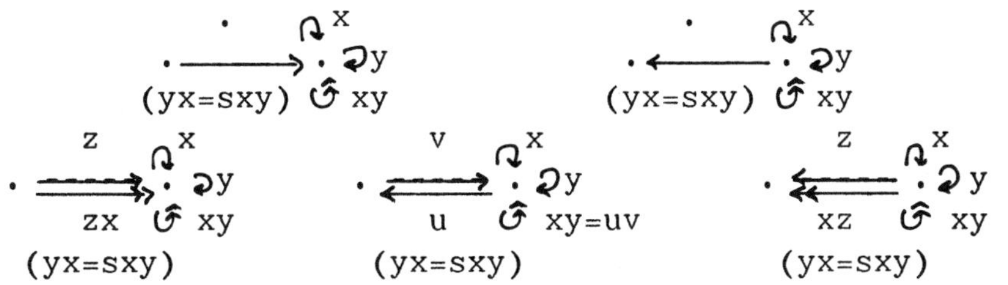

. BIBLIOGRAPHY .

[Ci] C. Cibils: Rigid monomial relations algebras, Prepublication de l'Institut Fourier n°146, Grenoble (1990).

[DP] T. Dana-Picard: Classifying generic algebras, Ph.D. Thesis, Bar-Ilan University 1990.

[DP-Sc1] T. Dana-Picard-M. Schaps: Classifying generic algebras, to appear in Rocky Mount. J. of Math.

[DP-Sc2] T. Dana-Picard-M. Schaps:Classifying generic algebras II, Preprint, Bar-Ilan University.

[Fl] F. Flanigan: Which algebras deform into a total matrix algebra? J. of Alg. 29 (1974), 103-112.

[Ga] P.Gabriel: Finite representation type is open, in Representations of Algebras ,LNM 488, Springer Verlag 1974, 132-155.

[Ge] M. Gerstenhaber: On the deformation of rings and algebras, Ann. of Maths 79 n°1 (1964), 59-103.

[Ha] D. Happel: Deformations of five dimensional algebras with unit, in Ring Theory,LN in Pure and Applied Maths (1978), 453-494.

[Ja] N. Jacobson: Structure of rings, A.M.S. Colloq. Publ. vol 37, A.M.S., Providence R.I. (1964).

[Kr] H. Kraft: Geometric methods in representation theory,LNM 944, Springer Verlag 1980, 180-258.

[Ma] G. Mazzola: The algebraic and geometric classification of associative algebras of dimension 5, Manuscr. Math.27(1979), 81-101.

[Sc1] M. Schaps: Deformations of finite dimensional algebras and their idempotents, Trans. A.M.S. 307 n°2 (1988), 843-856.

[Sc2] M. Schaps: Deformations of algebras and Hochschild cohomology II, in Perspectives in Ring Theory, NATO ASI Series 1988.

Department of Mathematics and Computer Science
Ben-Gurion University of the Negev
P.O.B. 653
Beer-Sheva
Israel
e-mail: dana@bengus.bitnet

Canadian Mathematical Society
Conference Proceedings
Volume 11 (1991)

THE WEYL GROUP OF A WILD GRAPH

J. A. de la Peña

ABSTRACT. Let A be a nxn Cartan matrix and Δ the associated Coxeter (valued) graph. Let W be the Weyl group of A. If Δ is a wild graph (that is, Δ is neither Dynkin nor affine), and $n \geq 3$, then Tits [13] has shown that W has a non-abelian free subgroup. We give an elementary proof of this theorem using some representation theory arguments. We also show that, in this case, for every faithful representation $\sigma: W \to GL(V)$, there exist a homomorphism $\mathbb{C} \to k$ and an element $w \in W$ such that $\tilde{\sigma}(w)$ is a hyperbolic matrix, where $\tilde{\sigma}: W \to GL(V \otimes_{\mathbb{C}} k)$ is the induced representation.

Let $A = (a_{ij})$ be a nxn *Cartan matrix*, that is, $a_{ii} = 2$ for $1 \leq i \leq n$, $a_{ij} = 0$ if and only if $a_{ji} = 0$, and a_{ij} is a non positive integer for $i \neq j$. Associated with A there is a *Coxeter* (valued) *graph* Δ that we assume to be connected.

Let $E = \mathbb{R}^n$ and $e_1, \ldots, e_n$ be a set of linearly independent vectors in E. Define (-,-) the (non-symmetric) bilinear form in E given by $(e_i, e_j) = a_{ij}$. The *reflections* $s_i \in GL(E)$ are defined by $v \cdot s_i = v - (v, e_i)e_i$. The subgroup W of $GL(E)$ generated by $s_1, \ldots, s_n$ is called the *Weyl group* of A.

Weyl groups play a fundamental role in the theroy of Lie and Kac-Moody algebras. In representation theory of algebras appear in the study of the root systems of hereditary algebras [9]. When Δ is a Dynkin or affine diagram, the structure of the group W is well known. If Δ is a *wild graph* (that is, Δ is

---

AMS subject classification: 16A64, 17B67, 20G99.

neither Dynkin nor affine), then there is a beautiful result of J. Tits [13] (see also the works of P. de la Harpe [6,7]).

Theorem 1: *If Δ is a wild graph and $n \geq 3$, then W has a non-abelian free subgroup.*

One of the purposes of this work is to present a self-contained elementary proof of Theorem 1. In certain cases we indicate how to construct the generators of a free non-abelian subgroup of W.

A matrix $g \in GL(V)$ is *hyperbolic* if it has two eigenvalues λ, μ with $\|\lambda\| \neq \|\mu\|$. We show the following result.

Theorem 2: *Let A be a nxn Cartan matrix with Coxeter diagram Δ and $n \geq 3$. The following conditions are equivalent:*

i) *Δ is a wild diagram.*

ii) *For every faithful complex representation $\sigma: W \to GL(V)$, there exists a locally compact field k, a homomorphism $\mathbb{C} \to k$ and an element $w \in W$ such that $\tilde{\sigma}(w)$ is hyperbolic, where $\tilde{\sigma}: W \to GL(V \otimes_{\mathbb{C}} k)$ is the induced representation.*

The implication ii) $\Rightarrow$ i) is contained in [7]. For the sake of completeness we sketch some proofs of [7,13].

1. Weyl groups and Coxeter elements.

1.1. Let W be the Weyl group of the Cartan matrix $A = (a_{ij})$. Then W is a *Coxeter group* with generators $s_1, \ldots, s_n$ and defining relations

$$s_i^2 = 1 \quad (i = 1, \ldots, n) \qquad \text{and } (s_i s_j)^{m_{ij}} = 1 \quad (i, j = 1, \ldots, n),$$

where the integers m_{ij} are given in the following table

$a_{ij}a_{ji}$	0	1	2	3	≥ 4
m_{ij}	2	3	4	6	∞

See [3,Chap.IV,1.5] and [10,3.12].

1.2. The Cartan matrix A satisfies one and only one of the following three properties:

(Finite)	$\det A \neq 0$; there exists $u > 0$ such that $u \cdot A > 0$.
(Affine)	$\det A = 0$; there exists $u > 0$ such that $u \cdot A = 0$.
(Indefinite)	there exists $u > 0$ such that $u \cdot A < 0$.

With A we associate the *Coxeter valued graph* Δ in the following way: Δ has vertices $\Delta_0 = \{1, \dots, n\}$, if $i < j$ and $a_{ij} \neq 0$, then there is an edge in Δ joining i and j with valuation $(|a_{ij}|, |a_{ji}|)$.

It turns out that A is of finite type if and only if Δ is a *Dynkin diagram*. In this case the Weyl group W is finite.

The matrix A is of affine type if and oly if Δ is an *affine* (also called *Euclidean*) diagram. In this case W is a semidirect product $\overset{\circ}{W} \ltimes T$, where $\overset{\circ}{W}$ is a finite Weyl group and T is an abelian group.

In case $n = 2$, the Weyl group is dihedral and hence solvable.

1.3. The element $c = s_1 s_2 \dots s_n \in W$ is called a *Coxeter transformation*. In general, the conjugation class of a Coxeter transformation depends on the order chosen for the indices.

If A is of finite type, all Coxeter transformations are conjugate and therefore have the same order h (=*Coxeter number* of A). The characteristic polynomial $P(x)$ of c has the form

$$P(x) = \Pi_{j=1}^{n}(x - \exp(2i\pi m_j/h)),$$

where $m_1, \dots, m_n$ are integers such that $1 = m_1 < m_2 \leq \cdots \leq m_{n-1} < m_n = h - 1$. Moreover, the order of W is $(m_1 + 1)(m_2 + 1) \cdots (m_n + 1)$. See [3,Chap.V].

If A is of affine type, then c has infinite order and

$$P(x) = (x - 1)^2 \cdot \Pi_{j=1}^{n-2}(x - \exp(2i\pi m_j/h)),$$

where $1 \leq m_1 \leq \cdots \leq m_{n-2} \leq h-1$ are integers and h is the Coxeter number of a Cartan matrix of finite type (canonically associated with A), [2,4].

1.4. Given a matrix $g \in GL(E)$, the number $\rho(g) = \max\{\|\lambda\| : \lambda$ is an eigenvalue of $g\}$ is called the ***spectral radius*** of g. In case A is of finite or affine type, we have seen that $\rho(c) = 1$ for a Coxeter element $c \in W$.

An element $g \in GL(E)$ is said to be ***hyperbolic*** if there are two eigenvalues λ, μ of g such that $\|\lambda\| \neq \|\mu\|$. Since an element $w \in W$ has $\det(w) = \pm 1$, then w is hyperbolic if and only if $\rho(w) > 1$.

Following [6], we say that $g \in GL(E)$ is ***sharp*** if g has a simple eigenvalue λ such that $\|\mu\| < \|\lambda\| \quad (= \rho(g))$ for any other eigenvalue μ of g. We say that g is ***very sharp*** if both g and g^{-1} are sharp.

1.5. Proposition. *Let A be a Cartan matrix and c be a Coxeter transformation. Then*

a) *If A is not of finite type, then $\rho(c)$ is an eigenvalue of c* [5].

b) *If A is of indefinite type, then c is hyperbolic* [8]. □

If the Coxeter graph Δ associated with A is a *tree* of wild type then c is very sharp [1]. It is an *open problem* if this property holds for any Coxeter graph of wild type.

1.6. We say that a Cartan matrix A is ***bipartite*** if it is ***symmetrizable*** (that is, there exists a diagonal matrix $D = \text{diag}\,(f_1, \ldots, f_n)$ with f_i positive integers $(1 \leq i \leq n)$ such that $A \cdot D$ is symmetric) and the Coxeter graph Δ of A admits an orientation $\vec{\Delta}$ such that every vertex is either a source or a sink.

If A is bipartite, we will always ***assume*** that the vertices of Δ are ordered in such a way that $\{1, \ldots, m\}$ are the sinks of $\vec{\Delta}$ and $\{m+1, \ldots, n\}$ are the sources. Let $c = s_1, \ldots, s_n$ be the corresponding Coxeter transformation.

Proposition [2,11,15]. *Let A be a bipartite Cartan matrix of indefinite type. Then*

a) *The Coxeter transformation is very sharp.*

b) *There exists a vector $y^+ >> 0$ such that $y^+ \cdot c = \rho(c) y^+$.* $\square$

2. Free subgroups of Weyl Groups.

2.1. Let $g, h \in GL(E)$ be two very sharp elements. Let y^+ (resp. y^-, z^+, z^-) be an eigenvector of g (resp. g^{-1}, h, h^{-1}) with eigenvalue λ of norm $\|\lambda\| = \rho(g)$ (resp. $\rho(g^{-1}), \rho(h), \rho(h^{-1})$). Then g and h are said to be *in general position* if y^+ and y^- (resp. z^+ and z^-) are not eigenvectores of h (resp. of g).

Proposition [6,13]. *If $g, h \in GL(E)$ are two very sharp elements in general position, there exists a number N such that for all $m \geq N$, the subgroup of $GL(E)$ generated by g^m and h^m is free.*

Sketch of proof: Let y^+, y^-, z^+, z^- be as above and assume that these vectors have euclidean norm 1. Consider the images $\bar{g}, \bar{h}$ of g, h in $PGL(E) = GL(E)/\mathbb{R}^*$. Hence $y^+, y^- \in \mathbb{S}^n$ are fixed points of $\bar{g}$. Moreover y^+ is an attracting point and y^- is a repulsing point of $\bar{g}$. That is, for any neighborhood U of y^+ in $\mathbb{S}^n$ and for any compact subset K of $\mathbb{S}^n \setminus \{y^-\}$, we have $\bar{g}^m(K) \subset U$ for any integer m with $|m|$ big enough. By symmetry, we may find disjoint compact sets K and K' with $y^+, y^- \in K$, $z^+, z^- \in K'$ and such that $\bar{g}^m(K') \subset K$, $\bar{h}^m(K) \subset K'$ for $|m| \geq N$.

Let $m \in \mathbb{Z}$ be such that $|m| \geq N$ and let $g_1 = g^m$, $h_1 = h^m$. We show that the group generated by g_1 and h_1 is free. This is an application of an argument of Klein that de la Harpe names properly "ping-pong criterium". Assume that $g_1^{a_1} \cdot h_1^{b_1} \cdot g_1^{a_2} \cdots h_1^{b_{s-1}} \cdot g_1^{a_s} = 1$. If $a_1 \neq 0 \neq a_s$, then

$$\bar{g}_1^{a_1} \cdot \bar{h}_1^{b_1} \cdots \bar{h}_1^{b_{s-1}} \cdot \bar{g}_1^{a_s}(K') \subset \bar{g}_1^{a_1} \cdot \bar{h}_1^{b_1} \cdots \bar{h}_1^{b_{s-1}}(K) \subset \cdots \subset \bar{g}_1^{a_1}(K') \subset K,$$

a contradiction. If $a_1 = 0 < a_s$, then $g_1^{-1} \cdot h_1^{b_1} \cdots h_1^{b_s - 1} \cdots g_1^{a_s} \cdot g_1 = 1$ and we are in the first situation. The remaining cases can be treated similarly. □

2.2. Let A be a *symmetrizable* Cartan matrix. Let $D = \text{diag}\,(f_1, \ldots, f_n)$ be a diagonal matrix with f_i positive integers ($i = 1, \ldots, n$) and $A \cdot D$ be symmetric. Let $\vec{\Delta}$ be an oriented graph whose underlying graph is the Coxeter valued graph Δ of A. We assume that the vertices of $\vec{\Delta}$ form a $(+)$-*accesible sequence* (that is, 1 is a sink of $\vec{\Delta}$, 2 is a sink of $\vec{\Delta} \setminus \{1\} \ldots$).

Fix division rings F_i of dimension f_i over a central field k and for $i < j$, fix $F_i - F_j$-bimodules ${}_iM_j$ of dimension $\dim_{F_i}({}_iM_j) = |a_{ji}|$, $\dim\,({}_iM_j)_{F_j} = |a_{ij}|$. Recall that a *representation* X of the valued graph Δ associates to each vertex i a right F_i-module $X(i)$ and to each arrow $\alpha: i \to j$, a F_j-map $X(\alpha): X(i) \otimes_{F_i} {}_iM_j \to X(j)$. The *dimension vector* of X is $\underline{\dim}\, X = (\dim\; X(i)_{F_i})_i \in \mathbb{Z}^n$.

For each vertex i, we denote by P_i (resp. I_i) the indecomposable projective (resp. injective) representation associated with i. Then $\underline{\dim}\; P_i$ has j-th coordinate $\sum_s |a_{ix_{s1}} a_{x_{s1}x_{s2}} \cdots a_{x_{sm_s}j}|$, with $\{i \to x_{s1} \to x_{s2} \to \cdots \to x_{sm_s} \to j\}_s$ the set of all paths from i to j in $\vec{\Delta}$. Similarly we may describe $\underline{\dim}\; I_i$.

Let P (resp. I) be the matrix whose i-th row is the vector $\underline{\dim}\; P_i$ (resp. $\underline{\dim}\; I_i$). Then P is invertible and we may form the matrix $-P^{-1}I$. The following is well known.

Lemma. *Let c be the Coxeter transformation corresponding to the symmetrizable Cartan matrix A. Then $c = -P^{-1}I$.*

Proof: Check that $(\underline{\dim}\; P_i) \cdot c = -\underline{\dim}\; I_i$. □

2.3. Proposition. *Let A be a bipartite Cartan matrix of indefinite type and with $n > 2$. Let $c \in W$ be the corresponding Coxeter transformation. Then there exists $i \in \{1, \ldots, n\}$ such that c and $s_i c s_i$ are two very sharp elements in general position.*

Proof: By (1.6), c and $s_i c s_i$ $(i = 1, \dots, n)$ are very sharp. Let ρ be the spectral radius of c and $y^+, y^- \gg 0$ be eigenvectors of c with $y^+ \cdot c = \rho y^+$ and $y^- \cdot c = \rho^{-1} y^-$ (observe that in this case $\rho(c^{-1}) = \rho(c)$).

We distinguish two cases.

a) There are at least two sinks in $\vec{\Delta}$.

Assume that y^+ is an eigenvector of $s_1 c s_1$. There exists a number $\mu \in C$ such that $y^+ \cdot s_1 c = \mu y^+ \cdot s_1$. First we remark that the vectors y^+ and $y^+ \cdot s_1$ are different. Otherwise, ρ is an eigenvalue of $c' = s_2 \dots s_n$. Let $\bar{A}$ be the Cartan matrix obtained by deleting the first column and the first row of A, then the Coxeter transformation $\bar{c}$ of $\bar{A}$ satisfies

$$c' = \begin{pmatrix} 1 & * \\ 0 & \bar{c} \end{pmatrix}.$$

Hence ρ is an eigenvalue of $\bar{c}$. But [11,15] show that $\rho(\bar{c}) < \rho(c)$, a contradiction.

Therefore $y^+ \cdot s_1 = y^+ + \alpha e_1$ with $\alpha \neq 0$. Since $e_1 = \underline{\dim}\, P_1$, $(\rho - \mu) y^+ = \alpha \underline{\dim}\, I_1 + \mu \alpha e_1$.

Thus $\mu \neq 0$. Since $y^+ \gg 0$, I_1 has all its coordinates different from 0. This implies that 1 is the unique sink of $\vec{\Delta}$. A contradiction.

Similar arguments complete the proof in this case.

b) The vertex 1 is the unique sink of $\vec{\Delta}$. Then Δ is a valued graph of the form

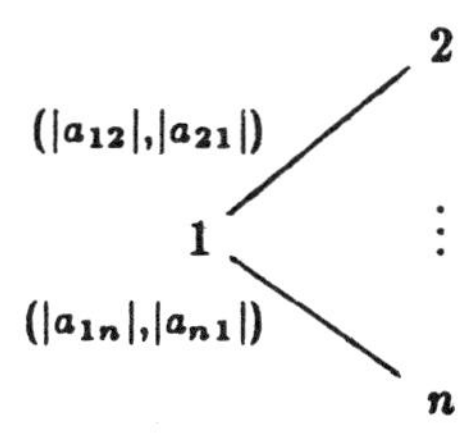

Simple calculations show the following:

$$c = \begin{pmatrix} -1 & a_{12} & \cdots & & a_{1n} \\ -a_{21} & a_{12}a_{21}-1 & & & \\ \vdots & & \ddots & a_{1i}a_{j1} & \vdots \\ & a_{1i}a_{j1} & & & \\ -a_{n1} & & \cdots & & a_{1n}a_{n1}-1 \end{pmatrix}$$

$\rho = 1/2(d-2+\sqrt{(d-2)^2-4})$, where $d = \sum a_{1i}a_{i1}$ and $y^+ = \rho/(\rho+1)(1+1/\rho, -a_{12}, \ldots, -a_{1n})$. We show that c, s_2cs_2 are in general position. Indeed, $y^+ \cdot s_2 = y^+ + \alpha e_2$, where $\alpha = (\rho-1)a_{12}/(1+\rho)$. If y^+ is an eigenvector of s_2cs_2, then $y^+ \cdot s_2c = \mu y^+ \cdot s_2$ for some $\mu \in C$. Hence $(\rho-\mu)y^+ = -\alpha e_2 \cdot c + \alpha\mu e_2$, where $e_2 \cdot c = (-a_{21}, a_{12}a_{21}-1, a_{13}a_{31}, \ldots, a_{1n}a_{n1})$. Evaluating in the two first coordinates, we get: $a_{12}a_{21} = (1+\rho)^2/\rho = d$, which implies $n = 2$, against the hypothesis. Similar arguments complete the proof. □

2.4. A Cartan matrix A is said to be *minimal of indefinite type* if A is of indefinite type and every restriction A_i of A, obtained by deleting the i-th column and the i-th row of A, is of finite or affine type.

Lemma. *Let A be a Cartan matrix minimal of indefinite type with $n > 3$. Then there is a bijection $\sigma: \{1, \ldots, n\} \to \{1, \ldots, n\}$ such that the Coxeter elements $c = s_{\sigma(1)} \cdots s_{\sigma(n)}$ and s_1cs_1 are very sharp and in general position.*

Proof: If A is bipartite the result follows from (2.3). It is enough to check the claim for the non bipartite diagrams appearing in the list [14], a task that can be done with the help of a computer. □

2.5. *Proof of Theorem 1:* Let A be an indefinite nxn Cartan matrix with $n \geq 3$. If A is bipartite, by (2.1), (2.3), the Weyl group W of A has a free non-abelian subgroup. If A is non bipartite,the Coxeter graph Δ of A has an induced valued subgraph Δ' such that the corresponding mxm Cartan matrix A' is of indefinite type, $m \geq 3$ and m is minimal with these properties. Let W' be the Weyl group of A', then W' is a subgroup of W. It is enough to show the result for A'.

It A' is minimal of indefinite type, by (2.4), (2.1) we get the result. Otherwise, A' has one of the following shapes

with $a_{12}a_{21} > 4$. Then W' has a quotient isomorphic to a Weyl group $\bar{W}$ of a 3×3 Cartan matrix $\bar{A}$ minimal of indefinite type. Since $\bar{W}$ has a free non-abelian subgroup, therefore W' has also one. □

Remark. Of course, (2.4) can be shown for all minimal indefinite Cartan matrices by a direct calculation. Nevertheless, when it applies (2.3) yields a more explicit construction.

3. Representation of Weyl groups.

3.1. Let W be a group and k be a field. A *representation* $\sigma: W \to GL(V)$, where V is a k-vector space, is called *completely irreducible* if for every subgroup of finite index W_1 of W the restriction $\sigma: W_1 \to GL(V)$ is irreducible.

Let A be a nxn Cartan matrix. Let $E = \mathbb{C}^n$ and $(e_i)_i$ be the canonical basis. Denote by $B: E \times E \to \mathbb{C}$ the symmetric bilinear form given by

$$B(e_i, e_j) = \begin{cases} -\cos(\pi/m_{ij}) & \text{if } a_{ij}a_{ji} < 4 \\ -1 & \text{if } a_{ij}a_{ji} \geq 4. \end{cases}$$

Let $\sigma_0: W \to GL(E)$ be the canonical inclusion. If B is not degenerated, then σ_0 is irreducible. If B is degenerated, set $\bar{E} = E/E^0$, where E^0 is the radical of B. Then the induced representation $\bar{\sigma}_0: W \to GL(\bar{E})$ is irreducible.

Proposition [7]. a) *If A is not of finite type and B is not degenerated, then σ_0 is completely irreducible.*

b) *If A is not of affine type and B is degenerated, then $\bar{\sigma}_0$ is completely irreducible.* □

3.2. If A is a Cartan matrix of indefinite type, we know (1.5) that the Coxeter element $c \in W$ is hyperbolic. The proof of Theorem 1 as given in [7] consists in using the hyperbolic element c to construct two very sharp elements in general position.

Proposition [7,13]. *Let A be a nxn Cartan matrix with $n \geq 3$ and let W be its Weyl group. Assume that $w \in W$ is such that $\sigma_0(w) \in GL(E)$ is a hyperbolic matrix. Then W has a non-abelian free subgroup.*

Sketch of the proof: Assume that B is not degenerated (the other case is similar).

a) There is a subgroup of finite index W_1 of W such that $\sigma_0(W_1)$ is contained in $SL(E)$. Since $\sigma_0(w)$ is hyperbolic, it has infinite order, hence there is some number k such that $w^k \in W_1$. Moreover, $\sigma_0(w^k)$ is hyperbolic. Let $\mu_1, \ldots, \mu_n$ be the set of eigenvalues (with multiplicities) of $\sigma_0(w^k)$. Assume $\|\mu_1\| = \|\mu_2\| = \cdots = \|\mu_m\| > \|\mu_{m+1}\| \geq \cdots \geq \|\mu_n\|$. Consider the induced representation $\tilde{\sigma}: W_1 \to SL(\Lambda^m E)$. Then $\tilde{\sigma}(w^k)$ is sharp with spectral radius $\tilde{\rho} = \|\mu_1\|^m$. Moreover, the Zariski closure of $\sigma_0(W_1)$ in $SL(E)$ is semisimple. Therefore $\tilde{\sigma}$ is completely reducible. Let $\Lambda^m E = \oplus_1^s F_i$ be a decomposition in irreducible W_1-spaces. An eigenvector of $\tilde{\sigma}(w^k)$ with eigenvalue $\tilde{\rho}$ belong to some F_j. Hence $\tilde{\sigma}_j: W_1 \to SL(F_j)$ is an irreducible representation such that $\tilde{\sigma}_j(w^k)$ is sharp.

b) Let $\sigma: W_1 \to GL(F)$ be an irreducible representation such that the Zariski closure of $\sigma(W_1)$ is connected. Let $w_1 \in W_1$ be such that $\sigma(w_1)$ is sharp. Arguments of Tits [13] or Dixon [7] show that there exist elements $x, y \in W_1$ such that $\sigma(x \cdot w_1^s \cdot y \cdot w_1^{-s})$ is very sharp for some $s > 0$.

c) Let $w_2 \in W_1$ be an element such that $\sigma(w_2)$ is very sharp. An easy topological argument shows that there is some $g \in W_1$ such that $\sigma(w_2)$ and $\sigma(g \cdot w_2 \cdot g^{-1})$ are very sharp in general position. The result then follows from (2.1). □

3.3. *Proof of Theorem 2:* Assume that (ii) is satisfied. Then (3.2) and (1.2) imply that A is of indefinite type. Alternatively, if A is of finite or affine type, it is easy to see that for any $w \in W$, the eigenvalues of $\sigma_0(w)$ are roots of unity.

Assume that (i) holds. Let $\sigma: W \to GL(V)$ be a complex faithful representation. Suppose that $\sigma(w)$ is not hyperbolic for all $w \in W$.

Since $\det \sigma(w) = \pm 1$ then $\|\lambda\| = 1$ for any $\lambda \in \operatorname{Spec} \sigma(w) = \{\lambda \in \mathbb{C}: \lambda \text{eigenvalue of } \sigma(w)\}$.

a) Let U be the set of roots of unity in $\mathbb{C}$. Assume that $\operatorname{Spec} \sigma(w) \subset U$ for every $w \in W$

Let $0 = V_0 \subset V_1 \subset \cdots \subset V_s = V$ be a Jordan-Holder sequence of W-modules that is, the induced representations $\sigma_i: W \to GL(V_i/V_{i-1})$ are irreducible, $i = 1, \ldots, s$.

We claim that for all $1 \leq i \leq s$, $w \in W$, we have $\operatorname{Spec} \sigma_i(w) \subset U$ and $\sigma_i(W)$ is a finite group. By induction on i. For $i = 1$, $\sigma_1(w)$ acts as w on V_1, hence $\operatorname{Spec} \sigma_1(w) \subset U$ and by Theorem (3.4) below, $\sigma_1(W)$ is a finite group, say of order m_1. Let $w \in W$ and suppose $\mu \in \operatorname{Spec} \sigma_2(w)$. Let $0 \neq v_2 \in V_2, v_1 \in V_1$ be such that $v_2 \cdot w = \mu v_2 + v_1$. Then $(v_2 \cdot w^{m_1} - v_2) \cdot w = \mu(v_2 \cdot w^{m_1} - v_2)$. Hence either $\mu \in \operatorname{Spec} \sigma(w) \subset U$ or $v_2 = v_2 \cdot w^{m_1} = \mu^{m_1} v_2 + (\mu^{m_1 - 1} v_1 + \mu^{m_1 - 2} v_1 \cdot w + \cdots + v_1 \cdot w^{m_1 - 1})$. In the last case, passing to V_1/V_2 we get $\mu^{m_1} = 1$. Again (3.4) implies that $\sigma_2(W)$ is a finite group. The induction can be continued in this way.

On the other hand, the representation $\bar{\sigma} = (\sigma_i)_i: W \to \Pi_i GL(V_i/V_{i-1})$ has nilpotent kernel (of index $\leq s$). Then W has a nilpotent subgroup of finite index. By Theorem 1, A is of finite or affine type a contradiction.

b) Assume that there is some $w \in W$ and some $\lambda \in \operatorname{Spec} \sigma(w)$ such that λ is not a root of unity. By [13,4.1], there is a locally compact field k endowed with an absolute value ω and a homomorphism $\alpha: \mathbb{C} \to k$ such that $\omega(\alpha(\lambda))) \neq 1$.

The induced representation $\tilde{\sigma}: W \rightarrow GL(V \otimes_{\mathbb{C}} k)$ has an eigenvalue $\alpha(\lambda)$. Since $\det \sigma(w) = \pm 1$, then $\tilde{\sigma}(w)$ is hyperbolic. □

3.4. Theorem (Schur's first theorem) [12]. *Let $\sigma: W \rightarrow GL(V)$ be an irreducible complex representation. If for every $w \in W$, Spec $\sigma(w)$ is formed bv roots of unity, then $\sigma(W)$ is finite.*

Proof: Let $n = \dim V$. If $\lambda \in$ Spec $\sigma(w)$, for some $w \in W$, then λ is a root of unity and a root of a polynomial of degree n. There are only finitely many numbers with this property. Hence the traces of $\sigma(w)$, for $w \in W$ take values in a finite set $\{\alpha_1, \ldots, \alpha_t\} \subset \mathbb{C}$. By an argument of Burnside, $\sigma(W)$ is finite of order at most t^{n^2}. □

References.

[1] N. A'Campo: Sur les valeurs propes de la transformation de Coxeter. Invent. math. 33 (1976), 61-67.

[2] S. Berman, Y.S. Lee and R.V. Moody: The spectrum of a Coxeter transformation, Affine Coxeter Transformations and defect map. J. Alg. 111 (1989), 339-357.

[3] N. Bourbaki: Groupes et algèbres de Lie. Chap 4,5 et 6. Hermann (1968).

[4] A.J. Coleman: Killing and the Coxeter transformation of Kac-Moody algebras. Invent. math. 95 (1989), 447-477.

[5] V. Dlab and C.M. Ringel: Eigenvalues of Coxeter transformations and the Gelfand-Kirillov dimension of the preprojective algebras. Proc. AMS 83 (1981), 228-232.

[6] P. de la Harpe: Free groups and M-matrices. Bull. London Math. Soc. 13 (1982), 127-131.

[7] P. de la Harpe: Groupes de Coxeter infinis non affines. Expo. math. 5 (1987), 91-96.

[8] R.B. Howlett: Coxeter groups and M-matrices. Bull. London Math. Soc. 13 (1982), 127-131.

[9] V. Kac: Infinite root systems, representations of graphs and invariant theory. Invent. math. 56 (1980), 57-92.

[10] V. Kac: Infinite Dimensional Lie Algebras. Birkhäuser (1983).

[11] J.A. de la Peña and M. Takane: Spectral properties of Coxeter transformations and applications. Arch. Math. 55 (1990), 110-124.

[12] D.A. Supranenko: Matrix Groups. Vol. 45 Translations of Mathematical Monographs. AMS (1976).

[13] J. Tits: Free subgroups in linear groups. J. Alg. 20 (1972), 250-270.

[14] Ch. Xi: The wild hereditary algebras with the small growth number. Preprint. Bielefeld.

[15] Y. Zhang: Eigenvalues of Coxeter transformations and the structure of regular components of the Auslander-Reiten quiver. Comm. Alg.

Instituto de Matemáticas, UNAM

México 04510, D. F.

MEXICO

Canadian Mathematical Society
Conference Proceedings
Volume 11 (1991)

ALGEBRAS WITH A LARGE DUAL

Gabriella D'Este

Let A be an algebra over the field K, and let $D({}_K A_A)$ denote the dual of A as a vector space over K, that is let $D({}_K A_A)$ denote the abelian group $\mathrm{Hom}_K({}_K A_A, {}_K K)$ with its usual structure as a left A-module ([1] Proposition 4.4). Under these hypotheses, let L be a subfield of K. Then comparing the A-modules $D({}_K A_A)$ and $D({}_L A_A)$, it is easy to see ([2] Proposition 4) that the following facts hold:

1. If $[K:L]$ is finite, then $D({}_K A_A)$ is isomorphic to $D({}_L A_A)$.
2. If $[K:L]$ is infinite, then the direct sum of $|L|^{[K:L]}$ copies of $D({}_K A_A)$ is isomorphic to a submodule of $D({}_L A_A)$.
3. If A admits a simple left A-module of finite dimension over K, then $D({}_K A_A)$ is isomorphic to $D({}_L A_A)$ if and only if $[K:L]$ is finite.

We shall say, for brevity, that the K-algebra A *has a large dual*, if there is a subfield L of K with $[K:L]$ infinite such that $D({}_K A_A)$ is isomorphic to $D({}_L A_A)$. In this

---

1980 Mathematics Subject Classification (1985 Revision) : 16A52 , 16A64 .

case, we know from 2. that $D({}_K A_A)$ has the pathological property of having a submodule isomorphic to the direct sum of $|L|^{[K:L]}$ copies of $D({}_K A_A)$. The aim of this note is to show that there exist algebras with a large dual. In this way, we solve an open problem of [2]. As we shall see, the existence of algebras with a large dual follows from the existence of fields with a large dual, and from the existence of small algebras with a large dual. Indeed, if A and B are algebras over the same field such that $A \leq B$ and A has a large dual, then B , too, has a large dual. On the other hand, if K is a reasonably large field, that is a field of infinite degree over its prime subfield, then any K-algebra A can be embedded in a K-algebra B such that B has a large dual.

In the following, the symbols $\mathbb{Q}$, $\mathbb{R}$ and $\mathbb{C}$ denote the fields of rational numbers, real numbers and complex numbers respectively. Finally, the symbol ω denotes the least infinite cardinal.

We begin with a result on fields.

THEOREM 1 Let L , K , E be fields with $L \leq K \leq E$ such that $[K:L]$ and $[E:K]$ are infinite. Then the following facts hold:

(i) If $|D({}_K E_E)| = |D({}_L E_E)|$, then $D({}_K E_E)$ is isomorphic to $D({}_L E_E)$.

(ii) If $[E:K] \geq [K:L]$, then $D({}_K E_E)$ is isomorphic to $D({}_L E_E)$.

PROOF (i) Let $m = |K|^{[E:K]}$ and let $m' = |L|^{[E:L]}$. Since ${}_E E$ is the only simple left E-module and the only indecomposable injective left E-module, we deduce from ([2]

Theorem 3) that

$$(1) \quad D({}_K E_E) \simeq \bigoplus_m E_E \quad ; \quad D({}_L E_E) \simeq \bigoplus_{m'} E_E \quad .$$

On the other hand, we clearly have

$$(2) \quad m = |K|^{[E:K]} \geq |E| \quad ; \quad m' = |L|^{[E:L]} \geq |E| \quad .$$

Combining (1) and (2) with the hypothesis that $|D({}_K E_E)| = |D({}_L E_E)|$, we obtain $m = m'$. Hence $D({}_K E_E)$ is isomorphic to $D({}_L E_E)$, and so (i) holds.

(ii) Our assumptions on $[E:K]$ and $[K:L]$ guarantee that $[E:K] = [E:L]$. Therefore, we obviously have

$$(3) \quad |D({}_K E_E)| = |K|^{[E:K]} \geq |L|^{[E:K]} = |L|^{[E:L]} = |D({}_L E_E)|.$$

On the other hand, by ([2] Proposition 4), we also have $|D({}_K E_E)| \leq |D({}_L E_E)|$; consequently $|D({}_K E_E)| = |D({}_L E_E)|$. Threfore, by (i), $D({}_K E_E)$ is isomorphic to $D({}_L E_E)$. Hence (ii) holds, and the theorem is proved. //

REMARK The following example shows that the converse of condition (ii) of Theorem 1 does not hold. In fact, under the hypotheses of Theorem 1, there are examples where L, K and E satisfy the following condition:

$$(*) \quad D({}_K E_E) \simeq D({}_L E_E) \quad \text{and} \quad [E:K] < [K:L] \quad .$$

For instance, let L,K,E be fields with $L \leq K \leq E$ such that $|L| = 2^{2^\omega}$, $[K:L] = 2^\omega$ and $[E:K] = \omega$. Since $|K| = 2^{2^\omega}$ and $[E:L] = 2^\omega$, it follows that

$$|D({}_K E_E)| = |K|^{[E:K]} = 2^{2^\omega} = |L|^{[E:L]} = |D({}_L E_E)| \quad .$$

This result and condition (i) of Theorem 1 guarantee that $D({}_K E_E)$ is isomorphic to $D({}_L E_E)$. On the other hand, we clearly have $[E:K] < [K:L]$. Hence L,K,E satisfy (*).

The next corollary shows that we cannot replace finite by countable in remark 1. of the introduction. In fact, duals of algebras and duals of fields, with respect to fields quite closed to each other, behave in a completely different way.

COROLLARY 2 Let K *and* L *be fields with* $L \leq K$ *and* $[K:L] = \omega$. *Then the following facts hold:*

(i) There is an infinite dimensional K-algebra A *such that* $D({}_K A_A)$ *is not isomorphic to* $D({}_L A_A)$.

(ii) If E *is a field containing* K, *then* $D({}_K E_E)$ *is isomorphic to* $D({}_L E_E)$ *if and only if* $[E:K]$ *is infinite.*

PROOF (i) Let $A = K[x]$; then A admits simple modules of dimension one over K . Consequently, by remark 3. of the introduction, $D({}_K A_A)$ is not isomorphic to $D({}_L A_A)$.

(ii) Suppose first that $[E:K]$ is finite. Then we deduce from remark 3. of the introduction that $D({}_K E_E)$ is not isomorphic to $D({}_L E_E)$. Assume next that $[E:K]$ is infinite. Since $[K:L] = \omega$, it follows that $[E:K] \geq [K:L]$. Therefore, by condition (ii) of Theorem 1 , $D({}_K E_E)$ is isomorphic to $D({}_L E_E)$. This completes the proof of (ii). //

We now prove that there are "enough" algebras with a large dual.

COROLLARY 3 Let K *and* L *be fields with* $L \leq K$ *and* $[K:L]$ *infinite, and let* A *be a K-algebra. Then the following facts hold:*

(i) *If* $D({}_K A_A)$ *is isomorphic to* $D({}_L A_A)$ *, then* $D({}_K B_B)$ *is isomorphic to* $D({}_L B_B)$ *for any K-algebra* B *such that* $A \leq B$ *.*

(ii) *If* $D({}_K A_A)$ *is not isomorphic to* $D({}_L A_A)$ *, then* A *can be embedded in a K-algebra* B *such that* $D({}_K B_B)$ *is isomorphic to* $D({}_L B_B)$ *.*

PROOF (i) The hypothesis $D({}_K A_A) \simeq D({}_L A_A)$ and the adjoint isomorphism ([3] Proposition 2.10.9) imply that the following B-modules are isomorphic:

(1) $$D({}_K B_B) = \mathrm{Hom}_K({}_K B_B, {}_K K) \simeq \mathrm{Hom}_K({}_K A_A \otimes_A {}_A B_B, {}_K K) \simeq$$

$$\simeq \mathrm{Hom}_A({}_A B_B, \mathrm{Hom}_K({}_K A_A, {}_K K)) \simeq \mathrm{Hom}_A({}_A B_B, \mathrm{Hom}_L({}_L A_A, {}_L L)) \simeq$$

$$\simeq \mathrm{Hom}_L({}_L A_A \otimes_A {}_A B_B, {}_L L) \simeq \mathrm{Hom}_L({}_L B_B, {}_L L) = D({}_L B_B) \quad .$$

Hence (i) holds.

(ii) Let E be a field such that $K \leq E$ and $[E:K] = [K:L]$. Then the choice of E and condition (ii) of Theorem 1 guarantee that there is an isomorphism of E-vector spaces

(2) $$D({}_K E_E) \simeq D({}_L E_E) \quad .$$

Next, let A' be the K-algebra obtained from A by extending the base field to E , that is let $A' = E \otimes_K A$. Since A is a vector space over K, we deduce from ([1] Proposition 19.16) that there is a monomorphism of K-algebras

(3) $$A \simeq K \otimes_K A \longrightarrow E \otimes_K A = A' \quad .$$

On the other hand, A' is a non-zero vector space over E . Hence, by (i) and (2), there exists an isomorphism of A'-

modules

(4) $D({}_K A'_{A'}) \simeq D({}_L A'_{A'})$.

Consequently, by (3) and (4), we may choose $B = A'$. Hence (ii) holds, and the corollary is proved. //

The next statement shows that the cardinality of a large dual must be quite large.

COROLLARY 4 Let K *and* L *be fields with* $L \leq K$ *and* $[K:L]$ *infinite. If* $[K:L] = |D({}_K A_A)|$ *, then* $D({}_K A_A)$ *is not isomorphic to* $D({}_L A_A)$.

PROOF The hypothesis $[K:L] = |D({}_K A_A)|$ implies that $[K:L] = |K| = |D({}_K A_A)|$. Consequently, we clearly have

$$|D({}_K A_A)| = |K| < 2^{|K|\cdot \dim_K A} \leq |L|^{[K:L]\cdot \dim_K A} =$$

$$= |L|^{\dim_L A} = |D({}_L A_A)|$$

Therefore, $D({}_K A_A)$ cannot be isomorphic to $D({}_L A_A)$. //

REMARK Let L,K,E be as in the hypotheses of Theorem 1. Then we deduce from ([2] Proposition 4) that

(*) $[K:L] \leq |K| \leq |D({}_K E_E)| \leq |D({}_L E_E)|$.

On the other hand, by Corollary 4 and by condition (i) of Theorem 1 , we have

(**) $[K:L] < |D({}_L E_E)|$.

We give some examples where exactly one strict inequality occurs in (*) .

Example 1 Let $L = \mathbf{Q}$, let $K = \mathbf{R}$, and let $E = \mathbf{R}(x)$. In this case, we have

$$|D({}_K E_E)| = |K|^{[E:K]} = 2^\omega < 2^{2^\omega} = |L|^{[E:L]} = |D({}_L E_E)| .$$

Consequently, we obtain

$$[K:L] = |K| = |D({}_K E_E)| < |D({}_L E_E)| .$$

Example 2 Let $L = \mathbf{Q}$, let K be the field of all algebraic numbers, and let $E = \mathbf{C}$. In this case, we have $[E:K] = 2^\omega > \omega = [K:L]$. Hence, by Theorem 1, $D({}_K E_E)$ is isomorphic to $D({}_L E_E)$. Since $|D({}_K E_E)| = 2^{2^\omega}$, we get

$$[K:L] = |K| < |D({}_K E_E)| = |D({}_L E_E)| .$$

Example 3 Let L,K,E be as in the example considered in the remark after Theorem 1. In this case, by the preceding observations, we have $D({}_K E_E) \simeq D({}_L E_E)$ and

$$[K:L] < |K| = |D({}_K E_E)| = |D({}_L E_E)| .$$

REFERENCES

[1] ANDERSON F.W. - FULLER K.R. *Rings and categories of modules*, Graduate Texts in Mathematics 13, Springer, New York - Heidelberg - Berlin (1973).

[2] D'ESTE G. *Simple submodules and multiplicities*, to appear in Forum Mathematicum.

[3] ROWEN L.H. *Ring Theory*, Pure and applied math. 127, Academic Press (1988).

Istituto di Matematica
Università di Salerno
84100 Salerno, Italy

Canadian Mathematical Society
Conference Proceedings
Volume 11 (1991)

A family of infinite-dimensional non-selfextending bricks for wild hereditary algebras

DIETER HAPPEL AND LUISE UNGER

Abstract. Let A be a finite-dimensional wild hereditary algebra. We will construct here infinitely many infinite-dimensional A-modules without selfextension whose endomorphism algebra is a division ring

0. INTRODUCTION.

Let A be a tame hereditary finite-dimensional algebra over an algebraically closed field k. Then Ringel [R1] has shown that there exists a unique (up to isomorphism) infinite-dimensional A-module Q satisfying $\operatorname{End} Q = D$ is a division ring such that Q is finite-dimensional when considered as module over D and $\operatorname{Ext}^1_A(Q,Q) = 0$. We refer to [R2] for further investigations.

If A is a wild hereditary finite-dimensional k-algebra there has been the feeling that there should be many such modules, but they do not seem to have appeared in the literature.

If X is an A-module whose endomorphism ring is a division ring we will say that X is a brick.

The purpose of this article is to exhibit quite explicitly infinite-dimensional non-selfextending bricks for a generalized Kronecker algebra (see 1.1. below). A standard argument then implies the existence of such modules for an arbitrary wild hereditary finite-dimensional k-algebra (see 2.4. below).

In the final section we will show that our results can be applied to show that certain systems of equations in the free associative algebra and its universal field of fractions (see 1.2. below) have a solution, while they do not have a solution in the polynomial ring and its field of fractions, the field of rational functions.

If V is a k-vectorspace we denote the identity map on V by 1. If s is a natural number we denote by V^s the direct sum of s copies of V. The elements of V^s will be considered as row vectors with coefficients in V. If P is any matrix, then P^T denotes the transpose of P. The composition of two morphisms $\phi : L \longrightarrow M$ and $\psi : M \longrightarrow N$ in a given category $\mathcal{C}$ is denoted by $\phi\psi$.

We would like to thank Claus Ringel for sharing his insight with us and in particular for drawing our attention to the universal field of fractions of Cohn.

1980 *Mathematics Subject Classification* (1985*Revision*):16A64,16A46,12E15,16A06

1. PRELIMINARIES.

1.1. Let k be an algebraically closed field and $r \geq 1$ be a natural number. Then the generalized Kronecker algebra $\mathcal{K}_r$ is defined as the path algebra over k of the quiver

$$\vec{\Delta}_r =$$

with $r+1$ arrows which we label $\alpha_0, \dots, \alpha_r$.

By $\operatorname{Mod} \mathcal{K}_r$ we denote the category of representations of $\mathcal{K}_r$.

Thus an object $X = (X(1), X(2); X(\alpha_0), \dots, X(\alpha_r))$ is given by a pair of k-vectorspaces $X(1), X(2)$ together with $r+1$ k-linear maps $X(\alpha_i)$ from $X(1)$ to $X(2)$. A morphism $f = (f(1), f(2)) : X \to Y = (Y(1), Y(2); Y(\alpha_0), \dots, Y(\alpha_r))$ is given by two k-linear maps $f(i) : X(i) \to Y(i)$ for $1 \leq i \leq 2$ such that

$$f(1)Y(\alpha_j) = X(\alpha_j)f(2) \quad \text{for} \quad 0 \leq j \leq r.$$

The full subcategory with objects $X = (X(1), X(2); X(\alpha_0), \dots, X(\alpha_r))$ satisfying $\dim_k X(i) < \infty$ for $1 \leq i \leq 2$ is denoted by $\operatorname{mod} \mathcal{K}_r$.

For $X = (X(1), X(2); X(\alpha_0), \dots, X(\alpha_r)) \in \operatorname{mod} \mathcal{K}_r$ the dimension vector is defined by $\mathbf{dim}\, X = (\dim_k X(1), \dim_k X(2)) = (x(1), x(2))$.

We also will use the non-symmetric bilinear form on $\mathbb{Z}^2$ defined by

$$\langle (a,b), (c,d) \rangle = ac + bd - (r+1)ad \quad \text{for} \quad (a,b) \text{ and } (c,d) \in \mathbb{Z}^2.$$

For $X, Y \in \operatorname{mod} \mathcal{K}_r$ it holds that

$$\langle \mathbf{dim}\, X, \mathbf{dim}\, Y \rangle = \dim_k \operatorname{Hom}_{\mathcal{K}_r}(X, Y) - \dim_k \operatorname{Ext}^1_{\mathcal{K}_r}(X, Y).$$

For the last result we refer to [R3].

We identify $\operatorname{Mod} \mathcal{K}_r$ with the category of right $\mathcal{K}_r$-modules and $\operatorname{mod} \mathcal{K}_r$ with the category of finite-dimensional right $\mathcal{K}_r$-modules.

1.2. Let $t \geq 1$ be a natural number. Let $F_t = k < x_1, \dots, x_t >$ be the free associative algebra in t generators. In the language above F_t is just the path algebra over k of the quiver with one vertex and t loops. Thus a right F_t-module is given by a k-vectorspace V together with t linear transformations on V. If $\phi_1, \dots, \phi_t$ are linear transformations of a vectorspace V we denote by $(V; \phi_1, \dots \phi_t)$ the corresponding F_t-module. The category of F_t-modules is denoted by $\operatorname{Mod} F_t$ and the full subcategory of finite-dimensional F_t-modules is denoted by $\operatorname{mod}^f F_t$.

By $U_t = k \,\langle\!\langle\, x_1, \dots, x_t \,\rangle\!\rangle$ we denote the universal (skew) field of fractions (see [C]). According to [S1],(chapter 5) U_t is the universal localization of F_t, and therefore by [S1], (Theorem 4.7) we know that $\operatorname{Ext}^1_{U_t}(M, N) = \operatorname{Ext}^1_{F_t}(M, N)$ for all U_t-modules M and N. In particular we infer that $\operatorname{Ext}^1_{F_t}(U_t, U_t) = 0$, where we consider U_t as F_t-module. This module is given by the k-vectorspace U_t together with the linear transformations $r(x_1), \dots, r(x_t)$, where for a generator x_i of U_t the linear transformation $r(x_i) : U_t \to U_t$ is defined by right multiplication, thus for $w \in U_t$ we have that $wr(x_i) = wx_i$. This follows from the fact that the epimorphism of rings $F_t \to U_t$ induced by the localization is injective [C].

Finally we denote by $k[x_1, \dots, x_t]$ the polynomial ring in t variables and by $k(x_1, \dots, x_t)$ its quotient field, the field of rational functions.

1.3. For a brick $X \in \operatorname{mod} \mathcal{K}_r$ (in this case we even have $\operatorname{End} X = k$) the right perpendicular category $X^\perp$ of X is the full subcategory of $\operatorname{Mod} \mathcal{K}_r$ containing those modules Y which satisfy $\operatorname{Hom}(X, Y) = \operatorname{Ext}^1(X, Y) = 0$. Perpendicular categories have been extensively studied in [GL] and [S2]. We will use the following facts. $X^\perp$ is an abelian category and the inclusion of $X^\perp \subset \operatorname{Mod} \mathcal{K}_r$ is exact and closed under extensions.

Thus for $Y, Z \in X^\perp$ we infer that $\operatorname{Ext}^1_{X^\perp}(Y, Z) = \operatorname{Ext}^1_{\mathcal{K}_r}(Y, Z)$.

1.4. We will use the reflection functors σ, σ^- from $\operatorname{Mod} \mathcal{K}_r$ to $\operatorname{Mod} \mathcal{K}_r$ as introduced in [BGP]. For the convenience of the reader we recall the definition of σ and the properties we will need later.

Let $Y = (Y(1), Y(2); Y(\alpha_0), \dots, Y(\alpha_r))$ be a representation of $\overrightarrow{\Delta}_r$. Let K be the kernel of the linear map

$$(Y(\alpha_0), \dots, Y(\alpha_r))^T : Y(1)^{r+1} \to Y(2).$$

By μ we denote the inclusion of K into $Y(1)^{r+1}$. For $0 \le i \le r$ we denote by π_i the projection from $Y(1)^{r+1}$ onto the i-th summand. The representation σY is then defined by $(K, Y(1); \mu\pi_0, \dots, \mu\pi_r)$.

We denote by $\mathcal{M}$ (respectively $\mathcal{M}'$) the full subcategory of $\operatorname{Mod} \mathcal{K}_r$ whose objects have no non-zero direct summand of the form $(0, W; 0, \dots, 0)$ (respectively $(V, 0; 0, \dots, 0)$. The restriction of σ to $\mathcal{M}$ is an equivalence from $\mathcal{M}$ to $\mathcal{M}'$ with inverse given by σ^-.

For $Y = (Y(1), Y(2); Y(\alpha_0), \dots, Y(\alpha_r)) \in \mathcal{M} \cap \operatorname{mod} \mathcal{K}_r$ with $\mathbf{dim}\, Y = (y(1), y(2))$ we infer that $\mathbf{dim}\, \sigma Y = ((r+1)y(1) - y(2), y(1))$. Indeed, if $Y \in \mathcal{M}$ it follows that the linear map $(Y(\alpha_0), \dots, Y(\alpha_r))^T : Y(1)^{r+1} \to Y(2)$ is surjective.

2. THE CONSTRUCTION.

Throughout this section X will denote the representation

$$X = (k, k^r; 0, \mu_1, \dots, \mu_r),$$

where μ_i is the i-th inclusion map.

Then X is a brick and thus it follows from 1.1. that $\dim_k \operatorname{Ext}^1_{\mathcal{K}_r}(X, X) = r$.

2.1. LEMMA. *The right perpendicular category $X^\perp$ is the full subcategory of $\operatorname{Mod}\mathcal{K}_r$ whose objects $Y = (Y(1), Y(2); Y(\alpha_0), \dots, Y(\alpha_r))$ satisfy that $Y(\alpha_0)$ is bijective.*

PROOF: For this we construct a minimal projective resolution of X. The indecomposable projective modules in $\operatorname{mod}\mathcal{K}_r$ have the following description as representations of $\overrightarrow{\Delta}_r$:
$P_1 = (k, k^{r+1}; \mu_0, \dots, \mu_r)$ (μ_i the canonical inclusion) and $P_2 = (0, k; 0, \dots, 0)$.
Then a minimal projective resolution of X is given by:

$$0 \to P_2 \xrightarrow{\alpha} P_1 \xrightarrow{\pi} X \to 0, \text{ where}$$

$$\alpha = (0, \mu_1), \text{ and } \pi = (1, \begin{pmatrix} 0 & 0 & \dots & 0 \\ 1 & 0 & \dots & 0 \\ 0 & 1 & \dots & 0 \\ \vdots & & \ddots & \vdots \\ 0 & 0 & \dots & 1 \end{pmatrix}).$$

Applying $\operatorname{Hom}_{\mathcal{K}_r}(-, Y)$ to the exact sequence above and the definition of $X^\perp$ show that $Y = (Y(1), Y(2); Y(\alpha_0), \dots, Y(\alpha_r)) \in X^\perp$ if and only if the induced map

$$\operatorname{Hom}(\alpha, Y) : \operatorname{Hom}(P_1, Y) \to \operatorname{Hom}(P_2, Y)$$

is bijective. Now $\operatorname{Hom}(P_1, Y)$ is isomorphic to $Y(1)$ and $\operatorname{Hom}(P_2, Y)$ is isomorphic to $Y(2)$. With these identifications we see that $\operatorname{Hom}(\alpha, Y)$ is identified with $Y(\alpha_0)$. Thus the assertion.

2.2. LEMMA. *The right perpendicular category $X^\perp$ is equivalent to $\operatorname{Mod} F_r$.*

PROOF: We define a functor $G : X^\perp \to \operatorname{Mod} F_r$ as follows:
For $Y = (Y(1), Y(2); Y(\alpha_0), \dots, Y(\alpha_r)) \in X^\perp$ let GY be the representation defined by $GY = (Y(1); Y(\alpha_1)Y(\alpha_0)^{-1}, \dots, Y(\alpha_r)Y(\alpha_0)^{-1})$.
For a morphism $f = (f(1), f(2))$ from Y to Y' in $X^\perp$ let $Gf = f(1)$.

It can be easily checked that G defines a functor with inverse given by G' where $G'(V; \phi_1, \dots, \phi_r) = (V, V; 1, \phi_1, \dots, \phi_r)$ and $G'f = (f, f)$ for a morphism f in $\operatorname{Mod} F_t$.
Hence $X^\perp$ and $\operatorname{Mod} F_t$ are equivalent.

Also note that the restriction of the functor G to $X^\perp \cap \operatorname{mod}\mathcal{K}_r$ gives an equivalence with $\operatorname{mod}^f F_r$.

2.3. THEOREM. *The representation $Q_r = (U_r, U_r; 1, r(x_1), \dots, r(x_r))$ of $\overrightarrow{\Delta}_r$ has no selfextensions. Furthermore, Q_r is infinite-dimensional over k with $\operatorname{End} Q_r = U_r$ and finite-dimensional when considered as module over its endomorphism ring.*

PROOF: We consider U_r as F_r-module. In the notation introduced in 1.2. this gives $(U_r; r(x_1), \dots, r(x_r))$. By the remarks in 1.2. we infer that this module has no selfextensions and is finite-dimensional when considered as module over its endomorphism ring U_r. By 2.2. we see that $G'U_r = Q_r$. By 2.1. we infer that the image of G' is $X^\perp$. So the assertion follows from 1.3.

2.4. Note that Q_1 is the unique indecomposable torsionfree divisible module studied in [R1].

2.5. For the rest of this section we assume that $r > 1$. We now prove that the existence of $Q = Q_r$ yields an infinite family of representations of $\overrightarrow{\Delta}_r$ which are infinite-dimensional, having no selfextensions and whose endomorphism ring is a division ring.

First we observe that the right perpendicular category $X^\perp$ is contained in $\mathcal{M}$ (compare 1.4.). Indeed for $Y = (0, W; 0, \dots, 0) \in \operatorname{Mod} \mathcal{K}_r$ we obtain by the Auslander-Reiten-formula that $\operatorname{Ext}^1(X, Y) \simeq D\operatorname{Hom}(Y, \tau X)$, where τ denotes the Auslander-Reiten-translation in $\operatorname{mod} \mathcal{K}_r$. Since by assumption $\operatorname{Ext}^1(X, X) \neq 0$ we know that X is a regular module, in particular we know that $(\tau X)(2) \neq 0$. Thus $\operatorname{Hom}(Y, \tau X) \neq 0$.

Let G' be the functor defined in 2.2. By the previous remark we can consider the composition G^1 of G' with σ. Then G^1 is a functor from $\operatorname{Mod} F_r$ to $\mathcal{M}'$ with image $(\sigma X)^\perp$, and G^1 is an equivalence to $\operatorname{Mod} F_r$ with $(\sigma X)^\perp$ with inverse given by the composition of the restriction of σ^- to $(\sigma X)^\perp$ with the functor G defined in 2.2.

Inductively we obtain for all $i \in \mathbb{N}$ equivalences $G^i : \operatorname{Mod} F_r \longrightarrow (\sigma^i X)^\perp$. Let $Q^i = \sigma^i Q$ for $i \geq 0$. Then Q^i is infinite-dimensional with $\operatorname{End} Q^i = U_r$ and satisfies $\operatorname{Ext}^1(Q^i, Q^i) = 0$. Moreover, Q^i is finite-dimensional when considered as module over its endomorphism ring.

THEOREM. *For all $i \geq 0$ the modules Q^i are pairwise non-isomorphic, infinite-dimensional non-selfextending bricks.*

PROOF: By the previous remarks it is enough to show that these modules are pairwise non-isomorphic. Assume to the contrary that $Q^i \simeq Q^j$ for some $i > j$. Let $s = i - j > 0$. Thus $Q = Q^0 \in (\sigma^s X)^\perp$. Let $M \in \operatorname{mod} \mathcal{K}_r$ be the representation defined by $(k, k; 1, \beta_1, \dots, \beta_r)$ with all $\beta_j \neq 0$. We then obtain a surjective morphism $f = (g, g)$ from Q to M where g maps a generator x_j of U_r to β_j. This gives an exact sequence $0 \longrightarrow \ker f \longrightarrow Q \longrightarrow M \longrightarrow 0$. Application of $\operatorname{Hom}(\sigma^s X, -)$ yields $\operatorname{Ext}^1(\sigma^s X, M) = 0$, for $\operatorname{Ext}^1(\sigma^s X, Q) = 0$ by assumption. Using the notation introduced in 1.1. we see that $\langle \operatorname{\mathbf{dim}} \sigma^s X, \operatorname{\mathbf{dim}} M \rangle \geq 0$. On the other hand it can be shown by induction on s using the formula in 1.4. that $\langle \operatorname{\mathbf{dim}} \sigma^s X, \operatorname{\mathbf{dim}} M \rangle < 0$. Hence $Q^0 \notin (\sigma^s X)^\perp$ for all $s > 0$, implying that $Q^i \not\simeq Q^j$ for $i \neq j$.

2.6. COROLLARY. *A finite-dimensional, wild hereditary k-algebra has infinitely many pairwise non-isomorphic modules, which are non-selfextending bricks of infinite dimension.*

PROOF: Let A be a finite-dimensional, wild hereditary k-algebra and denote by n the number of isomorphism classes of simple A-modules. If $n = 2$, then the assertion follows from 2.5. If $n > 2$, then it follows from [K] that there is some $r \geq 2$ and a full and faithful exact functor F from $\operatorname{Mod} \mathcal{K}_r$ to $\operatorname{Mod} A$ (the category of all right A-modules) whose image is an extension-closed subcategory.

In fact, the functor from mod $\mathcal{K}_r$ to mod A (the category of finite-dimensional right A-modules) given in [K] extends to a functor from Mod $\mathcal{K}_r$ to Mod A having the required properties. Thus the infinite family constructed in 2.5. yields an infinite family of A-modules satisfying the assertions by considering the family $F(Q^i)$ for $i \geq 0$.

2.7. We want to show without proof that there exist more modules which have the properties of the theorem in 2.3.

We consider the case $r = 2$. Let Q be the following representation of $\overrightarrow{\Delta}_2$:

$$(U_6^3, U_6^2; \begin{pmatrix} r(x_1) & 0 \\ r(x_2) & r(x_3) \\ 0 & 1 \end{pmatrix}, \begin{pmatrix} r(x_4) & r(x_5) \\ 0 & r(x_6) \\ 1 & 1 \end{pmatrix}, \begin{pmatrix} 1 & 0 \\ 0 & 1 \\ 0 & 0 \end{pmatrix}).$$

This module can be constructed by following the procedure in the proof of 2.1. and 2.2. if we start with the brick X defined by:

$$(k^2, k^3; \begin{pmatrix} 1 & 0 & 0 \\ 0 & 1 & 0 \end{pmatrix}, \begin{pmatrix} 0 & 1 & 0 \\ 0 & 0 & 1 \end{pmatrix}, \begin{pmatrix} 0 & 0 & 0 \\ 0 & 0 & 0 \end{pmatrix}).$$

3. EQUATIONS IN THE UNIVERSAL FIELD OF FRACTIONS.

We keep the notation from the previous sections. For a k-algebra K and $x \in K$ we denote by $r(x) : K \to K$ the k-linear map defined by $\lambda r(x) = \lambda x$ for $\lambda \in K$.

Let K be a k-algebra and let $x_1, \ldots, x_r$ be elements of K. Then we can define a representation X_K of $\overrightarrow{\Delta}_r$ as follows:

$$X_K = (K, K; 1, r(x_1), \ldots, r(x_r)).$$

Note that X_K coincides with the module Q_r considered in 2.3. in case $K = U_r$, if the chosen elements are the generators of U_r.

3.1. LEMMA. *$Ext^1_{\mathcal{K}_r}(X_K, X_K) = 0$ if and only if for $f_1, \ldots, f_r \in Hom_k(K, K)$ there is $\alpha \in Hom_k(K, K)$ such that $f_i = r(x_i)\alpha - \alpha r(x_i)$ for all $1 \leq i \leq r$.*

PROOF: We consider the following representations U, V of $\overrightarrow{\Delta}_r$:

$$U = (0, K; 0, \ldots, 0) \quad \text{and} \quad V = (K, 0; 0, \ldots, 0).$$

We have an exact sequence

$$0 \to U \to X_K \to V \to 0.$$

Application of $\operatorname{Hom}(X_K, -)$ yields the following exact sequence:

$$\operatorname{Hom}(X_K, V) \xrightarrow{\delta} \operatorname{Ext}^1(X_K, U) \to \operatorname{Ext}^1(X_K, X_K) \to \operatorname{Ext}^1(X_K, V) \to 0.$$

Thus $\operatorname{Ext}^1(X_K, X_K) = 0$ if and only if $\operatorname{Ext}^1(X_K, V) = 0$ and δ is surjective.
We will show next that $\operatorname{Ext}^1(X_K, V) = 0$.
In fact, let $\eta \in \operatorname{Ext}^1(X_K, V)$. Thus η is of the form:

$$0 \to V \xrightarrow{f} E \xrightarrow{g} X_K \to 0,$$

where $E = (K^2, K; \alpha_0, \ldots, \alpha_r)$ with $\alpha_0, \ldots, \alpha_r \in \operatorname{Hom}_k(K^2, K)$ and $f = (\mu, 0)$ for some $\mu \in \operatorname{Hom}_k(K, K^2)$. As a map of k-vectorspaces μ is a section. Thus there exists $\pi : K^2 \to K$ such that $\mu\pi = 1$. Also $f' = (\pi, 0)$ is a morphism from E to V such that $ff' = 1$, hence η is split exact.

Now let $\eta \in \operatorname{Ext}^1(X_K, U)$. Then η is of the form:

$$0 \to U \xrightarrow{f} E \xrightarrow{g} X_K \to 0,$$

where $E = (K, K^2; \alpha_o, \ldots, \alpha_r)$ with $\alpha_0, \ldots, \alpha_r \in \operatorname{Hom}_k(K, K^2)$, $f = (0, \mu)$ for some $\mu \in \operatorname{Hom}_k(K, K^2)$ and $g = (\beta, \pi)$ for some $\beta \in \operatorname{Hom}_k(K, K)$ and some $\pi \in \operatorname{Hom}_k(K^2, K)$.
We may assume that $\mu = (1, 0), \pi = (0, 1)^T$ and $\beta = 1$.
This yields that $\alpha_0 = (\beta_0, 1)$ and $\alpha_i = (\beta_i, r(x_i))$ for $1 \leq i \leq r$ for some $\beta_0, \ldots, \beta_r \in \operatorname{Hom}_k(K, K)$.

Now consider $\eta' \in \operatorname{Ext}^1(X_K, U)$ with η' of the form:

$$0 \to U \xrightarrow{f'} E' \xrightarrow{g'} X_K \to 0,$$

where $E' = (K, K^2; (0, 1), (f_1, r(x_1)), \ldots, (f_r, r(x_r)))$ with $f_i = \beta_i - r(x_i)\beta_0$. Also $f' = (0, (1, 0))$ and $g' = (1, (0, 1)^T)$. Now it follows that the morphism $\phi : E \to E'$ given by $(1, \begin{pmatrix} 1 & 0 \\ -\beta_0 & 1 \end{pmatrix})$ induces an isomorphism of η with η'.

Hence we can assume that an extension η has the normal form:

$$0 \to U \xrightarrow{f} E \xrightarrow{g} X_K \to 0,$$

where $E = (K, K^2; (0, 1), (f_1, r(x_1)), \ldots, (f_r, r(x_r)))$ for some k-linear maps $f_1, \ldots, f_r \in \operatorname{Hom}_k(K, K)$. Also $f = (0, (1, 0))$ and $g = (1, (0, 1)^T)$.

By definition of the connecting homomorphism δ we know that δ is surjective, if and only if for all $\eta \in \operatorname{Ext}^1(X_K, U)$ in normal form there exist k-linear maps $\gamma_1, \gamma_2, \gamma_3, \gamma_4 \in \operatorname{Hom}_k(K, K)$ such that the following diagram of exact sequences is commutative:

$$\begin{array}{ccccccccc}
0 & \longrightarrow & U & \xrightarrow{f} & E & \xrightarrow{g} & X_K & \longrightarrow & 0 \\
 & & \| & & \gamma\downarrow & & \downarrow (\gamma_2, 0) & & \\
0 & \longrightarrow & U & \xrightarrow{(0,1)} & X_K & \xrightarrow{(1,0)} & V & \longrightarrow & 0
\end{array}$$

where $\gamma = (\gamma_1, (\gamma_3\ \gamma_4)^T)$. This yields the following equations:

$$\gamma_1 = \gamma_2, \quad \gamma_3 = 1, \quad \gamma_1 = (0,1)\begin{pmatrix}\gamma_3\\ \gamma_4\end{pmatrix} \quad \gamma_1 r(x_i) = (f_i, r(x_i))\begin{pmatrix}\gamma_3\\ \gamma_4\end{pmatrix}$$

for $1 \le i \le r$.

This shows that $\gamma_3 = 1, \quad \gamma_1 = \gamma_4, \quad \gamma_1 r(x_i) = f_i + r(x_i)\gamma_1$ for $1 \le i \le r$.

Thus we have shown that the connecting homomorphism δ is surjective if and only if for all $f_1, \ldots, f_r \in \mathrm{Hom}_k(K, K)$ there exists $\alpha \in \mathrm{Hom}_k(K, K)$ such that $f_i = \alpha r(x_i) - r(x_i)\alpha$ for $1 \le i \le r$. This shows the assertion.

3.2. COROLLARY. *Let U_r be the universal field of fractions of F_r. There exists $\alpha \in Hom_k(U_r, U_r)$ such that $r(x_i)\alpha - \alpha r(x_i) = 0$ for all $1 \le i \le r$.*

PROOF: This follows from 2.3 and 3.1 by choosing $f_1 = \cdots = f_r = 0$.

3.3. COROLLARY. *Let $r \ge 2$ and suppose that K is a commutative k-algebra generated by the elements $x_1, \ldots, x_r$. Assume that K admits a non-trivial k-derivation δ. Then $Ext^1_{K_r}(X_K, X_K) \neq 0$.*

PROOF: By assumption there is a non-trivial k-derivation δ. Thus we have a k-linear map $\delta : K \to K$ such that $(\lambda\mu)\delta = \lambda(\mu)\delta + (\lambda)\delta\mu$ for all $\lambda, \mu \in K$.

We may assume that $x_1\delta \neq 0$.

This implies that $\delta r(x_1) - r(x_1)\delta \neq 0$. In fact, assume that $\delta r(x_1) - r(x_1)\delta = 0$. Thus for all $\lambda \in K$ we have that $\lambda\delta r(x_1) - (\lambda r(x_1))\delta = -\lambda(x_1)\delta = 0$, a contradiction.

Now choose $f_1 = 0, f_2 = \delta, f_3 = \cdots = f_r = 0$ and assume that there exists $\alpha \in \mathrm{Hom}_k(K, K)$ such that $f_i = r(x_i)\alpha - \alpha r(x_i)$ for all $1 \le i \le r$. This gives the following two equations:

$$r(x_2)\alpha r(x_1) - \alpha r(x_2) r(x_1) = f_2 r(x_1) \quad \text{and}$$
$$r(x_1) r(x_2)\alpha - r(x_1)\alpha r(x_2) = r(x_1) f_2.$$

Using $r(x_1)\alpha - \alpha r(x_1) = f_1 = 0$ and $r(x_1)r(x_2) = r(x_2)r(x_1)$, for K is commutative, we infer that:

$$\begin{aligned} r(x_1) f_2 &= r(x_1) r(x_2)\alpha - r(x_1)\alpha r(x_2) \\ &= r(x_2) r(x_1)\alpha - \alpha r(x_1) r(x_2) \\ &= r(x_2)\alpha r(x_1) - \alpha r(x_2) r(x_1) \\ &= f_2 r(x_1), \end{aligned}$$

a contradiction to the choice of f_2.

If we choose the k-algebra K to be the field of rational functions $k(x_1, \ldots, x_r)$ we can construct an infinite-dimensional selfextending brick, if $r \ge 2$. Just choose the generators of $k(x_1, \ldots, x_r)$ to define X_K. Note that in this case $\mathrm{End}_{K_r} X_K$ is isomorphic to $k(x_1, \ldots, x_r)$.

3.4. Now suppose that the k-algebra K is the free associative algebra F_r. Again we choose the generators of F_r to define X_K. Note that in this case $X_K = G'F_r$, where G' is the functor defined in 2.2. As F_r is a projective F_r-module we infer that $\mathrm{Ext}^1_{\mathcal{K}_r}(X_K, X_K) = 0$. So we know from 3.1. that for all $f_1, \ldots, f_r \in \mathrm{Hom}_k(K, K)$ there is $\alpha \in \mathrm{Hom}_k(K, K)$ such that $f_i = r(x_i)\alpha - \alpha r(x_i)$ for all $1 \leq i \leq r$.

We will now give a direct argument of this fact. Note that in this way we obtain a different proof that $\mathrm{Ext}^1_{\mathcal{K}_r}(X_K, X_K) = 0$.

Suppose that we have given $f_1, \ldots, f_r \in \mathrm{Hom}_k(K, K)$. We will define a k-linear map $\alpha \in \mathrm{Hom}_k(K, K)$ inductively on the reduced length $\ell(w)$ of a word $w \in F_r$. If $\ell(w) = 0$ then $w = 1$, then we set $1\alpha = 1$. If $\ell(w) > 0$ then $w = w'x_i$ for some word $w' \in F_r$ with $\ell(w') < \ell(w)$ and some generator x_i. Then we set $w\alpha = w'f_i + w'\alpha x_i$. Using induction on the reduced length $\ell(w)$ of a word $w \in F_r$ it is now easy to see that $f_i = r(x_i)\alpha - \alpha r(x_i)$ for all $1 \leq i \leq r$.

REFERENCES

[BGP] I.N. Bernstein, I.M. Gelfand, V.A. Ponomarev, *Coxeterfunctors and Gabriel's theorem*, Russ. Math. Surveys 28(1973), 17-32.

[C] P.M. Cohn, *Free Rings and their relation*,Acedemic Press (New York 1971).

[GL] W. Geigle, H. Lenzing, *Perpendicular categories with applications to representations and sheaves*, J. of Algebra, to appear

[K] O. Kerner *Preprojective components of wild tilted algebras*, Manus. math. 61(1988), 429-445.

[R1] C.M. Ringel, *Infinite dimensional representations of finite dimensional hereditary algebras*, Ist. Naz. Alta Mat., Symp. Math. 23(1979), 321-412.

[R2] C.M. Ringel, *The spectrum of a finite dimensional algebra*, Proc. Conf. on Ring theory, Antwerp 1978, Marcel Dekker, New York 1979, 535-598.

[R3] C.M. Ringel, *Tame algebras and integral quadratic forms*, Springer Lecture Notes 1099, Heidelberg 1984.

[S1] A. Schofield, *Representations of rings over skew fields*, London Math. Soc. Lect. Notes 92, Cambridge 1985

[S2] A. Schofield, *Universal localization for hereditary rings and quivers*, Springer Lecture Notes 1197, Heidelberg 1986, 149-164.

Dieter Happel
Fakultät für Mathematik
Universität Bielefeld
Postfach 8640
W-4800 Bielefeld 1
Rép. Féd. d'Allemagne

Luise Unger
Fachbereich Mathematik
Universität Paderborn
Warburgerstr.100
W-4790 Paderborn
Rép. Féd. d'Allemagne

Canadian Mathematical Society
Conference Proceedings
Volume 11 (1991)

Regular modules over wild hereditary algebras

Otto Kerner and Frank Lukas

Abstract

If A is a wild hereditary algebra over an algebraically closed field, if r is any natural number then there exists a quasi-simple regular module X such that $Hom(X, \tau^{-r}X)$ is nonzero.

The results of [1] and [8] on maps between regular modules over wild hereditary algebras are normally summarized by the sentence that in the regular part of such an algebra there exist a lot of maps in τ-direction and only few maps in τ^--direction. There are convincing arguments that in this case the τ-direction really points to the preinjective component. The expression "a lot of maps" can be made more precise in the following way:
Firstly, as it was shown essentially already in [1], for non-zero regular modules X and Y, the dimensions of the spaces $Hom(X, \tau^n Y)$ grow exponentially with n.
Secondly, for a regular component $\mathcal{C}$ in [11] there was defined the quasi-rank $rk(\mathcal{C})$ of this component by $rk(\mathcal{C}) = min\{m \mid rad(X, \tau^l X) \neq 0 \text{ for all } l \geq m\}$ where X is some quasi-simple module in $\mathcal{C}$, and it was shown there, that

$$\bar{t}(A) = max\{rk(\mathcal{C}) \mid \mathcal{C} \in \Omega(A)\}$$

is finite; hereby $\Omega(A)$ denotes the set of regular components of A.
In this paper we study maps between regular modules in τ^--direction. Surprisingly also in this direction, there exist a lot of maps. More precise our main result is

Theorem: *Let A be a finite dimensional wild hereditary algebra over an algebraically closed field. Then we have*

$$inf\{rk(\mathcal{C}) \mid \mathcal{C} \in \Omega(A)\} = -\infty$$

1980 Mathematics Subject Classification (1985 Revision). 16A46,16A64

As the problem of classification all indecomposables is considered as hopeless in the wild situation one might try to classify smaller parts, for example exact, full abelian subcategories of $A - \boldsymbol{mod}$. In part 4 we will show that full exact abelian subcategories of $\boldsymbol{add}\ \mathcal{R}$, the category of regular modules of a wild hereditary algebra A, are completely contained in the additive category generated by $\bigcup_{t(C) > -\bar{t}(A)} \mathcal{C}$, so our main result also means that a big part of $\boldsymbol{add}\ \mathcal{R}$ never is touched by full exact abelian subcategories of $\boldsymbol{add}\ \mathcal{R}$. The proof of the main theorem is done in two steps.

First we show that for any natural number p there exist algebras with (infinitely many) components of quasi-rank $-p$. It should be mentioned that the only example of components of negative quasi-rank known in the literature was given in [13]: The path-algebra of the quiver $\circ \leftleftarrows \circ \leftarrow \circ$ has infinitely many components of quasi-rank -1.

For the construction of components of type $-p$ we use the concept of ***simplification***, introduced by Ringel [12]: If $\mathcal{X}$ is any set of pairwise orthogonal bricks (that is modules with trivial endomorphism-rings) in $A - \boldsymbol{mod}$, then we consider the category $\mathcal{E}(\mathcal{X})$, the smallest full, exact, extension closed abelian subcategory containing $\mathcal{X}$. This category is hereditary, that is $Ext^2_{\mathcal{E}(\mathcal{X})}(-,-) = 0$ and $\mathcal{X}$ is the set of simple objects of $\mathcal{E}(\mathcal{X})$. If all elements of $\mathcal{X}$ have self-extensions, the category $\mathcal{E}(\mathcal{X})$ has amazing properties.

For example any regular component contains at most one indecomposable object of $\mathcal{E}(\mathcal{X})$ which additionally normally is quasi-simple. Especially if $\mathcal{X} = \{X\}$ where X is a brick with self-extensions of quasi-length $p + 1$ then the indecomposable objects $Y \not\cong X$ of $\mathcal{E}(X)$ are quasi-simple modules in components of quasi-rank $-p$.

In the second step we use tilting theory: By [10] there exist for any two wild hereditary algebras A and B bijections

$$\mu : \Omega(A) \longrightarrow \Omega(B)$$

induced by tilting modules. We use these bijections to compare the quasi-rank of components between different algebras and derive the general statement from the first partial result.

Notations The word algebra always denotes a finite dimensional, unitary, basic algebra over some field k. In the first two paragraphs k may be arbitrary, in the remainder we assume k to be algebraically closed. The letter A normally is reserved for wild hereditary,

connected algebras. If A is such an algebra, $A - mod$ denotes the category of finitely generated left A-modules and $\mathcal{R}$ denotes the class of indecomposable regular A-modules.
We call a module X a ***brick*** if $End(X)$ is a division ring. An indecomposable partial tilting module, that is a brick without self-extensions (and projective dimension at most 1) is called a ***stone***.
By q_A we denote the Tits-form which coincides in the hereditary case with the Euler-quadratic form, that is

$$q_A(\underline{dim}\, X) = dim_k End(X) - dim_k Ext^1(X, X).$$

By $\Gamma(A)$ we denote the Auslander-Reiten quiver of A. The vertices of $\Gamma(A)$ are isomorphism-classes of indecomposable A-modules; if the context is clear, we will not distinguish between an indecomposable module X and its class $[X]$.
By $\Omega(A)$ we denote the set of regular components of $\Gamma(A)$. If $\mathcal{C}$ is a regular component – it is of type $\mathbb{Z}A_\infty$ by [13] – and X is quasi-simple in $\mathcal{C}$ then the quasi-rank $rk(\mathcal{C})$ of $\mathcal{C}$ is defined by

$$rk(\mathcal{C}) = min\{l \mid rad(X, \tau^s X) \neq 0 \text{ for all } s \geq l\}.$$

By $\bar{t}(A)$ we mean $max\{rk(\mathcal{C}) \mid \mathcal{C} \in \Omega(A)\}$. If U is a quasi-simple module and r a natural number, we denote by $U(r)$ ($[r]U$)the indecomposable regular module with quasi-length r and quasi-socle (quasi-top) U.
From the concept of perpendicular categories we use the following result (see [4] or [16]): If A is an algebra and X a square free partial tilting module, then the right perpendicular category $X^\perp$, defined by the objects $\{Y \mid Hom(X,Y) = Ext^1(X,Y) = 0\}$ is an abelian subcategory of $A - mod$ which is equivalent to a module category $B - mod$. We have $B = End(Z)$, where Z is defined by Bongartz's construction (see [2]) such that $X \oplus Z$ is a square-free tilting module.
The Auslander-Reiten translations will be denoted by τ andτ^-; to emphasise the algebra, we sometimes add a subscript, for instance $\tau_A, \tau_{\mathcal{C}}^-$. In general we follow the notations used in [14].

1 Simplification à la Ringel

In this part A is a connected wild hereditary k-algebra, k some field.

For a class $\mathcal{X}$ of pairwise orthogonal bricks C.M. Ringel defined in [12, §1] the full subcategory $\mathcal{E}(\mathcal{X})$ of all modules Y in $A - mod$ with $\mathcal{X}$–filtration, that is with a chain

$$0 = Y_0 \subset Y_1 \subset \ldots \subset Y_{n-1} \subset Y_n = Y$$

with $Y_i/Y_{i-1} \in \mathcal{X}$ for $1 \le i \le n$.

He proved that $\mathcal{E}(\mathcal{X})$ is an exact abelian subcategory of $A - mod$, closed under extensions, and $\mathcal{X}$ is the class of all simple objects in $\mathcal{E}(\mathcal{X})$.

Lemma 1.1 *Let $\mathcal{X}$ be a class of pairwise orthogonal bricks with self-extensions.*

(i) All nonzero modules in $\mathcal{E}(\mathcal{X})$ have self-extensions. Especially $\mathcal{E}(\mathcal{X})$ is contained in add $\mathcal{R}$.

(ii) Any regular τ-orbit contains at most one module of $\mathcal{E}(\mathcal{X})$.

Proof: (a) We consider the quadratic form

$$q(\underline{dim}\, Y) = dim_k Hom(Y,Y) - dim_k Ext(Y,Y)$$

Clearly for any brick X with self-extensions we have $q(\underline{dim}\, X) \le 0$, since $Ext(X,X)$ is an $End(X)$– vectorspace.

If $0 \ne Y \in \mathcal{E}(\mathcal{X})$, then there exist finitely many pairwise orthogonal bricks $X_1, \ldots, X_r \in \mathcal{X}$ such that $\underline{dim}\, Y = \sum_{i=1}^{r} m_i \cdot \underline{dim}\, X_i$ with $m_i \in \mathbb{N}$. Thus we get

$$\begin{aligned} q(\underline{dim}\, Y) &= \textstyle\sum_{i=1}^{r} m_i^2 \cdot q(\underline{dim}\, X_i) + \sum_{i \ne j} m_i m_j \cdot dim_k Hom(X_i, X_j) \\ &\quad - \textstyle\sum_{i \ne j} m_i m_j \cdot dim_k Ext^1(X_i, X_j) \le 0 \end{aligned}$$

that is $Ext(Y,Y) \ne 0$.

(b) Suppose Y and $\tau^n Y$ are in $\mathcal{E}(\mathcal{X})$ for some $n \ge 1$ and some indecomposable module Y. Since the $\mathcal{E}(\mathcal{X})$-socle of Y is nonzero we have a monomorphism $X \to Y$ for some $X \in \mathcal{X}$. Applying the functor τ^n to this map yields an injective map $\tau^n X \to \tau^n Y$. By [11, 1.7] we know that $Hom(X, \tau^n X)$ is nonzero. Since the composition $X \to \tau^n X \to \tau^n Y$ is nonzero for every nonzero map $X \to \tau^n X$ and X is simple in $\mathcal{E}(\mathcal{X})$ all nonzero maps from X to $\tau^n X$ have to be injective. But the cokernel of this injective map is a submodule of the regular module $\tau^n Y/X \in \mathcal{E}(\mathcal{X})$ and therefore regular, a contradiction.

Lemma 1.2 *Let $\mathcal{X}$ be a class of pairwise orthogonal bricks and $Y = U(r)$ be indecomposable in $\mathcal{E}(\mathcal{X})$ with regular length $r \geq 2$ and quasi-socle U. If $Hom(X,U)$ is nonzero for some $X \in \mathcal{X}$ then the modules $U, \tau^- U(r-1)$ are in $\mathcal{E}(\mathcal{X})$.*

Proof: If $\iota : U \hookrightarrow Y$ denotes the embedding and $0 \neq f \in Hom(X,U)$, then $f \circ \iota : X \to Y$ is nonzero and injective, as $\mathcal{E}(\mathcal{X})$ is a full abelian subcategory and X is simple in $\mathcal{E}(\mathcal{X})$. If f is an isomorphism then we have $U \in \mathcal{X} \subset \mathcal{E}(\mathcal{X})$ and therefore $Y/U \cong \tau^- U(r-1) \in \mathcal{E}(\mathcal{X})$. If f is not an isomorphism, we consider the irreducible epimorphism $\pi : Y \to Y/U$. This map has the factorisation

$$\pi = \beta \circ \alpha : Y \xrightarrow{\beta} Y/f(X) \xrightarrow{\alpha} Y/U$$

The irreducibility of π implies that α is a split epimorphism, that is $Y/f(X) \cong Y/U \oplus U/f(X)$. Since $\mathcal{E}(\mathcal{X})$ is a full abelian category we get from $Y/f(X) \in \mathcal{E}(\mathcal{X})$ that also Y/U and $U/f(X)$ are objects of $\mathcal{E}(\mathcal{X})$. Especially U is in $\mathcal{E}(\mathcal{X})$, too.

Lemma 1.3 *Let $Y = U(r)$ be a regular module with U quasi-simple and $r \geq 2$. Then Y is a brick if and only if $Hom(Y,U) = 0$.*

Proof: If $Hom(Y,U)$ is nonzero, then the wing $\mathcal{W}(Y)$ is not a standard wing and thus Y is not a brick, see [11, 1.1].
If $Hom(Y,U)$ is zero we also get $Hom(Y/U,U) = 0$. But this implies by [11, 1.3] that $Hom(\tau^{-i}U,U) = 0$ for $i = 1,\ldots,r-1$. So U is a stone, $\mathcal{W}(Y)$ is a standard wing and consequently Y is a brick, see [11, 1.1].

Proposition 1.4 *Let $\mathcal{X}$ be a class of pairwise orthogonal bricks with self-extensions.*

(a) Every indecomposable non-brick in $\mathcal{E}(\mathcal{X})$ is quasi-simple in $A-mod$.

(b) Every regular component of the Auslander-Reiten quiver contains at most one module of $\mathcal{E}(\mathcal{X})$.

Proof: (a) Take $Y \in \mathcal{E}(\mathcal{X})$ indecomposable, of quasi-length r, say $Y = U(r)$ with U quasi-simple. If Y is not a brick we get $Hom(Y,U) \neq 0$ from (1.3) and therefore we have $Hom(X,U) \neq 0$ for some $X \in \mathcal{X}$. We then have $U, \tau^- U(r-1) \in \mathcal{E}(\mathcal{X})$. By (1.1(i)) the module U therefore has self-extensions and consequently $U(2),\ldots,U(r)$ are non-bricks and the above argument can be repeated with $\tau^{-}U(r-1)$. In this way we see that the wing $\mathcal{W}(Y)$ is contained in $\mathcal{E}(\mathcal{X})$. For $r \geq 2$ this is a contradiction to (1.1(ii)).

(b) For a regular component $\mathcal{C}$ of $\Gamma(A)$ containing a module $Y \in \mathcal{E}(\mathcal{X})$ we have to consider two cases:

First, if Y is not a brick, Y is quasi-simple and the component $\mathcal{C}$ does not contain any brick. This means that every indecomposable module in $\mathcal{E}(\mathcal{X}) \cap \mathcal{C}$ has to be quasi-simple and the assertion follows from (1.1(ii)).

In the second case $Y = U(r)$ is a brick. By [11, 1.1] the modules $U(i)$ $1 \leq i \leq r-1$ are stones and the modules $U(i)$ for $i > r$ have nontrivial endomorphisms. So indecomposable elements of $\mathcal{E}(\mathcal{X})$ can sit at most in the τ-orbit of Y and again we can apply (1.1(ii)).

Remarks: (1) Following Gabriel [3, §7] for a regular brick X with $End(X) = D$ and $dim_D Ext^1(X,X) = t$ the length-category $\mathcal{E}(X)$ is equivalent to the category of finite-dimensional modules over the power-series ring $D\langle\langle X_1, \ldots, X_t\rangle\rangle$ in non-commuting variables $X_1, \ldots, X_t$. Therefore for $t \geq 1$ the category $\mathcal{E}(X)$ is always of infinite representation type; moreover for $t \geq 2$ the category $\mathcal{E}(X)$ even is wild.

(2) Note that (1.4) also says that for a component $\mathcal{C}$ there exists at most one τ-orbit, which possibly contains a module $Y \in \mathcal{E}(\mathcal{X})$ where $\mathcal{X}$ is a family of pairwise orthogonal self-extending bricks. If there are stones in $\mathcal{C}$ it is the (unique) τ-orbit of bricks with self-extensions, otherwise it is the τ-orbit of quasi-simple modules.

In the sequel the following corollary plays an important role:

Corollary 1.5 *Let X be a regular brick of quasi-length $r+1$ $(r \geq 1)$ with self-extensions. Then there exists infinitely many regular components $\mathcal{C}$ with $rk(\mathcal{C}) = -r$.*

Proof: Clearly X is the only brick in $\mathcal{E}(X)$. Therefore any indecomposable non-simple object in $\mathcal{E}(X)$ is quasi-simple in $A-mod$ by (1.4). Since X is a brick of quasi-length $r+1$ we get from [10] $Hom(X, \tau^{-k}X) = 0$ if and only if $k \geq r+1$. One easily checks that $Hom(Y, \tau^{-k}Y) = 0$ if and only if $Hom(X, \tau^{-k}X) = 0$ for all nonzero objects Y in $\mathcal{E}(X)$ and all $k \in \mathbb{N}_0$. The result follows, as $\mathcal{E}(X)$ contains infinitely many indecomposable objects.

2 Constructions

If A is connected wild hereditary over some field k and T a tilting module without preinjective direct summand, then the tilted algebra $B = End(T)$ has exactly one preprojective component which is the preprojective component of a (connected) wild concealed algebra C, see [16]. By [10], the tilting module T induces a bijection $\mu_T = \mu$ from the set of regular Auslander-Reiten-components of $\Gamma(A)$ to the set $\Omega(C)$ of regular Auslander-Reiten-components of $\Gamma(C)$:

If $\mathcal{C}$ is a component in the torsionfree part

$$\mathcal{Y}(T) = \{Y \in B - mod \mid Tor_1^B(T, Y) = 0\}$$

different from the preprojective and connecting component then there exist indecomposable (quasi-simple) modules X and Z in $\mathcal{C}$ such that on the one hand $(\rightarrow X) = F(\rightarrow G(X))$ and on the other hand $(Z \rightarrow)$ is in $C - mod$. Hereby $(\rightarrow X)$ $((Z \rightarrow))$ denotes the full subquiver of $\mathcal{C}$ of predecessors of X (successors of Z), and F or G the tilting functors $F = Hom_A(T, -)$ or $G = T \otimes_B -$.

$(\rightarrow G(X))$ is part of a regular component $\mathcal{C}$ in $\Gamma(A)$, the cone $(Z \rightarrow)$ is part of a regular component $\mathcal{D}$ in $\Gamma(C)$ and the bijection

$$\mu : \Omega(A) \longrightarrow \Omega(C)$$

is defined via $\mu(\mathcal{C}) = \mathcal{D}$.

This fact also can be expressed in the terms $\tau_C^{-r} Z = \tau_B^{-r} Z$ for all $r \in \mathbb{N}_0$ and $\tau_B^r X = F(\tau_A^r G(X))$. For details we refer to [10] and [11].

Using that τ_B, restricted to $\mathcal{Y}(T)$ is a full (but not necessarily faithfull) functor, is was shown in [11]:

Lemma 2.1 *Under the above notations for any component $\mathcal{C} \in \Omega(A)$ we have*

$$rk(\mathcal{C}) \geq rk(\mu(\mathcal{C})).$$

Suppose now that in $A - mod$ where A is wild connected and hereditary, there is a brick X with self-extensions of quasi-length $r + 1$. Then by (1.5) we get infinitely many regular components $\mathcal{C}$ with $rk(\mathcal{C}) = -r$, each of them containing an object of $\mathcal{E}(X)$. If T is a

tilting module without preinjective direct summand, if $B = End(T)$ is the tilted algebra with preprojective component defined by the concealed algebra C, in the set $\Omega(C)$ by (2.1) we then have infinitely many components $\mathcal{D} = \mu(\mathcal{C})$ with $rk(\mathcal{D}) \leq -r$.
Denoting by $\tilde{\tilde{A}}_{p,q}$ and K_r $(r \geq 3)$ the quivers

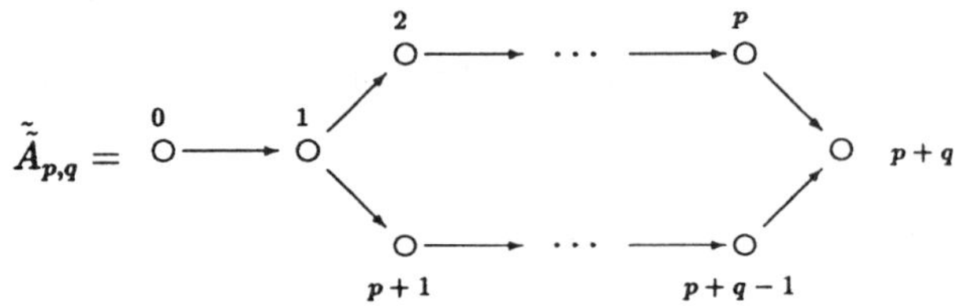

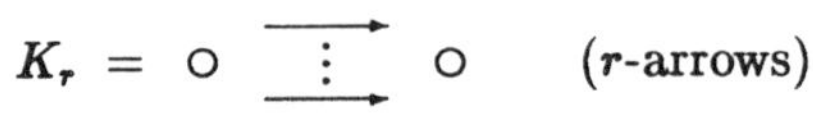

we get

Proposition 2.2 *Let A be the path algebra of $\tilde{\tilde{A}}_{p,q}$ $(p \geq 1, q \geq 1)$ or K_r $(r \geq 3)$. Then we have*

$$\underline{t}(A) = inf\ \{rk(\mathcal{C}) \mid \mathcal{C} \text{ regular component in } \Gamma(A)\} = -\infty$$

Proof: Let A' be $\tilde{\tilde{A}}_{p+1,q}$. Consider the tube $\mathcal{T}$ of period $p+1$ of the Euclidean algebra $\tilde{A}_{p+1,q}$. By [11] this tube $\mathcal{T}$ defines in A' an exceptional component $\mathcal{C}$ of quasi-rank $(p+1, t)$: $\mathcal{C}$ is the component containing the quasi-simple module

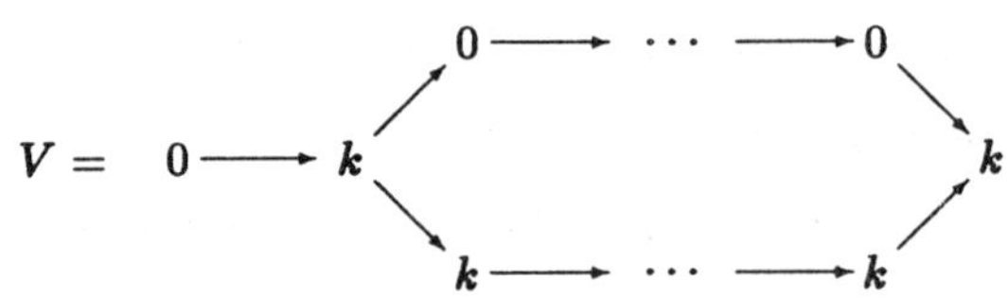

Especially $\mathcal{C}$ contains a brick X with self-extensions of quasi-length $p+1$ and thus by (1.5) there are in $\Omega(A')$ infinitely many components $\mathcal{D}$ of quasi-rank $rk(\mathcal{D}) = -p$.
Consider now $T = V \oplus \tilde{V}$ with $\tilde{V} = \bigoplus_{\substack{i=0 \\ i\neq 2}}^{p+q} P(i)$, where $P(i)$ denotes the projective module corresponding to the vertex i. It can be checked easily that T is a tilting module, $\tilde{V}$ is the projective generator of $V^{\perp}$ and $A = End(\tilde{V})$ is the path-algebra of the quiver $\tilde{\tilde{A}}_{p,q}$. Thus by (2.1) A has infinitely many components $\mathcal{D}$ whose quasi-rank is at most $-p$ (and of course it

also has infinitely many components of quasi-rank $-(p-1)$ or $-(q-1)$).

Using moreover the chain ($q \geq 1$):

$$\dots \longrightarrow \tilde{\tilde{A}}_{p+2,q} \longrightarrow \tilde{\tilde{A}}_{p+1,q} \longrightarrow \tilde{\tilde{A}}_{p,q} \longrightarrow \dots \longrightarrow \tilde{\tilde{A}}_{1,q}$$

we immediately get by (2.1) $\underline{t}(A) = -\infty$ for all these algebras.

If $A' = \tilde{\tilde{A}}_{1,1}$ then it was shown in [9, 17] that for each $r \geq 3$ there exists a tilting module $T(r)$ (without preinjective direct summand in [9] and even regular in [17]), such that the preprojective component of the tilted algebra $B(r) = End(T(r))$ is the preprojective component of the hereditary algebra C of type K_r. Again by (2.1) we therefore have $\underline{t}(C) = -\infty$.

3 Proof of the main result

For the rest of the paper we assume that the field k is algebraically closed. The main result of this part is

Theorem 3.1 *If A is connected wild hereditary, then*

$$\underline{t}(A) = inf\{rk(\mathcal{C}) \mid \mathcal{C} \in \Omega(A)\} = -\infty$$

Before starting the proof we consider the following situation:

Let A be connected wild hereditary with $n \geq 3$ simple modules. Then there exist regular stones in $A-mod$, see for instance [15]. Let V be a quasi-simple regular stone such that $\tau^{-i}V$ is sincere for all $i \in \mathbb{N}_0$ and let V' be the minimal projective generator in $V^{\perp} \cong C - mod$ (with C connected wild hereditary, see [16]). Then $T = V \oplus V'$ is a tilting module without preinjective direct summand and the preprojective component in $\Gamma(B)$, with $B = End(T)$, is the preprojective component of $C - mod$.

If X is an indecomposable module in $\mathcal{G}(T)$ not Ext-projective, if t denotes the torsion-radical associated to the torsion class $\mathcal{G}(T)$ and

$$0 \longrightarrow \tau X \longrightarrow M \longrightarrow X \longrightarrow 0$$

is the Auslander-Reiten-sequence ending in X, then

$$0 \longrightarrow t(\tau X) \longrightarrow t(M) \longrightarrow X \longrightarrow 0$$

is the Auslander-Reiten sequence in $\mathcal{G}(T)$ ending in X, see f.e. [6]. Moreover by [9]

$$0 \longrightarrow t(\tau X) \longrightarrow \tau X \longrightarrow Q \longrightarrow 0$$

is the universal exact sequence in $Ext^1(\mathcal{F}(T), t(\tau X))$ and Q is Ext-injective that is $Q \in add(\tau T)$, as A is hereditary.

As V' is a projective generator in $V^{\perp} \cong C - mod$ and X is not Ext-projective we have $Hom(X, V') = 0$ that is $Q \in add(\tau V)$. So we get for $X \in \mathcal{G}(T)$, $X \notin add(T)$ the short exact sequence

$$0 \longrightarrow t(\tau X) \longrightarrow \tau X \longrightarrow (\tau V)^l \longrightarrow 0. \tag{1}$$

Applying the functor $Hom(-, \tau V)$, we check $l = dim_k Hom(X, V)$, whereas the universality of (1) implies that $l = dim_k Ext^1(\tau V, t\tau X)$. Denoting again by F or G the tilting-functors $F = Hom_A(T, -)$ or $G = T \otimes_B -$ we can note:

Lemma 3.2 *If A is connected wild hereditary with $n \geq 3$ simple modules, if V is a quasi-simple regular stone in $A - mod$, if $T = V \oplus V'$ a tilting module where V' is the minimal projective generator in $V^{\perp}$ and $B = End(T)$ then we have:*

If X is indecomposable in $\mathcal{G}(T)$, not preinjective and such that $\tau_B^r F(X) \neq 0$ $(r > 0)$ then there exists a short exact sequence

$$0 \longrightarrow \tau_A^{-r} G(\tau_B^r F(X)) \xrightarrow{\iota_r} X \xrightarrow{\pi_r} Q \longrightarrow 0 \tag{2}$$

where Q has a filtration

$$Q = Q_0 \supset Q_1 \supset Q_2 \supset \ldots \supset Q_{r-1} \supset Q_r = 0$$

with $Q_i / Q_{i+1} \in add(\tau_A^{-i} V)$.

Proof: Let us denote by X_i for $0 \leq i \leq r$ the indecomposable module $G(\tau_B^i F(X)) \in \mathcal{G}(T)$. None of the modules X_i $(0 \leq i < r)$ is Ext-projective in $\mathcal{G}(T)$ or preinjective in $A - mod$ by assumption and we have $X_{i+1} = t(\tau_A X_i)$. Thus we get by [9] the short exact sequences

$$0 \longrightarrow X_{i+1} \longrightarrow \tau X_i \longrightarrow (\tau V)^{l_i} \longrightarrow 0 \quad (i = 0, \ldots, r-1)$$

with $l_i = dim_k Ext^1(\tau V, X_{i+1})$.

As no preinjective module is involved in these sequences we can apply the functors $\tau^{-(i+1)}$ and get

$$0 \longrightarrow \tau_A^{-(i+1)} X_{i+1} \longrightarrow \tau_A^{-i} X_i \longrightarrow (\tau_A^{-i} V)^{l_i} \longrightarrow 0$$

which immediately gives the result.

As an immediate consequence we get the following necessary condition for preprojective B-modules

Corollary 3.3 *Under the above assertions we have: If X is indecomposable in $\mathcal{G}(T)$ such that $F(X) = \tau_B^{-r}F(T_1)$ is preprojective in $B - mod$ where T_1 is an indecomposable direct summand of V' then we have the following short exact sequence*

$$0 \longrightarrow \tau_A^{-r}T_1 \longrightarrow X \longrightarrow Q \longrightarrow 0$$

where Q has a filtration

$$Q = Q_0 \supset Q_1 \supset \ldots \supset Q_r = 0$$

with $Q_i/Q_{i+1} \in add(\tau^{-i}V)$.

Lemma 3.4 *Let A be connected wild hereditary with $n \geq 3$ simple modules, let V be a quasi-simple regular stone in $A - mod$ with $\tau^{-i}V$ sincere for all $i \geq 0$ and $T = V \oplus V'$ a tilting module where V' is the minimal projective generator of $V^{\perp}$. If B is the tilted algebra $B = End(T)$ and X is indecomposable in $\mathcal{G}(T)$, not preinjective such that $\tau_B^r F(X) \neq 0$ and (2) is the short exact sequence as in (3.2)*

$$0 \longrightarrow \tau_A^{-r}G(\tau_B^r F(X)) \xrightarrow{\iota_r} X \xrightarrow{\pi_r} Q \longrightarrow 0$$

then we have:

If Y is any A-module and $f : Y \to X$ is a linear map which has a factorisation through $add(V)$, then f is in the image of $Hom(Y, \iota_r)$.

Proof: Consider the commutative triangle

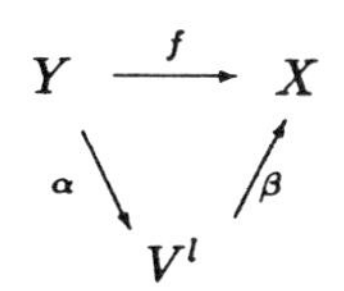

Assume f is not in the image of $Hom(Y, \iota_r)$. Then we have $f \circ \pi_r \neq 0$ and thus $\beta \circ \pi_r \neq 0$. By (3.2) we have the short exact sequence

$$0 \longrightarrow Q_1 \longrightarrow Q \xrightarrow{\pi'} V^{l'} \longrightarrow 0$$

where Q_1 has a filtration in $add(\bigoplus_{i=1}^{r-1} \tau^{-i}V)$. As V is a quasi-simple stone, we have $Hom(V, \tau^{-i}V) = 0$ for all $i > 0$ (see [10]), which implies $Hom(V, Q_1) = 0$. Therefore with $\beta \circ \pi_r \neq 0$ we get $0 \neq \beta \circ \pi_r \circ \pi' \in Hom(V^l, V^{l'})$. As V is a brick and X is indecomposable this means $V \cong X$ but $\tau_B F(X) \neq 0$ then gives a contradiction.

Proof of (3.1): We will use induction on the number n of simple A-modules. As any wild hereditary algebra A with two simples is of type K_r for some $r \geq 3$ by (2.2) we have $\underline{t}(A) = -\infty$.

For $n > 2$ we may chose a quasi-simple regular stone V in $A-mod$ such that $\tau^{-i}V$ is sincere for all $i \geq 0$. Consider the tilting module $T = V \oplus V'$ as in the beginning of this part. $C = End(V')$ is a wild, connected hereditary algebra with $(n-1)$ simple modules, thus we have $\underline{t}(C) = -\infty$.

Let r be any natural number. Then there exist infinitely many regular components $\mathcal{D}$ in $\Gamma(C)$ such that $rk(\mathcal{D}) \leq -r$. We use, as in part 2, the bijection $\mu = \mu_T$ from the set $\Omega(A)$ of regular components of $\Gamma(A)$ to those of $\Gamma(C)$. For $\mathcal{D} \in \Omega(C)$ with $rk(\mathcal{D}) \leq -r$ we chose $\mathcal{C} \in \Omega(A)$ with $\mu(\mathcal{C}) = \mathcal{D}$. Thus there exists a regular component $\mathcal{C}'$ in $\Gamma(B)$, where $B = End(T)$, with a quasi-simple module X such that $(X \rightarrow)$ is contained in $\mathcal{D}$ and $G(\rightarrow \tau_B^N X)$ is contained in $\mathcal{C}$, for some N.

By the choice of $\mathcal{D}$ there is a non-zero map $f : X \rightarrow \tau_B^{-r}X$. If $Hom(\tau_B^N X, \tau_B^{N-r}X) \neq 0$ then $Hom(\tau_B^{N+r}X, \tau_B^N X)$ is nonzero, that is $rk(\mathcal{C}) \leq -r$: Indeed, $\tau_B^{N+i+1}X = F(\tau_A G(\tau_B^{N+i}X))$ (for $i \geq 0$) just means that $Hom_B(\tau_B^{N+i}X, -) = \underline{Hom}_B(\tau_B^{N+i}X, -)$, where $\underline{Hom}$ denotes the projectively stable Hom-group.

If $Hom(\tau_B^N X, \tau_B^{N-r}X) = 0$, then there exists $Y' = \tau_B^{s'}X$ with $0 < s' < N$ such that $Hom(Y', \tau^{-r}Y') \neq 0$ but $\underline{Hom}(Y', \tau^{-r}Y) = 0$. Take some nonzero map $f' : Y' \rightarrow \tau^{-r}Y'$; as $Hom(Y', C) = 0$ the map f' factors through the projective $F(V)$, and we get the commutative triangle

$$\begin{array}{ccc} Y' & \xrightarrow{f'} & \tau_B^{-r}Y' \\ & \searrow \quad \nearrow & \\ & F(V^r) & \end{array}$$

Defining $X_i = G(\tau_B^{-i}Y')$ for $i = 0, \ldots, r$ this means

$$\begin{array}{ccc} X_0 & \xrightarrow{f} & X_{-r} \\ & {}_{\alpha}\searrow \quad \nearrow_{\beta} & \\ & V^r & \end{array}$$

By (3.2) we have the short exact sequence

$$0 \longrightarrow \tau^{-r}X_0 \xrightarrow{\iota} X_{-r} \xrightarrow{\pi} Q \longrightarrow 0$$

and by (3.4) we have $f = h \circ \iota$ with $0 \neq h \in Hom(X_0, \tau^{-r}X_0)$.

Clearly the module $X_0 = G(Y')$ is regular in $A - mod$. Let us finally show that X_0 is quasi-simple in $A - mod$:

If not, we have the Auslander-Reiten sequence

$$0 \longrightarrow \tau X_0 \longrightarrow U \oplus U' \longrightarrow X_0 \longrightarrow 0$$

where U is a submodule of X_0 and τX_0 is contained in U'. As Y' is quasi-simple in $B - mod$, we can consider the relative Auslander-Reiten sequence

$$0 \longrightarrow t\tau X_0 \longrightarrow t(U) \oplus t(U') \longrightarrow X_0 \longrightarrow 0,$$

t again the torsion radical, and see that $t(U) \oplus t(U')$ is indecomposable. From $Hom(t\tau X_0, U') \neq 0$ we deduce that U' is not in $\mathcal{F}(T)$, that is $t(U') \neq 0$. Thus we have $t(U) = 0$ which implies the condition $\tau_A^- U \in add(T)$. But this contradicts to the choice of Y', being an element of a regular component in $\Gamma(B)$. Therefore X_0 is a quasi-simple module in some component $\mathcal{C}_1 \in \Omega(A)$ with $rk(\mathcal{C}_1) \leq -r$.

4 Abelian subcategories

Let $\mathcal{A}$ be a full exact abelian subcategory of $add\ \mathcal{R}$ the category of regular A-modules, where A is a wild hereditary algebra over an algebraically closed field k. Then $\mathcal{A}$ is a length–category in the sense of [3]. Let $\mathcal{S}$ be the set of (representatives of) simple objects of $\mathcal{A}$. As $\mathcal{A}$ is full in $add(\mathcal{R})$, $\mathcal{S}$ is a set of pairwise orthogonal regular bricks and any object of $\mathcal{A}$ has an $\mathcal{S}$-filtration.

Lemma 4.1 *If $\mathcal{A}$ is a full exact abelian subcategory of $add\mathcal{R}$ and X is indecomposable in $\mathcal{A}$ with $Hom(X, \tau^{-l}X) \neq 0$ for some $l \in \mathbb{N}$, then there exist S_1, S_2 in $\mathcal{S}$ with $Hom(S_1, \tau^{-l}S_2)$ non-zero.*

Proof: Since $\tau : add\,\mathcal{R} \to add\,\mathcal{R}$ is an equivalence $\tau^{-l}\mathcal{A}$ again is abelian, with $\{\tau^{-l}S \mid S \in \mathcal{S}\}$ as set of simple objects. If $f : X \to \tau^{-l}X$ is nonzero, there exists a nonzero map g from a $\mathcal{S}$-composition-factor of X to a $\tau^{-l}\mathcal{S}$-composition-factor of $\tau^{-l}X$.
From [11] we know:

Lemma 4.2 *Let A be wild hereditary. If X and Y are regular A-modules with $Hom(X, Y) = 0$ then $Hom(X, \tau^{-l}Y) = 0$ for all $l \geq \bar{t}(A)$.*

Combing these two lemmas, we immediately get:

Corollary 4.3 *If $\mathcal{A}$ is a full abelian subcategory of $add\,\mathcal{R}$ and X, Y are indecomposable objects of $\mathcal{A}$ then we have $Hom(X, \tau^{-l}Y) = 0$ for all $l \geq \bar{t}$.*

Theorem 4.4 *If A is wild hereditary and $\mathcal{A}$ is a full exact abelian subcategory of $add\,\mathcal{R}$ then we have*

(a) If $\mathcal{C}$ is any regular component then there exists a wing $\mathcal{W}(Z)$ of length at most $\bar{t}$ such that $\mathcal{C} \cap \mathcal{A} \subset \mathcal{W}(Z)$.

(b) If $\mathcal{C}$ is any regular component containing an object $X \neq 0$ of $\mathcal{A}$, then $rk(\mathcal{C}) > -\bar{t}$.

Proof: If Y is an indecomposable object in $\mathcal{C}$ with quasi-length $s+1$ then $Hom(Y, \tau^{-s}Y) \neq 0$. Therefore all indecomposable objects of $\mathcal{A}$ have quasi-length at most $\bar{t}$ by (4.3). If there exists not a wing $\mathcal{W}(Z)$ of length at most $\bar{t}$ containing all indecomposable objects of $\mathcal{C} \cap \mathcal{A}$ then we easily find X_1, X_2 in $\mathcal{C} \cap \mathcal{A}$ such that $Hom(X_1, \tau^{-l}X_2) \neq 0$ for some $l \geq \bar{t}$, again a contradiction to (4.3).

5 Examples and applications

If A is wild connected and hereditary with (wild) quiver Q and Q contains an Euclidean subquiver Q' with corresponding path-algebra $A' = kQ'$ then we have a full exact embedding $A' - mod \hookrightarrow A - mod$.

If $\mathcal{S}$ is any set of homogenous quasi-simple A'-modules then of course $\mathcal{S}$ may be considered as set of pairwise orthogonal bricks in $A-mod$, but we have for $S, T \in \mathcal{S}$ with $S \not\cong T$ that $Ext(S,T) = Ext(T,S) = 0$, so $\mathcal{E}(\mathcal{S}) = \bigcup_{T\in\mathcal{S}} \mathcal{E}(T)$. If B is the radical-square zero algebra with underlying quiver (with n vertices)

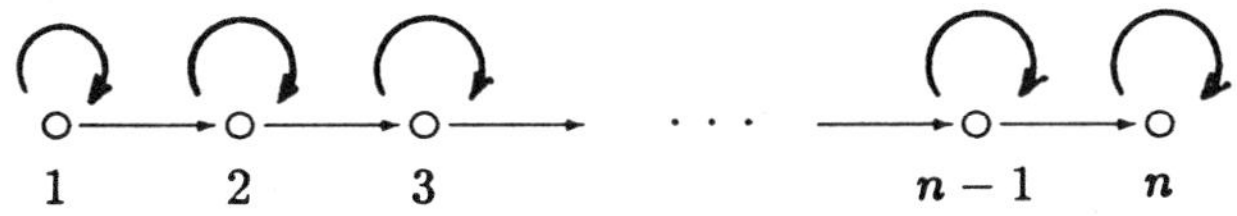

and if A is wild hereditary then there exists a full exact embedding (see [12])

$$E : B - mod \hookrightarrow A - mod$$

If $\{S_i \mid i = 1, \ldots, n\}$ is a set of simple B-modules then $\{E(S_i) \mid i = 1, \ldots, n\}$ is a set of pairwise orthogonal bricks with self-extensions satisfying the condition $Ext(E(S_{i+1}), E(S_i)) \neq 0$. As E normally will not be extension-closed, possibly more extension-groups are non-zero. We have mentioned above that for a brick X with $t = dim_k Ext^1(X,X)$ the category $\mathcal{E}(X)$ is equivalent to $K\langle\langle X_1, \ldots, X_t\rangle\rangle - mod$, the category of finite length modules over the power series ring in non-commuting variables $X_1, \ldots X_t$, especially $\mathcal{E}(X)$ is wild for $t \geq 2$. This implies

Proposition 5.1 *Let A be wild connected hereditary over an algebraically closed field with $n(A)$ simple modules*

(a) The full subcategory $\mathcal{S}_0$ of quasi-simple modules contained in components of quasi-rank 0 is wild.

(b) If $n(A) \geq 3$ then the category $\mathcal{S}_{-1}$ defined by the quasi-simple modules contained in components of quasi-rank -1 is wild.

Proof: (a) We have to find a quasi-simple brick X in $A - mod$ with $dim\ Ext^1(X,X) \geq 2$. Then all indecomposable modules of $\mathcal{E}(X)$, except its simple object X (which is harmless) are contained in $\mathcal{S}_0$. If B is the path-algebra of the quiver

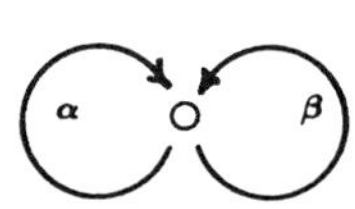

with relations $\alpha\beta = \beta\alpha = \alpha^2 = \beta^2 = 0$, and if S is the simple B-module then we have $dim_k Ext^1(S,S) = 2$. Again by [12] we consider the full exact embedding

$$E : B - mod \hookrightarrow A - mod$$

$E(S)$ then is a brick with $dim_k Ext(E(S), E(S)) = t \geq 2$. As there are only finitely many stones V with $dim\ V < dim\ E(S)$ and there are by [7] infinitely many bricks with the same dimension-vector as $E(S)$, we find a quasi-simple brick X with $\underline{dim}\, X = \underline{dim}\, E(S)$ and thus $dim_k Ext(X,X) = t$.

(b) Similarly to (a) we have to find a brick X of quasi-length 2 with $dim\ Ext^1(X,X) \geq 2$:

As $n(A) > 2$ we may suppose that A is a one-point extension of a representation-infinite algebra B, that is

$$A = \begin{pmatrix} B & M \\ 0 & k \end{pmatrix}$$

where M is a projective B-module. Since by [15] there are only finitely many preprojective B-modules non-sincere (in B-mod) and almost all preprojective B-modules are regular, when considered as A-modules, we find an indecomposable preprojective B-module Y, regular and quasi-simple (by [17]) in $A - mod$ with

$$d = dim\ Hom(M,Y) \cdot dim\ Hom(M, \tau_B Y) \geq dim\ Hom(\tau_B Y, Y) + 2$$

$X = [2]Y$ then is a brick and we get

$$q_A(\underline{dim}\, X) = 1 + dim\ Hom(\tau_B Y, Y) - d = 1 - dim\ Ext^1(X,X)$$

thus $dim\ Ext^1(X,X) \geq 2$. All indecomposable objects of $\mathcal{E}(X)$ except X are in $\mathcal{S}_{-1}$.

If the algebraically closed field in addition is uncountable then the set $\Gamma(A)$, which has the same cardinality as k (if k is infinite) is uncountable, too. If X is any brick with self-extensions, then by [7] there exist uncountable many bricks with the same dimension vector, all but finitely many of them being quasi-simple, so we have uncountable many components of quasi-rank $+1$.

If $dim\ Ext^1(X,X) = t$ then $\mathcal{E}(X)$ has countable many indecomposables, if t is 1 and uncountable many for $t > 1$. Thus we get

Corollary 5.2 *Let A be wild connected and hereditary over an uncountable algebraically closed field.*

(a) The set of components of quasi-rank 1 *is uncountable*

(b) The set of components of quasi-rank 0 *is uncountable*

(c) If $n(A) > 2$ *the set of components of quasi-rank* -1 *is uncountable*

(d) If $n(A) = 2$ *the set of components of quasi-rank* ≤ -1 *is uncountable*

Let us finally mention an application to tilting theory: If A is wild connected and hereditary with at least 3 simple modules, then by [15] there exist regular tilting modules. If T is a regular tilting module and $\mathcal{C}$ is any regular component, by [8] there exists a quasi-simple module X in $\mathcal{C}$ such that the cone $(X \rightarrow)$ is contained in $\mathcal{F}(T)$ and there exists a natural number N such that the cone $(\rightarrow \tau^N X)$ is in $\mathcal{G}(T)$. As $Hom(\mathcal{G}(T), \mathcal{F}(T)) = 0$ we see that $rk(\mathcal{C}) > -N$ holds, especially there is no upper bound for N. In contrast to this fact it was shown in [11, 6.1] that for indecomposable modules Y and Z in $\mathcal{C}$ with $Y \in \mathcal{G}(T)$ and $Z \in \mathcal{F}(T)$ we have $\tau^{\bar{t}} Y \in (\rightarrow \tau^N X)$ and $\tau^{-\bar{t}} Z \in (X \rightarrow)$, where again $\bar{t}$ is $max\ \{rk(\mathcal{C}) \mid \mathcal{C} \in \Omega(A)\}$.
So if $rk(\mathcal{C})$ is very small we have a big part of $\mathcal{C}$ which contains neither torsion nor torsion-free modules.

References

[1] BAER, D.: Wild hereditary Artin algebras and linear methods.
Manuscripta math. 55, 68-82 (1986)

[2] BONGARTZ, K.: Tilted Algebras. Proc. ICRA III (Puebla, 1980)
Lecture Notes in Mathematics 903, 17-32 (1981)

[3] GABRIEL, P.: Indecomposable Representations II.
Symposia Math. Inst. Naz. Alta Mat. Vol. XI, 81–104 (1973)

[4] GEIGLE, W. AND LENZING, H.: Perpendicular categories with applications to representations and sheaves. To appear in J. Algebra

[5] HAPPEL, D. AND RINGEL, C.M.: Tilted algebras. Trans. Amer. Math. Soc 274 (1982) 399-443

[6] HOSHINO, M.: On splitting torsion theories induced by tilting modules. Comm. Algebra 11, 427-441 (1983)

[7] KAC, V.G.: Infinite root systems, representations of graphs and invariant theory, II. J. Algebra 78, 144-162 (1982)

[8] KERNER.O.: Tilting wild algebras. J. London Math. Soc. (2) 39, 29-47 (1989)

[9] KERNER, O.: Universal exact sequences for torsion-theories. To appear in Topics in Algebra – Representations of algebras, Banach Centre Publications Vol. 26, PWN Warsow (1990)

[10] KERNER, O.: Stable components of wild tilted algebras. To appear in J. Algebra

[11] KERNER O.: Exceptional components of wild hereditary algebras. To appear in J. Algebra

[12] RINGEL C.M.: Representations of K-species and bimodules. J. Algebra 41, 269-302 (1976)

[13] RINGEL, C.M.: Finite dimensional algebras of wild representation type. Math. Z. 161, 235–255 (1978)

[14] RINGEL, C.M.: Tame algebras and integral quadratic forms. Lect. Notes in Math. 1099 (1984).

[15] RINGEL, C.M.: The regular components of the Auslander-Reiten quiver of a tilted algebra. Chinese Ann. Math. Ser. B.9, No. 1, 1-18 (1988)

[16] STRAUSS, H.: A reduction process with applications to tilts of wild hereditary algebras. To appear in J. Algebra

[17] UNGER, L.: On wild tilted algebras which are squids. Archiv Math. 55 (1990), 542-550

Mathematisches Institut
Heinrich-Heine-Universität Düsseldorf
Universitätsstraße 1
4000 Düsseldorf 1
Germany

Canadian Mathematical Society
Conference Proceedings
Volume 11 (1991)

On p-Radical Groups in Representation Theory of Finite Groups

SHIGEO KOSHITANI

1. Introduction and notation. Let F be an arbitrary field of nonzero characteristic, say, p, and let G be a finite group with a Sylow p-subgroup P. We write FG for the group algebra of G over F. Keeping in mind these notation we call G p-radical if the induced module $F_P{\uparrow}^G$ from the 1-dimensional trivial FP-module F_P is semi-simple (completely reducible) as a right FG-module, where $F_P{\uparrow}^G = F_P \bigotimes_{FP} FG$. This definition is equivalent to the original one given by Motose and Ninomiya [7] (see Feit's book [**1**, VI Corollary 6.4]). Since the property "p-radical" does not depend on the choice of the extension fields of F, we assume that F is algebraically closed from now on (see [**1**, p.265]).

Relating to p-radical groups there have been several papers. For instance, Motose and Ninomiya [**7**], Knörr [**4**], Tsushima [**9**] and Okuyama [**8**]. In his paper [**8**], Okuyama proved the following remarkable fact. That is, p-radical groups are always p-solvable. The converse of this fact is false in general as pointed out by several persons. One of the simplest counter examples for it, is SL(2,3) for $p = 3$. The group SL(2,3) is a semi-direct product of the quaternion group of order 8 by the cyclic group $Z(3)$ of order 3, so that SL(2,3) is a 3-nilpotent group. However, the induced module $F_{Z(3)}{\uparrow}^{\mathrm{SL}(2,3)}$ is not semi-simple. As a matter of fact, this module has Loewy structure of the form

$$F_{Z(3)}{\uparrow}^{\mathrm{SL}(2,3)} \quad \cong \quad F_{\mathrm{SL}(2,3)} \bigoplus \tfrac{2}{2} \bigoplus 3$$

where 2 and 3 are simple F[SL(2,3)]-modules of dimensions 2 and 3, respectively. This means that SL(2,3) is not 3-radical. Going back to the general case, it should be mentioned, by the way, that Tsushima in his paper [9] completely determined group-theoretical structure of G such that G is p-radical, under the assumption that G is p-nilpotent.

1980 *Mathematics Subject Classification* (1985 *Revision*). Primary 20C05, 20C20.
Partially supported by the Alexander von Humboldt Foundation.

Throughout this note, all groups here are finite and all modules here are finitely generated right modules. In particular, we fix a finite group G and an algebraically closed field F of characteristic $p > 0$. Let H be a subgroup of G, and let M and N be FG- and FH-modules, respectively. Then, we write $M_{\downarrow H}$ for the restriction of M to FH, and we write $N^{\uparrow G}$ for the induced module of N to FG, namely, $N^{\uparrow G} = N \bigotimes_{FH} FG$. Hence, $N^{\uparrow G}$ can be considered as a right FG-module by a canonical action of FG, say, $(n\otimes a)b = n\otimes ab$ for $n \in N$ and $a, b \in FG$. We write F_G for the 1-dimensional trivial FG-module. We write e_B for the block idempotent (centrally primitive idempotent) in FG corresponding to a p-block B of G. Further notation and terminology follow the book of Feit [**1**].

2. Lemmas. In this section we list a couple of lemmas for our purpose of this note in the next section. One of them has been obtained in our previous paper [**6**] as a main theorem there.

LEMMA 1. *Assume that for any simple (irreducible) FG-module S, the restriction $S_{\downarrow N_G(Q)}$ of S to the normalizer $N_G(Q)$ of a vertex Q of S is also simple as an $F[N_G(Q)]$-module. Then, G is p-radical.*

PROOF AND REMARK: This has been proved in [**6**, Theorem]. Perhaps we should mention that the above condition is equivalent to a condition that $Q \subseteq \operatorname{Ker} S$ for any simple FG-module S and a vertex Q of S.

The condition in Lemma 1 does not seem too strong as stated in [**6**]. After we had proved Lemma 1, Hida could generalize Lemma 1 to a p-block form. To state it, we need one more definition which is due to Tsushima [**9**, p.83]. Namely, let B be a p-block of G with its block idempotent e_B. Then, we call B p-radical if a module $(F_P{}^{\uparrow G})e_B$ is semi-simple as a right FG-module, where P is a Sylow p-subgroup of G.

LEMMA 2 (HIDA). *Let B be a p-block of G. Suppose that for any simple FG-module S in B, $S_{\downarrow N_G(Q)}$ is also simple as an $F[N_G(Q)]$-module where Q is a vertex of S. Then, B is p-radical.*

PROOF AND REMARK: This is a result of Hida in [**3**]. The proof there is completely different from that of Lemma 1.

3. Main result. In this section we state one proposition and one theorem.

PROPOSITION. *Let P be a Sylow p-subgroup of G as before, and let B be a p-block of G. If the assumption in Lemma 2 holds for B, then for any simple FG-module S in B there is a vertex Q of S such that Q is a normal subgroup of P and $S_{\downarrow P}$ has the form*

$$S_{\downarrow P} \;\cong\; F(P/Q) \bigoplus \cdots \bigoplus F(P/Q)$$

as FP-modules.

PROOF: We know from Lemma 2 that B is p-radical. Take any simple FG-module S in B. Then, by the result of Okuyama [**8**, Lemma 1],

$$S_{\downarrow P} \cong F_{Q_1}{}^{\uparrow P} \oplus \dots \oplus F_{Q_m}{}^{\uparrow P}$$

for a positive integer m and for vertices $Q_1,\dots,Q_m$ of S such that $Q_i \subseteq P$. Let $Q = Q_1$. Then, since all vertices of S are conjugate in G and Ker S is normal in G, we get from the assumption that

$$\begin{aligned} Q \subseteq \text{Ker } S \cap P &= \text{Ker}\,(S_{\downarrow P}) \subseteq \text{Ker}\,(F_{Q_i}{}^{\uparrow P}) \\ &= \cap_{y\in P}\ (\text{Ker } F_{Q_i})^y = \cap_{y \in P} Q_i{}^y \subseteq Q_i \end{aligned}$$

for all i (see [**1**, II Lemma 2.13]). Hence $Q = Q_i$ for all i. Moreover, by the above, $Q \subseteq \cap_{y\in P} Q^y$, so that $Q \subseteq Q^y$ for all $y \in P$. This means that Q is normal in P. Thus, we have $S_{\downarrow P} \cong m \times F(P/Q)$ as desired.

THEOREM. *Assume that G is p-nilpotent. Let B be a p-block of G with defect group D. Then, the following three conditions are equivalent.*

(1) *B is p-radical.*

(2) *The assumption in Lemma 2 for B holds, that is, $D \subseteq$ Ker S, where S is a unique simple FG-module in B. (Note that G is p-nilpotent).*

(3) *$(F_D{}^{\uparrow G})e_B$ is semi-simple.*

Furthermore, if one of (1)–(3) holds, then there is a Sylow p-subgroup P of G such that D is normal in P and

$$S_{\downarrow P} \cong F(P/D) \oplus \dots \oplus F(P/D)$$

where S is the same as in (2).

PROOF: The latter half is obtained by Proposition.

(2) $\Rightarrow$ (3): Since B has only one simple FG-module, we get it from the fact that (1) and (19) in [**5**, Theorem] are equivalent.

(3) $\Rightarrow$ (1): Let P be a Sylow p-subgroup of G which contains D. There is an FP-epimorphism $F_D{}^{\uparrow P} \longrightarrow F_P$ from Frobenius reciprocity, so that this induces an FG-epimorphism $(F_D{}^{\uparrow G})e_B \longrightarrow (F_P{}^{\uparrow G})e_B$.

(1) $\Rightarrow$ (2): Since G is p-nilpotent, we can write $G = KP$, where $K = O_{p'}(G)$ and P is a Sylow p-subgroup of G containing D. Let S be a unique simple FG-module in B, so that S has D as its vertex.

Case 1. $P = D$. Since P is a vertex of S and B is p-radical, it follows from the result of Okuyama [**8**, Lemma 1] that

$$S_{\downarrow P} \cong F_P \oplus \dots \oplus F_P$$

which shows $D = P \subseteq$ Ker S.

Case 2. *D is normal in P and $D \neq P$.* Let $H = KD$. Then H is normal in G. Let $b_1,\dots,b_t$ be all p-blocks of H which are covered by B. Since G/H is a p-group, it follows from [**1**, V Lemmas 3.3 and 3.5] and

Mackey decomposition that

$$[(F_P{}^{\uparrow G})e_B]_{\downarrow H} \cong \bigoplus_{i=1}^{t} (F_{P\cap H}{}^{\uparrow H})e_{b_i}$$

as FH-modules, so that $b_1,\ldots,b_t$ are all p-radical by Clifford's theorem since B is p-radical and $P \cap H$ is a Sylow p-subgroup of H. Now, by Clifford's theorem,

$$S_{\downarrow H} \cong m \times (T_1 \oplus \ldots \oplus T_\ell)$$

for positive integers m and ℓ, and for non-isomorphic simple FH-modules $T_1,\ldots,T_\ell$ such that $T_1,\ldots,T_\ell$ are all conjugate in G. Since H is also p-nilpotent, $\ell = t$. Hence we may assume that T_i belongs to b_i for each $i = 1,\ldots,t$. Let D_i be a defect group of b_i such that $D_i \subseteq D$ for each i. By [**1**, III Lemma 4.6], there is some j such that $D = D_j$. On the other hand, the F-dimensions of all $T_1,\ldots,T_t$ are the same. Thus, it follows from a theorem of Hamernik and Michler [**2**, Theorem 2.1] that all $D_1,\ldots,D_t$ have the same order. This implies $D = D_1 = \ldots = D_t$. So, by the proof of Case 1, $D \subseteq \text{Ker } T_i$ for all i, so that $D \subseteq \text{Ker } S$.

Case 3. General Case. Since D is subnormal in P, we obtain that $D \subseteq \text{Ker } S$ just by repeating the method in Case 2.

REMARK. Keep the notation as in Theorem. Theorem says that $(F_P{}^{\uparrow G})e_B$ is semi-simple if and only if $(F_D{}^{\uparrow G})e_B$ is semi-simple, under the assumption that G is p-nilpotent. Perhaps, the assumption that G is p-nilpotent could be weakened to get the equivalence above.

REFERENCES

1. W. Feit, "The Representation Theory of Finite Groups," North-Holland, Amsterdam, 1982.
2. W. Hamernik and G. Michler, *On vertices of simple modules in p-solvable groups*, Mitt. Math. Sem. Giessen **121** (1976), 147–162.
3. A. Hida, *On p-radical blocks of finite groups*, Proc. Amer. Math. Soc. (to appear).
4. R. Knörr, *Semisimplicity, induction, and restriction for modular representations of finite groups*, J. Algebra **48** (1977), 347–367.
5. S. Koshitani, *On the kernels of representations of finite groups II*, Glasgow Math. J. **32** (1990), 341–347.
6. ———, *A remark on p-radical groups*, J. Algebra **134** (1990), 491–496.
7. K. Motose and Y. Ninomiya, *On the subgroups H of a group G such that $J(KH)KG \supset J(KG)$*, Math. J. Okayama Univ. **17** (1975), 171–176.
8. T. Okuyama, *p-Radical groups are p-solvable*, Osaka J. Math. **23** (1986), 467–469.
9. Y. Tsushima, *On p-radical groups*, J. Algebra **103** (1986), 80–86.

DEPARTMENT OF MATHEMATICS, FACULTY OF SCIENCE, CHIBA UNIVERSITY, YAYOI-CHO, CHIBA-CITY, 260, JAPAN

Canadian Mathematical Society
Conference Proceedings
Volume 11 (1991)

EXTENSIONS OF FINITE DIMENSIONAL ALGEBRAS AND TILTING MODULES

Nikolaos Marmaridis

ABSTRACT. Let A be a finite dimensional k-algebra and M an A-A-bimodule finite dimensional over k. Let $\vartheta : M \otimes_A M \longrightarrow M$ be an associative nilpotent A-A-bimodule homomorphism and let R denote the ϑ-extension algebra of A by M. The following are equivalent : (i) The triple (B , X , A) is a tilting triple and the A-modules DM and $X \otimes_A M$ belong to the torsion class T_A .
(ii) The triple $(E , X \otimes_A R , R)$ is a tilting triple.

INTRODUCTION.

The starting impetus of the present work is the following result of Miyachi (cf. [7], Th. 1.4.) :
Let A be a finite dimensional k-algebra, M an A-A bimodule and R the trivial extension ring of A by M. The module $X \otimes_A R_R$ is a tilting R-module if and only if X is a tilting A-module and the A-modules $Hom_k(M , k)$ and $X \otimes_A M$ are generated by X.
We extend this result to the case of ϑ-extensions of

1980 Mathematics Subject Classification (1985 Rev.) 16A64

algebras (cf. Section 1.). This is a rather general class of algebras, which includes trivial extension algebras, triangular matrix algebras, graded algebras etc..

Our paper is divided into three sections.

In the first section we recall preliminary facts needed from the theory of ϑ-extensions developed in [5]. The propositions 1.1. and 1.2. form the key for the proof of the main theorem 2.3..

The second section is devoted to the proof of Theorem 2.3. Further, we show that the class of ϑ-extensions is closed under the tilting process as this is described in the above mentioned Theorem.

In the third section we apply our theorem to the case of triangular matrix algebras.

We close our paper presenting a special class of ϑ-extensions, namely that of polytrivial extensions of degree ω (cf. 3.2.), which is also closed under the above mentioned tilting process.

1. PRELIMINARIES

In the following, we recall definitions and facts concerning extensions of rings and categories of modules (cf. [5]).

Let A be a ring, M an A-A bimodule and $\vartheta : M \otimes_A M \to M$ an A-A bimodule homomorphism. The morphism ϑ is said to be an *associative A-A bimodule homomorphism*, if $\vartheta \circ \vartheta \otimes 1_M = \vartheta \circ 1_M \otimes \vartheta$.

Let R be the abelian group $A \oplus M$ and (a_1, m_1) , (a_2, m_2) be two elements of R. If ϑ is an associative A-A bimodule homomorphism, then the multiplication $(a_1, m_1)(a_2, m_2) := (a_1 a_2, a_1 m_2 + m_1 a_2 + \vartheta(m_1 \otimes m_2))$ defines an associative ring structure on R. We call R the ϑ-*extension ring* of A by M.

Let Mod-A (resp. Mod-R) denote the category of right A-modules (resp. R-modules) and let F be the functor $-\otimes_A M : \text{Mod-}A \to \text{Mod-}A$.

We denote by $\eta(\vartheta)_X$ the morphism $1_X \otimes_A \vartheta : X \otimes_A M \otimes_A M \to X \otimes_A M$, where X is an A-module.

The family $(\eta(\vartheta)_X)$, where X runs over the objects of Mod-A determines a natural transformation $\eta(\vartheta) : F^2 \stackrel{\cdot}{\to} F$.

Using the category Mod-A and the natural transformation $\eta(\vartheta)$, we introduce a new category $\underline{\underline{R}}(\vartheta)$ as follows:

The objects of $\underline{\underline{R}}(\vartheta)$ are pairs (X, f), where X is an A-module and f is an A-homomorphism from FX to X satisfying the condition $f \circ \eta_X = f \circ Ff$.

The morphisms $a : (X, f) \to (Y, g)$ of $\underline{\underline{R}}(\vartheta)$ are the A-module homomorphisms $a : X \to Y$ satisfying the condition $g \circ Fa = a \circ f$. The morphisms in $\underline{\underline{R}}(\vartheta)$ are composed in the same way as in Mod-A.

Note, that Mod-R is isomorphic the category $\underline{\underline{R}}(\vartheta)$, (cf. 5.2., [5]). Using this isomorphism, we identify Mod-R with $\underline{\underline{R}}(\vartheta)$.

In the present paper, the ring A is assumed to be a finite dimensional k-algebra over an algebraically closed field k, the A-A-bimodule M is assumed to be finite dimensional over k and the A-A-bimodule homomorphism ϑ is *i-nilpotent*, i.e. there is an $i \in \mathbb{N}$, such that $\vartheta \circ 1_M \otimes \vartheta \circ 1_M \otimes 1_M \otimes \vartheta \circ \ldots \circ 1_M \otimes 1_M \otimes \ldots 1_M \otimes \vartheta = 0$ (i factors).

Because A and M are finite dimensional, R is a finite dimensional k-algebra. Since our main interest, lies in representation theory, we consider the full subcategories mod-R of Mod-R and mod-A of Mod-A consisting only of the modules of finite length.

The category, mod-R is isomorphic to the full subcategory $\underline{\underline{RF}}(\vartheta)$ of $\underline{\underline{R}}(\vartheta)$ which has as objects the pairs (X, f), where X is an object of mod-A.

We consider the following functors between mod-A and mod-R:

$$T = - \otimes_A R_R : \text{mod-}A \longrightarrow \text{mod-}R,$$

$$U = Hom_R({}_A R_R, -) : \text{mod-}R \longrightarrow \text{mod-}A,$$

$$C = - \otimes_R A_A : \text{mod-}R \longrightarrow \text{mod-}A,$$

$$Z = Hom_A({}_R A_A, -) : \text{mod-}A \longrightarrow \text{mod-}R,$$

$$U' = - \otimes_R R_A : \text{mod-}R \longrightarrow \text{mod-}A,$$

$$H = Hom_A({}_R R_A, -) : \text{mod-}A \longrightarrow \text{mod-}R,$$

Obviously, (T, U), (C, Z) and (U', H) form adjoint pairs of functors.

The following two propositions are proved in [5](5.4 and 2.8.) using the identification of mod-R with $\underline{\underline{RF}}(\vartheta)$.

PROPOSITION 1.1. *An R-module (X, f) is simple if and only if X is a simple A-module and f is the zero morphism.*

PROPOSITION 1.2. *If X is an A-module with a minimal projective presentation of the form $Q \xrightarrow{\xi} P \xrightarrow{\omega} X \longrightarrow 0$, then the exact sequence $TQ \xrightarrow{T\xi} TP \xrightarrow{T\omega} TX \longrightarrow 0$ is a minimal projective presentation of the R-module TX.*

In the following we shall denote by D the duality functor $Hom_k(-, k)$ and by τ_R, τ^-_R (resp. τ_A, τ^-_A) the

Auslander-Reiten translates (cf. [1], [2], [4]) defined on mod-R (resp. mod-A).

Moreover, we shall use freely and without any other reference results from tilting theory such as can be found in [8].

If X is a tilting A-module with endomorphism ring B, then we call the triple (B, X, A) a *tilting triple*. We denote by (T_A, F_A) the torsion theory induced by (B, X, A), where T_A is the torsion part and F_A is the torsion-free part.

2. TILTING MODULES AND ϑ-EXTENSIONS.

The following proposition relates the Auslander-Reiten translates of mod-A and of mod-R, where R is the ϑ-extension of A by M.

PROPOSITION 2.1. *For any A-module X it holds:*

i. *The R-module $\tau_R(TX)$ is isomorphic to $H(\tau_A X)$ and the R-module $\tau^-_R(HX)$ is isomorphic to $T(\tau^-_A X)$.*

ii. *The group $Hom_R(TX, \tau_R(TX))$ is isomorphic to the group $Hom_A(X, \tau_A X) \oplus Hom_A(X \otimes_A M, \tau_A X)$.*

PROOF. One can repeat the proofs 1.2 and 1.3. of [7] using our 1.2..

In order to prove the next theorem we need the following:

LEMMA 2.2. i. *If X is an A-module and $X_1 \oplus X_2 \oplus \ldots \oplus X_n$ is a direct decomposition of X in indecomposable A-modules, then $TX_1 \oplus TX_2 \oplus \ldots \oplus TX_n$ is a decomposition of TX in indecomposable R-modules.*

ii. *The number of isomorphism classes of the indecomposable direct summands of* TX *equals the number of isomorphism classes of the indecomposable direct summands of* X.

PROOF. i. We claim that each of the direct summands TX_i of TX is indecomposable. If $N_1 \oplus N_2$ is a direct decomposion of TX_i into two non-zero R-modules, then $CTX_i = CN_1 \oplus CN_2$. Now, $CTX_i \cong X_i$ and CN_1 , CN_2 are non-zero A-modules (cf.2.2. and 2.3. of [5]). This is a contradiction.

ii. Since $CTX_i \cong X_i$ for $1 \leqq i \leqq n$, the claim is obvious.

Let B denote the endomorphism ring of the A-module X and E the endomorphism ring of the R-module TX.

THEOREM 2.3. *Equivalent are:*

i. *The triple* (E, TX, R) *is a tilting triple.*

ii. *The triple* (B, X, A) *is a tilting triple and the* A*-modules* DM *and* $X \otimes_A M$ *belong to the torsion class* T_A .

PROOF. "i.→ ii." Since, $pdim TX \leqq 1$, the group $DHom_R(TX, \tau_R(TX))$ is isomorphic to $Ext^1_R(TX, TX)$, which is the zero group, because TX is a tilting R-module. Using 2.1.ii. we obtain that $Hom_A(X, \tau_A X) = 0$ and $Hom_A(X \otimes_A M, \tau_A X) = 0$.

Hence,

$$Ext^1_A(X, X) = D\overline{Hom}_A(X, \tau_A X) = 0 \quad , \tag{1}$$

and

$$Ext^1_A(X, X \otimes_A M) = D\overline{Hom}_A(X \otimes_A M, \tau_A X) = 0 \quad . \tag{2}$$

If $Q \xrightarrow{\xi} P \xrightarrow{\omega} X \longrightarrow 0$ is a minimal projective presentation of X, then $TQ \xrightarrow{T\xi} TP \xrightarrow{T\omega} TX \longrightarrow 0$ is a minimal projective presentation of TX, because of 1.2.. Since $pdim TX \leqq 1$, the morphism $T\xi$ has to be a monomorphism. This implies that ξ is a monomorphism too. Hence, $pdim X \leqq 1$ (3).

The number $d(TX)$ of isomorphism classes of the indecomposable direct summands of TX equals the number $d(X)$ of isomorphism classes of the indecomposable direct summands of X, because of 2.2.. Since TX is a tilting R-module, $d(TX)$ equals the number $s(R)$ of isomorphism classes of the simple R-modules. Because of 1.1. $s(R)$ is equal to the number $s(A)$ of isomorphism classes of the simple A-modules. Hence, $d(X) = s(A)$ (4).

The properties (1),(2) and (4) imply that X is a tilting A-module. Because of (3), we obtain that $X \otimes_A M$ belongs to T_A.

It remains to prove, that the A-module DM is in T_A.

Let $0 \longrightarrow Q \xrightarrow{\xi} P \xrightarrow{\omega} X \longrightarrow 0$ be a minimal projective resolution of X.

Applying T on the short exact sequence above and because $pdim TX \leqq 1$, we obtain that

$0 \longrightarrow TQ \xrightarrow{T\xi} TP \xrightarrow{T\omega} TX \longrightarrow 0$ is again a short exact sequence.

Since $T = - \otimes_A R_R$, we have $Tor_1^A(X, R) = 0$ and since R as an A-module equals $A \oplus M$, we obtain $Tor_1^A(X, M) = 0$. Because the group $Tor_1^A(X, M)$ is isomorphic to $DExt_A^1(X, DM)$, we obtain $Ext_A^1(X, DM) = 0$. This finishes the first part of the proof.

"ii.→i." Since $pdimX \leqq 1$, these is a short exact sequence in mod-A of the form

$$0 \longrightarrow Q \xrightarrow{\xi} P \xrightarrow{\omega} X \longrightarrow 0 \quad ,$$

which is a minimal projective resolution of X.
The above sequence induces the exact sequence

$$Tor_1^A(X, M) \longrightarrow TQ \xrightarrow{T\xi} TP \xrightarrow{T\omega} TX \longrightarrow 0 \quad ,$$

with left end term equal zero, because $Tor_1^A(X, M)$ is isomorphic to $DExt_A^1(X, DM) = 0$.
Since this sequence is a minimal projective resolution of TX (cf. 1.2.), we obtain that $pdimTX \leqq 1$, (5).
The groups $Hom_R(TX, \tau_R(TX))$ and $Hom_A(X, \tau_A X) \oplus Hom_A(X \otimes_A M, \tau_A X)$ are isomorphic. The first direct summand $Hom_A(X, \tau_A X)$ equals zero, because it is isomorphic to $DExt_A^1(X, X)$ and X is a tilting A-module.

Since $pdimX \leqq 1$, we have that the second direct summand $Hom_A(X \otimes_A M, \tau_A X)$ is isomorphic to $DExt_A^1(X, X \otimes_A M)$. This last group is zero, because $X \otimes_A M$ is in T_A. Hence, $Hom_R(TX, \tau_R(TX))$ is equal to the zero group.
The group $Ext_R^1(TX, TX)$ is also zero, because it is isomorphic to $DHom_R(TX, \tau_R(TX))$, (6).
Similarly to the first part of the proof, the number of isomorphism classes of the indecomposable direct summands of TX equals the number of isomorphism classes of the simple R-modules (7).
Because of (5), (6) and (7), TX is a tilting R-module.

COROLLARY 2.4. i. *An R-module (Y, g) belongs to the torsion class T_R if and only if Y belongs to the tor-*

sion class T_A.

ii. *An* R*-module* (Y, g) *belongs to the torsion-free class* F_R *if and only if* Y *belongs to the torsion-free class* F_A.

PROOF. Similar to that of [7], 1.5..

DEFINITION 2.5. The tilting triple (B, X, A) is said to be *liftable*, if DM and $X \otimes_A M$ belong to the torsion class T_A. The corresponding tilting triple (E, TX, R) is said to be a *lifted tilting triple*.

In the following our aim is to show that if (E, TX, R) is a lifted tilting triple, then E is again a ψ-extension of B by some bimodule N, where $\psi: N \otimes N \longrightarrow N$ is a B-B bimodule homomorphism.

Let X be an A-module and B its endomorphism ring. The endomorphism ring E of TX is isomorphic to the group $B \oplus N$, where $N = Hom_A(X, X \otimes_A M)$.

This group N can be considered as a B-B-bimodule, where the left structure is defined by $gf = (g \otimes 1_M) \circ f$ and the right structure is defined by $fg = f \circ g$, where $g \in B$ and $f \in N$.

One verifies easily, that there is a map $\psi: N \otimes_B N \rightarrow N$ defined as

$$\psi(f \otimes g) = 1_X \otimes \vartheta \circ f \otimes 1_M \circ g \ , \text{ where } f, g \in N .$$

This map ψ is in fact a B-B-bimodule homomorphism. So, we may consider the ψ-extension Q of B by the bimodule N.

PROPOSITION 2.6. i. *The homomorphism* ψ *is associative*

ii. *The homomorphism* ψ *is* *i-nilpotent.*

iii. *The algebra* E *is isomorphic to the* ψ*-extension* Q *of* B *by* N.

PROOF. i. Firstly, we will prove the associativity of ψ. We claim for any three elements f, g and h of $Hom_A(X, X \otimes_A M)$ and any element x of X it holds:

$$(\psi \circ 1_N \otimes \psi)(f \otimes g \otimes h)(x) = (\psi \circ \psi \otimes 1_N)(f \otimes g \otimes h)(x) .$$

Let $h(x) = \sum_r x_r \otimes m_r$, $g(x_r) = \sum_s y_{rs} \otimes n_{rs}$ and

$f(y_{rs}) = \sum_t z_{rst} \otimes p_{rst}$.

A simple calculation shows that

$$(\psi \circ \psi \otimes 1_N)(f \otimes g \otimes h)(x) = \sum_{r,s,t} z_{rst} \otimes \vartheta(\vartheta(p_{rst} \otimes n_{rs}) \otimes m_r)$$

and that

(1) $(\psi \circ 1_N \otimes \psi)(f \otimes g \otimes h)(x) = \sum_{r,s,t} z_{rst} \otimes \vartheta(p_{rst} \otimes \vartheta(n_{rs} \otimes m_r))$

Because the bimodule homomorphism ϑ is associative, the right hand terms of the above equalities coincide.

ii. Let ω_n be the composed morphism

$$\psi \circ 1_N \otimes \psi \circ 1_N \otimes 1_N \otimes \psi \circ \ldots \circ 1_N \otimes \ldots \otimes 1_N \otimes 1_N \otimes \psi$$

(n factors , $n \geq 2$) .

Let $f_1, f_2, \ldots, f_{n+1}$ be elements of $Hom_A(X, X \otimes_A M)$ and let x be an element of X.

We shall prove, using induction on n, the following:

The element $\omega_n(f_1 \otimes f_2 \otimes \ldots \otimes f_{n+1})(x)$ is a finite sum of elements of $X \otimes M$ of the form

$y \otimes (\vartheta \circ \vartheta \otimes 1_M \circ \vartheta \otimes 1_M \otimes 1_M \circ \ldots \circ \vartheta \otimes 1_M \otimes 1_M \otimes \ldots \otimes 1_M)(m_1 \otimes m_2 \otimes \ldots \otimes m_{n+1})$

where the composed morphism

$\vartheta \circ \vartheta \otimes 1_M \circ \vartheta \otimes 1_M \otimes 1_M \circ \ldots \circ \vartheta \otimes 1_M \otimes 1_M \otimes \ldots \otimes 1_M$ consists of n factors, and where $y \in X$ and $m_s \in M$, $1 \leqq s \leqq n+1$.
The claim is true for $n = 2$, because it is the formula (1) obtained in the proof of part i. of the present proposition.
If our claim is true for $n = k$, then we shall prove it for $n = k+1$.
We have

$$\omega_{k+1}(f_1 \otimes f_2 \otimes \ldots \otimes f_{k+2})(x) =$$

$$= \psi\left[f_1 \otimes (\psi(f_2 \otimes (\psi(f_3 \otimes \ldots \psi(f_k \otimes \psi(f_{k+1} \otimes f_{k+2}))\ldots))))\right](x) =$$

$$= \left[1_X \otimes \vartheta \circ \omega_k(f_1 \otimes \ldots \otimes f_{k+1}) \otimes 1_M \circ f_{k+2}\right](x) .$$

Let $f_{k+2}(x)$ be equal to $\sum_{t=1}^{r} x_t \otimes m_t$, $x_t \in X$, $m_t \in M$, $1 \leqq t \leqq r$. We obtain:

$$\left[1_X \otimes \vartheta \circ \omega_k(f_1 \otimes \ldots \otimes f_{k+1}) \otimes 1_M \circ f_{k+2}\right](x) =$$

$$= \left[1_X \otimes \vartheta \circ \omega_k(f_1 \otimes \ldots \otimes f_{k+1}) \otimes 1_M\right]\left(\sum_{t=1}^{r} x_t \otimes m_t\right) =$$

$$= \sum_{t=1}^{r} 1_X \otimes \vartheta(\omega_k(f_1 \otimes \ldots \otimes f_{k+1})(x_t) \otimes m_t) .$$

So, in order to prove our claim it is enough to show that the element

$$1_X \otimes \vartheta(\omega_k(f_1 \otimes \ldots \otimes f_{k+1})(x_t) \otimes m_t)$$

is a finite sum of elements of $X \otimes M$ of the desired form for any t, $1 \leqq t \leqq r$.

Because of the inductive assumption, the element $\omega_k(f_1 \otimes \ldots \otimes f_{k+1})(x_t)$ is a finite sum of the form:

$$\Sigma\, y_t \otimes (\vartheta \circ \vartheta \otimes 1_M \circ \ldots \circ \vartheta \otimes 1_M \otimes 1_M \otimes \ldots \otimes 1_M)(m_{1t} \otimes m_{2t} \otimes \ldots \otimes m_{k+1,t}),$$

where the composed morphism $\vartheta \circ \vartheta \otimes 1_M \circ \ldots \circ \vartheta \otimes 1_M \otimes 1_M \otimes \ldots \otimes 1_M$ consists of k-factors, $y_t \in X$ and $m_{s,t} \in M$, $1 \leq s \leq k+1$. Hence,

$$1_X \otimes \vartheta(\omega_k(f_1 \otimes \ldots \otimes f_{k+1})(x_t) \otimes m_t) =$$

$$= 1_X \otimes \vartheta \{\Sigma\, y_t \otimes [(\vartheta \circ \vartheta \otimes 1_M \circ \ldots \circ \vartheta \otimes 1_M \otimes 1_M \otimes \ldots \otimes 1_M)$$

$$(m_{1t} \otimes m_{2t} \otimes \ldots \otimes m_{k+1,t})] \otimes m_t\} =$$

$$= \Sigma\, y_t \otimes \vartheta \{[(\vartheta \circ \vartheta \otimes 1_M \circ \ldots \circ \vartheta \otimes 1_M \otimes 1_M \otimes \ldots \otimes 1_M)$$

$$(m_{1t} \otimes m_{2t} \otimes \ldots \otimes m_{k+1,t})] \otimes m_t\}.$$

Since

$$\vartheta \{[(\vartheta \circ \vartheta \otimes 1_M \circ \ldots \circ \vartheta \otimes 1_M \otimes 1_M \otimes \ldots \otimes 1_M)$$

$$(m_{1t} \otimes m_{2t} \otimes \ldots \otimes m_{k+1,t})] \otimes m_t\},$$

is equal to

$$(\vartheta \circ \vartheta \otimes 1_M \circ \vartheta \otimes 1_M \otimes 1_M \circ \ldots \circ \vartheta \otimes 1_M \otimes 1_M \otimes \ldots \otimes 1_M)$$

$$(m_{1t} \otimes m_{2t} \otimes \ldots \otimes m_{k+1,t} \otimes m_t),$$

the proof is finished.

Because ϑ is i-nilpotent the truth of the claim above implies immediately, that ψ is also i-nilpotent.

iii. We identify the module TX with the object $(X \oplus X \otimes M, q_X)$, where the homomorphism q_X equals

$$\begin{pmatrix} 0 & 0 \\ 1_{X\otimes M} & 1_X \otimes \vartheta \end{pmatrix} : X\otimes_A M \oplus X\otimes_A M \otimes_A M \longrightarrow X \oplus X\otimes_A M \ ,$$ (cf.

section 2 of. [5]).

One sees easily, that the endomorphism ring E of TX consists of the A-endomorphisms of $X\oplus X\otimes_A M$ of the form

$$\varphi = \begin{pmatrix} f & 0 \\ g & f\otimes 1_M + 1_X \otimes \vartheta \circ g \otimes 1_M \end{pmatrix} .$$

Let $\xi : E \longrightarrow B \oplus N$ be the group isomorphism determined by the pair of adjoint functors (T, U), i.e. where $\xi(\varphi) = (f, g)$.

The composition of φ with the element

$$\chi = \begin{pmatrix} f' & 0 \\ g' & f'\otimes 1_M + 1_X \otimes\vartheta \circ g'\otimes 1_M \end{pmatrix}$$ of E is mapped under ξ to

the element $(f\circ f', f\otimes 1_M \circ g' + g\circ f' + 1_X \otimes \vartheta \circ g\otimes 1_M \circ g')$.

Since $1_X \otimes\vartheta \circ g\otimes 1_M \circ g' = \psi(g\otimes g')$, the morphism ξ extends to a k-algebra isomorphism from E to the ψ-extension Q of B by the B-B-bimodule N.

REMARK 2.7. The proposition above shows that if we are dealing with lifted tilting triples, then the class of ϑ-extensions is closed.

3. APPLICATIONS

The first application, which we like to present, is that of triangular matrix algebras.

We consider the triangular matrix algebra

$$R = \begin{pmatrix} R_1 & {}_1M_2 & {}_1M_3 & \cdots & {}_1M_n \\ 0 & R_2 & {}_2M_3 & \cdots & {}_2M_n \\ \cdot & \cdot & \cdot & \cdots & \cdot \\ 0 & 0 & 0 & \cdots & R_n \end{pmatrix} .$$

Let M be the abelian group $\bigoplus_{r,s} {}_rM_s$, $1 \leq r < s \leq n$ and A be the ring $R_1 \times R_2 \times \ldots \times R_n$.
The group M can be considered as an A-A-bimodule in the natural way.
For any $r, t, s \in \mathbb{N}$, $1 \leq r < s < t \leq n$, we consider the structure map $\vartheta(r, s, t) : {}_rM_s \otimes {}_sM_t \to {}_rM_t$, which is determined by the multiplication of R .
Let $\vartheta : M \otimes M \to M$ be the A-A-bimodule homomorphism, which is uniquely determined by the fact that its restriction on the direct summand ${}_rM_s \otimes {}_sM_t$ of $M \otimes M$ coincides with $\vartheta(r, s, t)$. One checks easily that the homomorphism ϑ is associative and n-nilpotent. Hence, the ring R can be considered as the ϑ-extension of A by M .
Using the above interpretation of triangular matrix algebras and Theorem 2.3. we obtain:

PROPOSITION 3.1. *Let* $X = X_1 \times X_2 \times \ldots \times X_n$ *be an* A*-module, where* X_s *denotes the corresponding* R_s*-module,* B *(resp.* B_t*) be the endomorphism ring of* X *(resp.* X_t*) and let* E *be the endomorphism ring of* TX .
Equivalent are:

i. *The triple* (E, TX, R) *is a tilting triple.*

ii. α. *For any* t , $1 \leq t \leq n$, *the triple* (B_t, X_t, R_t) *is a tilting triple.*

β. *The right* R_s*-module* D_sM_t *belongs to the torsion* *class* T_s *induced by* (B_s, X_s, R_s) , *for any pair* (s,t), $1 \leq s < t \leq n$.

γ. *The right* R_t*-module* $X_s \otimes_s M_t$ *belongs to the torsion class* T_t *induced by* (B_t, X_t, R_t) *for any pair* (s, t) , $1 \leq s < t \leq n$.

Note that in the case above, the endomorphism ring E is isomorphic to the triangular matrix algebra:

$$\begin{pmatrix} End_{R_1} X_1 & Hom_{R_2}(X_2, X_1 \otimes_1 M_2) & \dots & Hom_{R_n}(X_n, X_1 \otimes_1 M_n) \\ 0 & End_{R_2} X_2 & \dots & HomR_n(X_n, X_2 \otimes_2 M_n) \\ \cdot & \cdot & \dots & \cdot \\ 0 & 0 & \dots & R_n \end{pmatrix}$$

Hence, the class of triangular matrix algebras is closed under the tilting process, if the tilting triples which are used are lifted tilting triples.

There are various subclasses of ϑ-extensions, where the same phenomenon takes place, i.e. where the endomorphism ring of TX with X an A-module, is again a ϑ-extension of the same type.

Our second example describes also such a particular case. Let A be a finite-dimensional algebra and (M_s) , $s \in \mathbb{N}$ be a family of A-A-bimodules, such that M_s equals to the zero module for $s \geq q+1$.

Assume, that for any pair $(s, t) \in \mathbb{N} \times \mathbb{N}$, there is a A-A-bimodule homomorphism $\vartheta(s, t) : M_s \otimes_A M_t \rightarrow M_{s+t+\omega}$, where ω is a fixed non-negative integer. Moreover, assume, that for any $s, t, r \in \mathbb{N}$ the diagram below is commutative:

$$\begin{array}{ccc} M_s \otimes_A M_t \otimes_A M_r & \xrightarrow{\vartheta(s,t)\otimes 1_{M_r}} & M_{s+t+\omega} \otimes_A M_r \\ \Big\downarrow 1M_s \otimes_A \vartheta(t,r) & & \Big\downarrow \vartheta(s+t+\omega, r) \\ M_s \otimes_A M_{t+r+\omega} & \xrightarrow{\vartheta(s,t+r+\omega)} & M_{s+t+r+2\omega} \end{array} .$$

Let R_ω denote the direct sum $A \oplus (\bigoplus_{s\in \mathbb{N}} M_s)$. We define a multiplication between the elements of R_ω as follows: If $x = (a, m_1, m_2, \ldots, m_s, \ldots)$ and $y = (a', n_1, n_2, \ldots, n_s, \ldots)$ are elements of R_ω, then their product is defined to be the element $z = (z_0, z_1, \ldots, z_s, \ldots)$ of R_ω with $z_0 = aa'$ and

$$z_s = an_s + m_s a' + \sum_{t+r+\omega=s} \vartheta(t,r)(m_t \otimes n_r) \ , \text{ for } s \in \mathbb{N} \ .$$

Obviously, R_ω is a finite dimensional k-algebra.

Let M denote the A-A-bimodule $\bigoplus_{t\in\mathbb{N}} M$ and $\vartheta : M \otimes_A M \to M$ the unique bimodule map, which restricted on $M_t \otimes_A M_r$ coincides with $\vartheta(t, r)$.

One verifies easily, that ϑ is an associative and i-nilpotent, A-A-bimodule homomorphism, where $i = \min\{t \in \mathbb{N} \cup \{0\} / t\cdot(1+\omega) \geq q\}$.

The ϑ-extension ring of A by M coincides with the above constructed algebra R_ω .

DEFINITION 3.2. The k-algebra R_ω is said to be *poly-trivial ϑ-extension of A by M of degree* ω .

Note, that polytrivial extensions of degree 0 coincide with the graded k-algebras in the sense of [6], Chapter

VI, 3. If the bimodule homomorphism ϑ equals zero, then the polytrivial extensions of degree zero are just the well-known trivial extensions of a ring by a bimodule (cf. [3]).

Let R_ω be a polytrivial ϑ-extension of A by M of degree ω and X be an A-module. Let B denote the endomorphism ring of X and E denote the endomorphism ring of the R_ω-module TX.

We consider the B-B-bimodule homomorphism, defined in section 2, $\psi : N \otimes_B N \longrightarrow N$, where $N = Hom_A(X, X \otimes_A M)$.

One verifies the following:

PROPOSITION 3.3. *The algebra* E *is the polytrivial* ψ*-extension of* B *by* N *of degree* ω.

The proposition above implies that also the class of polytrivial extensions of degree ω remains closed under the tilting process, if the tilting triples which are used are lifted tilting triples.

We leave to the reader to restate Theorem 2.3. in the case of polytrivial extensions of degree ω.

REFERENCES

1. AUSLANDER,M. and REITEN,I. : Representation theory of Artin algebras III, Comm. Algebra 3 (1975), 239-294.

2. AUSLANDER,M. and REITEN,I. : Representation theory of Artin algebras IV, Comm. Algebra 5 (1977), 443-518.

3. FOSSUM,R.M. , GRIFFITH,RH.A. and REITEN,I. : Trivial extensions of abelian categories, Lecture Notes Math., Vol. 456, Springer, Berlin, 1975.

4. GABRIEL,P. ; Auslander-Reiten sequences and representation-finite algebras, Lecture Notes Math., Vol.831, pp. 1-71, Springer, Berlin, 1980.

5. MARMARIDIS,N. : On extensions of abelian categories with applications to ring theory, Ioannina 1990, 1-26 (Preprint).

6. McLANE,S.M. : "Homology" , Springer, Berlin, 1967.

7. MIYACHI,J.-I. : Tilting modules over trivial extensions of Artin algebras, Comm. in Algebra 13(6),(1985), 1319-1326.

8. RINGEL,C.M. : Tame algebras and integral quadratic forms, Lecture Notes Math., Vol. 1099, Springer, Berlin, 1984.

University of Ioannia
45110 Ioannia
Greece

Canadian Mathematical Society
Conference Proceedings
Volume 11 (1991)

ON AUSLANDER-REITEN SEQUENCES OF TRIANGULAR MATRIX ALGEBRAS

Héctor A. Merklen

Abstract. *Some Auslander-Rieten sequences of $\widetilde{\Lambda}$-modules are characterized, where $\widetilde{\Lambda}$ is a triangular matrix algebra of the form* $\begin{bmatrix} K & M \\ 0 & \Lambda \end{bmatrix}$. *These sequences may be indicated explicitly by using the more or less standard notation (U, f, X) for $\widetilde{\Lambda}$-modules: they are the A–R–sequences beginning (resp. ending) at a Λ-module, i.e. a $\widetilde{\Lambda}$-module of the form $(0, 0, X)$. The cokernel functor $(U, f, X) \mapsto \operatorname{cok} f$ is also introduced and some of its properties are applied to calculate, in certain cases, preprojective $\widetilde{\Lambda}$-modules of level* 1.

In this paper, by triangular matrix algebra we mean an algebra of the form

$$\widetilde{\Lambda} = \begin{bmatrix} K & M \\ 0 & \Lambda \end{bmatrix}$$

where K and Λ are artin algebras and M is a K-Λ-bimodule. These algebras, of course, have been studied by many people and for many reasons. For example, these algebras are important, in many situations, as a means for induction arguments in proving theorems, or as a good context to produce examples, and, in all cases, they are useful because of interesting connections that appear, depending mainly on the properties of M, between the categories mod $\widetilde{\Lambda}$, mod K and mod Λ. There are so many research articles that take profit of these algebras that it would be very hard to mention all of them and even to choose only some of them to indicate as a reference. But it must be mentioned that a very important particular case of triangular matrix algebras $\widetilde{\Lambda}$ are the one-point extensions, where K is a field and Λ is a K-algebra

1980 Mathematics Subject Classification (1985 Revision). 16A64.

of finite dimension.

In this paper we characterize some Auslander-Reiten sequences of $\tilde{\Lambda}$-modules. We choose to work with right-modules, unless left-modules appear in a natural way. If there is any possibility of misunderstanding, we will carefully indicate that the modules under consideration are left and not right-modules.

After an introduction in section 1, where we also point out the natural ways of embedding $\text{mod}\ \Lambda$ and $\text{mod}\ K$ in $\text{mod}\ \tilde{\Lambda}$, we study, in section 2, Auslander-Reiten sequences over $\tilde{\Lambda}$ which end at a Λ-module and, in section 3, Auslander-Reiten sequences over $\tilde{\Lambda}$ which begin at a Λ-module. Finally, in section 4, we introduce a functor from $\text{mod}\ \tilde{\Lambda}$ to $\text{mod}\ \Lambda$ that we call the *cokernel functor*. Then we show that this functor preserves Auslander-Reiten sequences beginning at Λ-modules and how this fact can be used to compute the preprojective modules of level 1. Section 5 is devoted to an application to the study of artin algebras which are equivalent to a hereditary algebra modulo projectives and preprojectives of level 1.

1. Introduction

Throughout this paper, K and Λ denote basic artin algebras, M a finitely generated K-Λ-bimodule and $\tilde{\Lambda}$ denotes the triangular matrix algebra

$$\begin{bmatrix} K & M \\ 0 & \Lambda \end{bmatrix}$$

consisting of matrices $\begin{pmatrix} a & m \\ 0 & b \end{pmatrix}$ $(a \in K,\ m \in M,\ b \in \Lambda)$, with the sum performed component-wise according to the sums of K, M, Λ, respectively, and with the obvious multiplication: $\begin{pmatrix} a & m \\ 0 & b \end{pmatrix}\begin{pmatrix} a' & m' \\ 0 & b' \end{pmatrix} = \begin{pmatrix} aa' & am' + mb' \\ 0 & bb' \end{pmatrix}$. All our modules will be finitely generated, right-modules. With T we will denote the functor $- \otimes_K M : \text{mod}\ K \to \text{mod}\ \Lambda$ and with S the functor $\Lambda(M, -) : \text{mod}\ \Lambda \to \text{mod}\ K$. It is well known that we can identify the category $\text{mod}\ \tilde{\Lambda}$ with anyone of the comma categories $(T \downarrow 1)$, $(1 \downarrow S)$ (see [McL] and [FGR] that we describe below mainly to fix our notations.

Category $(T \downarrow 1)$

Objects: (U, f, N) $(U$ in $\text{mod}\ K$, N in $\text{mod}\ \Lambda$, $f \in \Lambda(TU, N))$

Morphisms: $(\alpha, \beta) : (U, f, N) \to (U', f', N')$ if and only if $\alpha \in K(U, U')$, $\beta \in \Lambda(N, N')$ and $f'T\alpha = \beta f$.

Category $(1 \downarrow S)$

Objects: $(U, \tilde{f}, N)$ (U in mod K, N in mod Λ, $\tilde{f} \in K(U, SN)$)

Morphisms: $(\alpha, \beta) : (U, \tilde{f}, N) \to (U', \tilde{f}', N')$ if and only if $\alpha \in K(U, U')$, $\beta \in \Lambda(N, N')$ and $\tilde{f}'\alpha = S\beta \cdot \tilde{f}$.

The indecomposable projectives of $(T \downarrow 1)$ are $(P, 1, TP)$ (with P an indecomposable projective K-module) and $(0, 0, Q)$ (with Q an indecomposable projective Λ-module). And the indecomposable injectives are $(I, 0, 0)$ (with I indecomposable injective K-module) and (SJ, e, J) (with J indecomposable injective Λ-module and e the evaluation map, $e : \Lambda(M, J) \otimes M \to J$, sending $\varphi \otimes x$ into $\varphi(x)$). The simple $\tilde{\Lambda}$-modules are given by $(U, 0, 0)$ (U simple K-module), $(0, 0, L)$ (L simple Λ-module).

It is fairly easy to compute some objects related to our comma categories. In particular, we have formulas to help computing the duality functors D (ordinary duality) and $* = \Lambda(-, \Lambda)$ (see [HM$_1$]), which are necessary to compute the Nakayma functor. And, for these computations, one needs also the representation for left $\tilde{\Lambda}$-modules. But we feel that, because of limitation of space, it would be preferrable to skip here all these details. We will only state that left $\tilde{\Lambda}$-modules are represented by columns $\begin{bmatrix} U \\ f \\ N \end{bmatrix}$, where U is a left K-module, N is a left Λ-module and f is a left K-homomorphism from $M \otimes_\Lambda N$ into U. The left indecomposable projective $\tilde{\Lambda}$-modules are the $\begin{bmatrix} P \\ 0 \\ 0 \end{bmatrix}$ (with P an indecomposable, projective K-module) and the $\begin{bmatrix} M \otimes Q \\ 1 \\ Q \end{bmatrix}$ (with Q an indecomposable, projective Λ-module), whereas the indecomposable injective left $\tilde{\Lambda}$-modules are the $\begin{bmatrix} I \\ e \\ \Lambda(M, I) \end{bmatrix}$ (with I an indecomposable injective, left K-module and e the evaluation map) and the $\begin{bmatrix} 0 \\ 0 \\ J \end{bmatrix}$ (with J an indecomposable, injective Λ-module).

There are natural embeddings of mod K, and mod Λ into mod $\tilde{\Lambda}$. They are given, with obvious notation, by $U \to (U, 0, 0)$ and by $N \to (0, 0, N)$, respectively.

2. Auslander-Reiten sequences over $\widetilde{\Lambda}$ that end up at a Λ-module

In this section we are interested in computing $\widetilde{\Lambda}$ minimal right almost split morphisms of the form $(U, f, X) \to (0, 0, N)$ and Auslander-Reiten sequences ending at a non-projective $\widetilde{\Lambda}$-module of the form $(0, 0, N)$ (i.e. where N is an indecomposable, non-projective Λ-module).

It is very easy to see that, if P is an indecomposable projective Λ-module, then the minimal right almost split morphism ending at $(0, 0, P)$ is

$(0, 0, \text{rad } P) \xrightarrow{(0,\iota)} (0, 0, P)$, with ι the inclusion map. For the case where $(0, 0, N)$ is not projective, we have the following result.

Proposition 2.1. *Let N be an indecomposable, non-projective Λ-module. Then the following is an Auslander-Reiten sequence over $\widetilde{\Lambda}$.*

$$0 \longrightarrow (U, g, \tau N) \xrightarrow{(1,i)} (U, ig, E) \xrightarrow{(0,\nu)} (0, 0, N) \longrightarrow 0$$

here i and ν define the Auslander-Reiten sequence over Λ ending at $N : 0 \to \tau N \xrightarrow{i} E \xrightarrow{\nu} N \to 0$ and the translate $\tau(0, 0, N) = (U, g, \tau N)$ is characteried by the fact that it is an indecomposable $\widetilde{\Lambda}$-module with the following property.

For every morphism of the form $x : TX \to \tau N$ there is a morphism $a : X \to U$ and an automorphism ϕ of τN that is congruent to 1 modulo the radical, such that $x = \phi.g.Ta$.

As a matter of fact, such $(U, g, \tau N)$ is isomorphic to $(S\tau N, e, \tau N)$, e the evaluation map.

Proof. It is easy to see that if a morphism $(U, f, L) \xrightarrow{(0,\lambda)} (0, 0, N)$ is split epic, then λ is split epic. Hence, we deduce that if $(U, h, E) \xrightarrow{(0,\nu)} (0, 0, N)$ is right almost split, then ν is right almost split.

Let us assume that $(0, \nu)$ is right almost split. Since $\nu h = 0$, h factors through $i : \tau N \to E$, the kernel of $\nu : h = ig$, so that the kernel of $(0, \nu)$ is $(U, g, \tau N) \xrightarrow{(1,i)} (U, ig, E)$.

Now, given a morphism $x \in \Lambda(TX, \tau N)$ or, what amounts to the same, a $\widetilde{\Lambda}$-module $(X, x, \tau N)$ we have a morphism $(X, ix, E) \xrightarrow{(0,\nu)} (0, 0, N)$ which is not split epic. Therefore, $(0, \nu)$ factors through (U, ig, E) according to the following diagram.

$$\begin{array}{ccccc}
 & TX & \xrightarrow{x} & \tau N & \\
 & \downarrow{\scriptstyle T1} & & \downarrow{\scriptstyle i} & \\
 & TX & \xrightarrow{ix} & E & \\
 & \downarrow{\scriptstyle Ta} & & \downarrow{\scriptstyle 1} & \\
0 & TU & \xrightarrow{ig} & E & \nu \\
 & \downarrow & & \downarrow{\scriptstyle \nu} & \\
 & 0 & \longrightarrow & N &
\end{array}$$

Hence, we have $i.g.Ta = ix$ or $x = g.Ta$. This shows that the Auslander-Reiten sequence ending at $(0,0,N)$ has the prescribed form and that $(U,g,\tau N)$ has the stated properties.

To prove the converse, let us assume that $(U,g,\tau N)$ has the stated properties. It will be enough then to show that $(0,\nu)$ is right almost split. Given a morphism $(X,x,Y) \xrightarrow{(0,f)} (0,0,N)$, which is not split epic, we know that f is not split epic. Hence, f factors as $f = \nu f'$ and $fx = 0$ implies $\nu f'x = 0$, saying that there exists a morphism $Tx \xrightarrow{y} \tau N$ such that $f'x = iy$. Now we use the hypothesis to introduce a morphism $(X,y,\tau N) \xrightarrow{(a,\phi)} (U,g,\tau N)$ with ϕ of the form $\phi = 1+\delta$, $\delta \in \operatorname{rad} \Lambda(\tau N,\tau N)$. Using that i is minimal left almost split, we can rewrite δ in the form $\delta = \delta' i$ and put i.g. $Ta = i(1_{\tau N} + \delta' i)y = (1_E + i\delta')iy = (1_E + i\delta')f'x = f_1'x$, where $f_1' = (1_E + i\delta')f'$. Since $\nu f_1' = \nu f' = f$, we obtain a factorization $(0,f) = (0,\nu)(a,f_1')$ and the proof is complete.

Finally, to show that $(U,g,\tau N)$ is isomorphic to $(S\tau N,e,\tau N)$, we observe first that the stated property may be described by the following commutative diagram.

$$\begin{array}{ccc}
TX & \xrightarrow{\forall\, x} & \tau N \\
\exists\, Ta \searrow & & \nearrow g \\
 & TU &
\end{array}$$

Using the adjuntion $\sim$ between our functors T and S, we see that the property may also be described by the following commutative diagram.

$$\begin{array}{ccc} X & \xrightarrow{\forall \tilde{x}} & S(\tau N) \\ & a \searrow \quad \nearrow \tilde{g} & \\ & U & \end{array}$$

But, clearly, an obvious solution for this problem is $U = S(\tau N)$, $\tilde{g} = 1_{S(\tau N)}$.

Remarks.

1) It is not difficult to prove, in the above situation, that $(0,\nu)$ is *minimal* right almost split if and only if ν is also minimal.
2) It should be noticed that $\tau(0,0,N)$ is of the form $(0,0,\tau N)$ if and only if $S(\tau N) = \Lambda(M,\tau N) = 0$.
3) It is not difficult to compute directly $\tau(0,0,N)$ by using the Nakayama functor. We write down the results and we refer to [HM$_1$] for other details. Given N, an indecomposable, non projective Λ-module, we start from its minimal projective resolution $Q \xrightarrow{p} P \to N$ which gives the following minimal projective resolution of $(0,0,N)$.

$$(0,0,Q) \xrightarrow{(0,p)} (0,0,P) \longrightarrow (0,0,N)$$

If to this resolution we apply the $*$ duality, we obtain the transpose $\mathrm{Tr}(0,0,N)$ as follows.

$$\begin{bmatrix} M\otimes P^* \\ 1 \\ P^* \end{bmatrix} \xrightarrow{\binom{1\otimes p^*}{p^*}} \begin{bmatrix} M\otimes Q^* \\ 1 \\ Q^* \end{bmatrix} \longrightarrow \mathrm{Tr}(0,0,N) \longrightarrow 0$$

Since cok $p^* = \mathrm{Tr}N$, it follows that $\mathrm{Tr}(0,0,N) = \begin{bmatrix} M\otimes \mathrm{Tr}N \\ 1 \\ \mathrm{Tr}N \end{bmatrix}$. Finally, computing the ordinary dual D we obtain the Auslander-Reiten translate as follows.

$$\tau(0,0,N) = D\mathrm{Tr}(0,0,N) = D\begin{bmatrix} M\otimes \mathrm{Tr}N \\ 1 \\ \mathrm{Tr}N \end{bmatrix} = (D(M\otimes \mathrm{Tr}N), D1, \tau N) =$$

$$= (\Lambda(M,\tau N), e, \tau N).$$

3. Auslander-Reiten sequences over $\widetilde{\Lambda}$ that begin at a Λ-module

In this section we are interested in computing the minimal left almost split morphisms of $\widetilde{\Lambda}$ of the form $(0,0,L) \to (U,f,Y)$ and the Auslander-Reiten sequences beginning at a $\widetilde{\Lambda}$-module of the form $(0,0,L)$ (i.e. where L is an indecomposable Λ-module which is not injective or, if it is injective, has $\Lambda(M,L) = S(L) \neq 0$.

When $(0,0,L)$ is injective, it is easy to see taht the left minimal almost split map beginning at it is $(0,0,L) \xrightarrow{(0,\lambda)} (0,0,\frac{L}{\operatorname{soc} L})$, with λ the canonical epimorphism. For the cases where $(0,0,L)$ is not injective, we have the following results.

We observe first that *if L is an indecomposable, non-injective Λ-module such that $S(L) = 0$, then $0 \longrightarrow (0,0,L) \xrightarrow{(0,i)} (0,0,E) \xrightarrow{(0,p)} (0,0,\tau^{-1}L) \longrightarrow 0$ is an Auslander-Reiten sequence if and only if $0 \longrightarrow L \xrightarrow{i} E \xrightarrow{p} \tau^{-1}L \longrightarrow 0$ is.* This is an easy corollary of Prop. 2.1. Hence, the following statements assume that $S(L) \neq 0$.

Remark. If one has formulas for computing the duality functors, it is fairly easy to determine the inverse translate $\tau^{-1}(0,0,L)$. In case that L is not injective, we start with a projective resolution of the left, indecomposable, non-projective Λ-module DL as follows.

$$Q' \xrightarrow{p'} P' \xrightarrow{\alpha} DL\,.$$

This already gives a projective cover of $D(0,0,L) = \begin{bmatrix} 0 \\ 0 \\ DL \end{bmatrix}$ in the following form.

$$\begin{bmatrix} M \otimes P' \\ 1 \\ P' \end{bmatrix} \xrightarrow{\binom{0}{\alpha}} \begin{bmatrix} 0 \\ 0 \\ DL \end{bmatrix} \longrightarrow 0$$

Next, we proceed to complete this for a projective resolution. Using the exact sequence $M \otimes Q' \xrightarrow{1 \otimes p'} M \otimes P' \xrightarrow{1\otimes\alpha} M \otimes DL \longrightarrow 0$ and a projective cover $Q_1' \xrightarrow{p_1'} M \otimes DL$, assuming that p_1' factors through $M \otimes P'$ as $p_1' = (1\otimes\alpha).q'$, we get an epimorphism $Q_1' \oplus M \otimes Q' \xrightarrow{(q',1\otimes p')} M \otimes P'$ which we use to write the desired projective resolution in the following form (it should be kept in mind that P' and Q' are projective, left Λ-modules, whereas Q_1' is a projective, left K-module).

$$\begin{bmatrix} Q_1' \\ 0 \\ 0 \end{bmatrix} \oplus \begin{bmatrix} M \otimes Q' \\ 1 \\ Q' \end{bmatrix} \xrightarrow{\left(\binom{q'}{0},\binom{1\otimes p'}{p'}\right)} \begin{bmatrix} M \otimes P' \\ 1 \\ P' \end{bmatrix} \xrightarrow{\binom{0}{\alpha}} \begin{bmatrix} 0 \\ 0 \\ DL \end{bmatrix} \longrightarrow 0$$

In order that this be a minimal projective resolution it may be necessary to throw out some direct summand of Q_1' that might prove to be superfluous.

Finally, applying to that resolution the $*$ duality, we obtain the desired translate $\tau^{-1}(0,0,L)$ as a cokernel of the map linking the projectives.

$$0 \longrightarrow \begin{bmatrix} 0 \\ 0 \\ DL \end{bmatrix}^* \longrightarrow (0,0,P'^*) \longrightarrow (Q_1'^*, 1, TQ_1'^*) \oplus (0,0,Q'^*) \longrightarrow$$

$$\longrightarrow (P,h,H) \longrightarrow 0$$

In case that L is an indecomposable injective but $S(L) = 0$, we consider as above the projective cover $Q_1' \xrightarrow{p_1'} M \otimes DL$ and the desired minimal projective resolution is the following.

$$\begin{bmatrix} Q_1' \\ 0 \\ 0 \end{bmatrix} \xrightarrow{\binom{p_1'}{0}} \begin{bmatrix} M \otimes DL \\ 1 \\ DL \end{bmatrix} \longrightarrow \begin{bmatrix} 0 \\ 0 \\ DL \end{bmatrix}$$

Applying here the $*$ duality we obtain again $\tau^{-1}(0,0,L) = (P,h,H)$ as follows.

$$0 \longrightarrow \begin{bmatrix} 0 \\ 0 \\ DL \end{bmatrix}^* \longrightarrow (0,0,(DL)^*) \longrightarrow (Q_1'^*, 1, TQ_1'^*) \longrightarrow$$

$$\longrightarrow (P,h,H) \longrightarrow 0$$

We notice that, in this case, the map between the projectives has the form $(0,\nu)$, so that $P = Q_1'^*$ and h is an epimorphism.

The following proposition characterizes the $\widetilde{\Lambda}$ Auslander-Reiten sequence beginning at an indecomposable of the form $(0,0,L)$ such that $S(L)$ is not zero.

Proposition 3.1. *Let L be an indecomposable Λ-module such that $S(L) = \Lambda(M,L) \neq 0$. If $(0,\lambda)$ is the minimal left almost split map starting from $(0,0,L)$, then λ is split monic: $\lambda = \binom{1}{0}$, so that this map is of the following form.*

$$(0,0,L) \xrightarrow{(0,\binom{1}{0})} \left(P, \begin{pmatrix} g \\ h \end{pmatrix}, L \oplus H\right)$$

In order that this be minimal almost split it is necessary and sufficient that the following four conditions are satisfied.

1) g does not factor through h, i.e., there is no morphism d such that $g = dh$.

2) For every morphism $L \xrightarrow{\lambda} Y$ that is not split monic, there is a morphism $H \xrightarrow{\lambda'} Y$ such that $\lambda g = \lambda' h$.

3) Given morphisms $TX \xrightarrow{x} L$, $TX \xrightarrow{y} Y$ such that x does not factor through y, there exist morphisms $p \xrightarrow{a} X$, $H \xrightarrow{b} L$, $H \xrightarrow{c} Y$ such that

$$x.Ta = g + bh, \qquad y.Ta = ch.$$

4) (P, h, H) is an indecomposable $\widetilde{\Lambda}$-module.

Obviously, if those four conditions are satisfied, the following is an Auslander-Reiten sequence of $\widetilde{\Lambda}$-modules.

$$0 \longrightarrow (0,0,L) \xrightarrow{(0,\binom{1}{0})} \left(P, \begin{pmatrix} g \\ h \end{pmatrix}, L \oplus H\right) \xrightarrow{(1,(0,1))} (P,h,H) \longrightarrow 0$$

Proof. The fact that λ is split monic comes from the fact that $(0,0,L) \xrightarrow{(0,1)} (S(L), e, L)$ is not split monic. Property 1) comes from the fact that, if g factors through h, then $(0,0,L) \xrightarrow{(0,\binom{1}{0})} (P, \binom{g}{h}, L \oplus H)$ is split monic. In a similar way, the fact that x does not factor through y implies that $(0,0,L) \xrightarrow{(0,\binom{1}{0})} (X, \binom{x}{y}, L \oplus Y)$ is not split monic, so that if factors through the minimal left almost split map. This implies that there exists a morphism $\left(a, \begin{pmatrix} 1 & b \\ 0 & c \end{pmatrix}\right)$ from $(P, \binom{g}{h}, L \oplus H)$ to $(X, \binom{x}{y}, L \oplus Y)$. In other words, we have the equality $\begin{pmatrix} 1 & b \\ 0 & c \end{pmatrix} \binom{g}{h} = \binom{x}{y} Ta$ from which property 3) follows readily. As for the validity of 2), it is obtained in a similar way considering the morphism $(0,0,L) \xrightarrow{(0,\lambda)} (0,0,Y)$ which is also not split monic.

Conversely, let us assume that 1), 2), 3) and 4) are satisfied. Then, the sequence at the end of the statement is exact, non split and has indecomposable end terms. Let $(0,0,L) \xrightarrow{(0,\lambda)} (X,t,Z)$ be a morphism which is not split monic. If λ is not split monic, property 2) says that $(0,\lambda)$ factors through the proposed minimal left almost split map. On the other hand, if λ is split monic, it may be assumed to be of the form $L \xrightarrow{\binom{1}{0}} L \oplus Y$, meaning that $Z = L \oplus Y$ so that t may be assumed to be of the form $\begin{pmatrix} x \\ y \end{pmatrix}$. Then, since the given morphism is not split monic, we deduce that x does not factor through y. Hence, as it is easily seen, property 3) provides the desired factorization, and the proof is complete. ■

The following proposition completes the information contained in Prop. 3.1. Also, it characterizes completely the Auslander-Reiten sequence beginning at $(0,0,L)$, under the assumption that the inverse translate $\tau^{-1}(0,0,L) = (P,h,H)$ is already known.

Proposition 3.2. *Let L be an indecomposable, non-injective Λ-module such that $S(L) = \Lambda(M,L) \neq 0$. Then the $\widetilde{\Lambda}$ Auslander-Reiten sequence beginning at $(0,0,L)$ has the following form, where P is a projective K-module.*

$$0 \longrightarrow (0,0,L) \xrightarrow{(0,\binom{1}{0})} \left(P, \begin{pmatrix} g \\ h \end{pmatrix}, L \oplus H\right) \xrightarrow{(0,(0,1))} (P,h,H) \longrightarrow 0$$

If $0 \longrightarrow L \xrightarrow{i} E \xrightarrow{p} \tau^{-1}L \longrightarrow 0$ is the Λ Auslander-Reiten sequence beginning at L, g is determined by h through the following diagram, which is commutative and has exact horizontal rows.

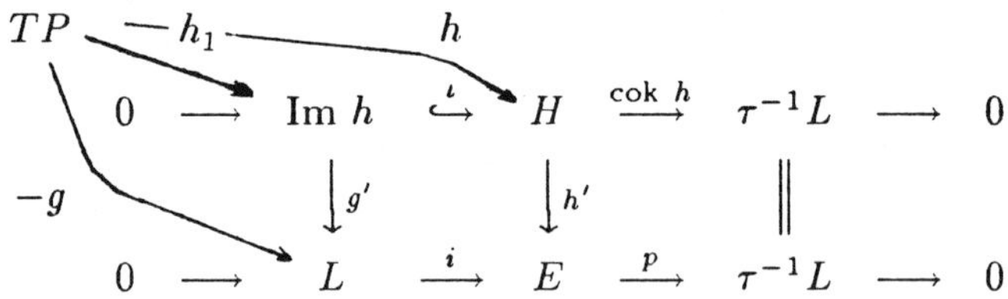

In other words, cok h is an epimorphism from H onto $\tau^{-1}L$ which does not split so that it has a factorization cok $h = ph'$. If g' is the passage of h' to the kernels (notice that Im h = ker(cok h)), g may be taken to be $-g'.h_1$, where h_1 is the same h considered as an epimorphism onto Im h.

Proof. Clearly, the first part of the statement is contained in Prop. 3.1. Assuming that P is not projective, let $X \xrightarrow{x} P$ be its projective cover. Then $(X, h.Tx, H) \xrightarrow{(x,1)} (P,h,H)$ is not split epic, so that it factors through the minimal right almost split map $(1,(0,1))$. This implies that $g.Tx$ factors through $h.Tx$ and, since T is right exact, g factors through h, which leads to a contradiction, as we saw in the proof of Prop. 3.1.

Given a Λ-morphism $L \xrightarrow{\lambda} X$ which is not split monic, since $(0,\lambda)$ is not split monic, it factors through our minimal left almost split map giving a $H \xrightarrow{\nu} X$ such taht $\lambda g + \nu h = 0$. In particular, taking $\lambda = i$, we see that g factors through $\mathrm{cok}(\mathrm{ker} h) = \mathrm{Im} h$ giving $g = -g'h_1$. Let us call, for the moment, E' the push-out of $(-g, h)$. Then we have the following commutative diagram with exact rows.

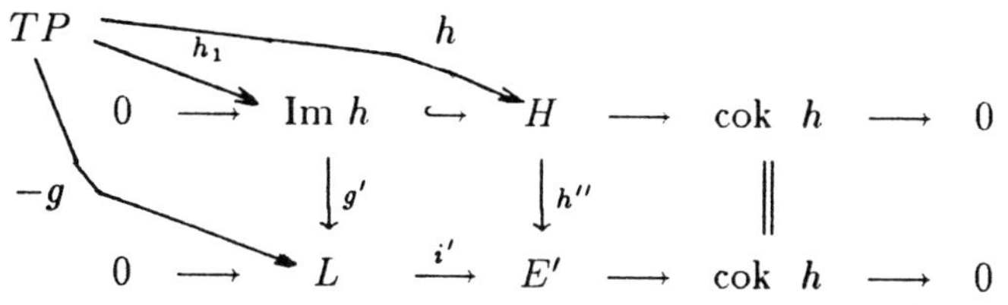

Now, using the above relation for $\lambda = i$: $ig + \nu h = 0$, we get the existence of a morphism $E' \xrightarrow{j} E$ satisfying $i = ji'$ which means that i' is split monic and that E' is of the form $E' = E \oplus X$. It follows also that cok h is of the form $\tau^{-1}L \oplus X$. The proof will be complete if we show that $X = 0$.

In order to do this, we redraw the above diagram putting $E \oplus X$ and $\tau^{-1}L \oplus X$ in place of E' and cok h, respectively, and introducing a new top exact row by means of (H_1, b_1, h_1'), the pull-back of $\left(\binom{b}{x}, \binom{1}{0}\right)$, and a new botton row by means of the Λ Auslander-Reiten sequence beginning at L.

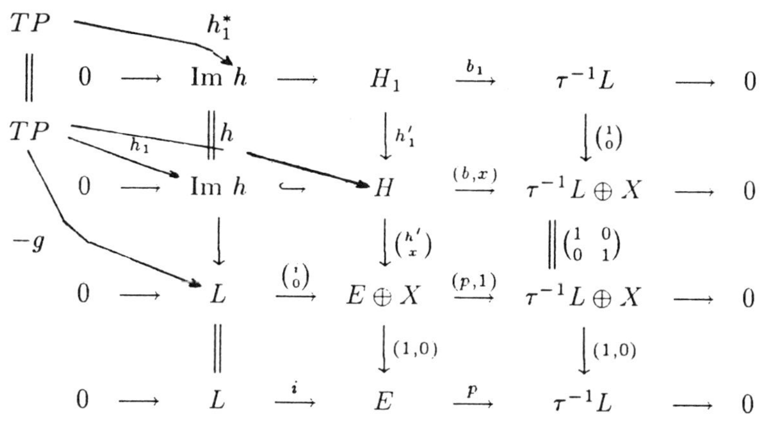

It follows from the diagram that g does not factor through h_1^* (otherwise, the upper row would split). Hence, given a morphism $\binom{z}{t} : TX \longrightarrow L \oplus Y$ such that z does not factor through t, it follows from Prop. 3.1 that there are morphisms a', b', c' such that $z.Ta' = g + b'h = g + (b'h_1')h_1^*$ and $t.Ta' = c'h = (c'h_1')h_1^*$. But then, applying Prop. 3.1 again we deduce that $(P, \binom{g}{h_1^*}, L \oplus H_1)$ is isomorphic to $(P, \binom{g}{h}, L \oplus H)$ and that $H_1 \cong H$ and, using the above diagram again, we see that the upper and botton rows give the diagram of our statement. ■

We end this section with the following propositions that treat the remaining case, when L is injective.

Proposition 3.3. *Let L be an indecomposable, injective Λ-module such that $S(L) = \Lambda(M, L) \neq 0$. In order that $(0,0,L) \xrightarrow{(0,\binom{1}{0})} (P, \binom{g}{h}, L \oplus H)$ be minimal left almost split, it is necessary and sufficient that there exist a K-module P_1, a Λ-module H_0 with simple top isomorphic to soc L and a morphism $H_0 \xrightarrow{k} TP_1$ satisfying the following property.*

Given a K-module X and Λ-modules N, N', such that

$$\operatorname{soc}\left(\frac{TX}{N}\right) \cong \frac{N + N'}{N} \cong \operatorname{soc} L$$

there is a K-morphism $P_1 \xrightarrow{a} X$ such that $\operatorname{Im}(Ta.k)$ is contained in N' but is not contained in N.

If the above mentioned minimal left almost split map is known, then k is isomorphic to the kernel of h. Conversely, if the conditions are satisfied, h may be taken to be the cokernel of k, and $-g$ to be any of the extensions of $H_0 \longrightarrow \operatorname{soc} L \hookrightarrow L$ to TP_1 (these extensions exist because of the factorization $H_0 \longrightarrow \operatorname{Im} k \longrightarrow \operatorname{soc} L$ and because L is assumed to be injective).

Proof. We apply proposition 3.1, keeping the corresponding notations. Let $H_0 \overset{k}{\hookrightarrow} TP$ be the kernel of h. By the exact sequence which defines (P, h, H) (see the remark at the beginning of this section), the kernel of h coincides with the image of $\nu : (DL)^* \longrightarrow TQ_1'^*$ and, since $(DL)^*$ is indecomposable projective, ker h has simple top. Given X, N, N' as in the property of the statement, let x be the composite $TX \longrightarrow \frac{TX}{N} \longrightarrow L$ and let y be the natural map $TX \longrightarrow \frac{TX}{N'}$. By property 3) of Prop. 3.1, there are morphisms a, b, c such that $x.Ta = g + bh$, $y.Ta = ch$. The last equality means that $\operatorname{Im}(Ta.k)$ is contained in N' and the first equality implies that $\operatorname{Im}(Ta.k)$ is not contained in N. Conversely, let us be given a morphism k with the properties of the statement, and let us call $TP_1 \xrightarrow{h_1} H_1$ the cokernel of k. Keeping again the notations of Prop. 3.1, let N be the kernel of g. Since

$L \supset TP/N \supset N + \ker h/\ker h \cong \ker h/N \cap \ker h$ and since $\operatorname{soc} L$ is essential, we can take a submodule N' of $\ker h$ such that $\operatorname{soc} TP/N \cong \operatorname{soc} \frac{N+\ker h}{N} \cong N+N'/N \cong \operatorname{soc} L$. Then, using the assumptions, we easily obtain a morphism $a_1 : P_1 \longrightarrow P$ such that cok $g.Ta_1.k = \phi_1$ is onto $\operatorname{soc} L$. If $-g_1$ is the extension of $H_0 \xrightarrow{\phi_1} \operatorname{soc} L \overset{i}{\hookrightarrow} L$ to TP_1, we have $-g_1 k = i\phi_1$ and we obtain a morphism $(P_1, \binom{g_1}{h_1}, L \oplus H_1) \xrightarrow{\left(a_1,\ \begin{pmatrix} 1 & b_1 \\ 0 & c_1 \end{pmatrix}\right)} (P, \binom{g}{h}, L \oplus H)$. But, applying 3) of Prop. 3.1 to the pair $\binom{g_1}{h_1}$, we also obtain a morphism $\left(a, \begin{pmatrix} 1 & b \\ 0 & c \end{pmatrix}\right)$ in the reverse direction. It follows that $\left(P_1, \binom{g_1}{h_1}, L \oplus H_1\right)$ is a direct summand of $\left(P, \binom{g}{h}, L \oplus H\right)$ and, since (P, h, H) is indecomposable, the two $\widetilde{\Lambda}$-modules are isomorphic. The proof is complete. ■

Proposition 3.4. *Let Λ be a finitely dimensional algebra over an algebraically closed field and let L be an indecomposable injective Λ-module such that $S(L) = \Lambda(M, L) \neq 0$. If the inverse translate $\tau^{-1}(0,0,L) = (P, h, H)$ is already known, then the morphism $TP \xrightarrow{g} L$ needed to complete the construction of the $\widetilde{\Lambda}$ Auslander-Reiten sequence beginning at $(0,0,L)$ is determined by h through the following commutative diagram with exact rows.*

$$\begin{array}{ccccccccc} 0 & \longrightarrow & H_0 & \xrightarrow{k} & TP & \xrightarrow{h} & H & \longrightarrow & 0 \\ & & \downarrow{\scriptstyle \phi} & & \downarrow{\scriptstyle -g} & & \downarrow{\scriptstyle h'} & & \\ 0 & \longrightarrow & \operatorname{soc} L & \overset{i}{\hookrightarrow} & L & \xrightarrow{\pi} & \frac{L}{\operatorname{soc} L} & \longrightarrow & 0 \end{array}$$

Proof. Let π be the minimal left almost split map starting at L. In the same way as in the proof of Prop. 3.2, we deduce that there is a morphism h' such that $\pi g = -h'h$, implying the $g(\ker h) \doteq \ker \pi = \operatorname{soc} L$. As we saw in the proof of Prop. 3.3, $H_0 = \ker h$ has simple top. Hence, there is an epimorphism $H_0 \xrightarrow{\phi} \operatorname{soc} L \hookrightarrow L$ and its extension to TP gives $-g$. ■

4. The cokernel functor

The results in section 3 lead to an interesting fact concerning the obvious cokernel functor of a triangular matrix algebra. Keeping our general notations, the cokernel functor associated to $\widetilde{\Lambda}$ is the functor that to each $\widetilde{\Lambda}$-module (U, f, n) associates the Λ-module cok N and, to each morphism (α, g) of

$\widetilde{\Lambda}$–modules, its passage to the cokernels.

$$\begin{array}{ccccc} TU & \xrightarrow{f} & N & \xrightarrow{\text{cok } f} & \overline{N} \\ \downarrow{\scriptstyle T\alpha} & & \downarrow{\scriptstyle g} & & \downarrow{\scriptstyle \overline{g}} \\ TU' & \xrightarrow{f'} & N' & \xrightarrow{\text{cok } f'} & \overline{N}' \end{array}$$

We will denote this right exact functor by cok_M or, simply, by cok. Referring to the above diagram, we have: $\text{cok}_M(U, f, N) = \overline{N}$ and $\text{cok}_M(\alpha, g) = \overline{g}$. It should be observed that cok_M ahs a left inverse: the embedding mod $\Lambda \longrightarrow$ mod $\widetilde{\Lambda}$ which sends each Λ–module N to the $\widetilde{\Lambda}$–module $(0, 0, N)$.

The following proposition summarizes some of the information contained in the results of section 3.

Proposition 4.1. *The following case describe the behavior of* cok_M *on Auslander-Reiten sequences beginning at a* Λ*–module.* Notations are the same as in section 3.

1) *Let* L *be an indecomposable, non-injective* Λ*–module and let* i, p *be morphisms defining the Auslander-Reiten sequence beginning at* L.
If $\Lambda(M, L) = 0$:

$$\begin{array}{ccccccccccc} 0 & \longrightarrow & (0,0,L) & \longrightarrow & (0,0,E) & \longrightarrow & (0,0,\tau^{-1}L) & \longrightarrow & 0 & & \binom{\widetilde{\Lambda}}{\text{ARS}} \\ \downarrow{\scriptstyle \text{cok}_M} & & \downarrow & & \downarrow & & \downarrow & & & & \\ 0 & \longrightarrow & L & \xrightarrow{i} & E & \xrightarrow{p} & \tau^{-1}L & \longrightarrow & 0 & & \binom{\Lambda}{\text{ARS}} \end{array}$$

If $\Lambda(M, L) \neq 0$:

$$\begin{array}{ccccccccccc} 0 & \longrightarrow & (0,0,L) & \longrightarrow & (P, \binom{g}{h}, L \oplus H) & \longrightarrow & (P, h, H) & \longrightarrow & 0 & & \binom{\widetilde{\Lambda}}{\text{ARS}} \\ \downarrow{\scriptstyle \text{cok}_M} & & \downarrow & & \downarrow & & \downarrow & & & & \\ 0 & \longrightarrow & L & \xrightarrow{i} & E & \xrightarrow{p} & \tau^{-1}L & \longrightarrow & 0 & & \binom{\Lambda}{\text{ARS}} \end{array}$$

2) *Let* L *be an indecomposable injective* Λ*–module. If* $\Lambda(M, L) = 0$:

$$\begin{array}{ccccc} & (0,0,L) & \longrightarrow & (0,0,\frac{L}{\text{soc}L}) & \binom{\widetilde{\Lambda} \text{ min}}{\ell.\text{a.s.}} \\ \downarrow{\scriptstyle \text{cok}_M} & \downarrow & & \downarrow & \\ & L & \longrightarrow & \frac{L}{\text{soc}L} & \binom{\Lambda \text{ min}}{\ell\text{a.s.}} \end{array}$$

If $\Lambda(M,L) \neq 0$:

$$\begin{array}{ccccccccccc}
 & 0 & \longrightarrow & (0,0,L) & \longrightarrow & (P,\binom{g}{h},L\oplus H) & \longrightarrow & (P,h,H) & \longrightarrow & 0 & \binom{\widetilde{\Lambda}}{\text{ARS}} \\
\big\downarrow \text{cok}_M & & & \big\downarrow & & \big\downarrow & & \big\downarrow & & & \\
 & & & L & \longrightarrow & \frac{L}{\text{soc}L} & \longrightarrow & 0 & & & \binom{\Lambda\text{min}}{\ell\text{a.s.}}
\end{array}$$

Proof. The first parts of 1) and 2) follow from Prop. 2.1. The second part of 1) is essentially contained in Prop. 3.2. The second part of 2) is essentially contained in Prop. 3.4.

Remarks.

1) $\tau^{-1}(0,0,L)$ will be also in $\operatorname{mod}\Lambda$ if and only if $\Lambda(M,L)=0$.
2) h will be an epimorphism if and only if L is injective.
3) Prop. 4.1 gives an important tool for the computation of $\underline{\underline{P}}_1(\widetilde{\Lambda})$, i.e. of the indecomposable preprojectives of level 1. In fact, it is well known that those are the indecomposables of the form $\tau^{-1}\widetilde{L}$, where $\widetilde{L}$ is an indecomposable component of the radical of an indecomposable projective. Hence, the result applies especially when the given projective comes from $\operatorname{mod}\Lambda$. In other words, the projectives coming from $\operatorname{mod}\Lambda$ are of the form $(0,0,Q)$ (Q an indecomposable projective Λ–module) and $\operatorname{rad}(0,0,Q)$ is $(0,0,\operatorname{rad}Q)$. Hence our results apply to preprojectives of level 1 of the form $(P,h,H)=\tau^{-1}(0,0,L)$, where L is an indecomposable direct summand of $\operatorname{rad}Q$. In this case it is clear that L is not injective.

In the particular case of one-point extensions, $K=k$ is a field and $\operatorname{mod}\widetilde{\Lambda}$ has only one indecomposable projective not coming from $\operatorname{mod}\Lambda$. It is given by $(k,1,M)$ and it is linked to preprojectives of level 1 of the form $(P,h,L)=\tau^{-1}(0,0,L)$, where L is an indecomposable direct summand of M. In this case, L may be injective.

We see that, in the case of one-point extensions, there are three kinds of preprojectives of level 1. The first kind are of the form (P,h,H), where $\operatorname{cok} h$ is in $\underline{\underline{P}}_1(\Lambda)$ and their Auslander-Reiten translate is of the form $(0,0,L)$, $L \mid \operatorname{rad}Q$, Q Λ–projective. The second class (if not empty) is formed by inverse translates of non-injective components L of M. They are of the form (P,h,H) and $\operatorname{cok} h=\tau^{-1}L$. The third class (if not empty) is formed by modules of the form $\tau^{-1}(0,0,L)=(P,h,H)$, with $\operatorname{cok} h=0$, and L injective.

5. An application

In [HM$_2$] and [HM$_3$] we have studied conditions for an algebra to be equivalent, modulo preprojectives up to the level n, to a hereditary algebra when a so called $*$–condition is satisfied. Our results require also the assumption that, for the algebra under consideration, there are no injective indecomposables which are preprojective of level less than or equal to n. Our results of sections 3 and 4 enable us to extend our theorems to the case when this hypothesis is not satisfied and even to the case when the corresponding hereditary algebras may be of finite representation type (cf. [HM$_3$]). We can do this generalization for the case $n = 1$ by showing that the problem may be reduced to an algebra which is a quotient of a hereditary algebra and where there is an injective–projective which is uniserial, all its submodules being projectives. This study will be the object of a separate publication.

Presently we are going to present the computations of some pre-projective modules of level 1 for the particular case which will be needed for the above mentioned study.

In what follows we consider a field k and a finite dimensional k–algebra $\tilde{\Lambda}$ of the form

$$\tilde{\Lambda} = \begin{bmatrix} k & S \\ 0 & \Lambda \end{bmatrix}$$

We assume that S is the socle of an injective–projective Λ–module which is uniserial and all of whose submodules are projective. We will use the notation I_0 for this module and P_j for its jth radical, i.e., $P_j = \operatorname{rad}^j I_0$. If r is its Loewy length, $P_r = S$. On the other hand, we will denote by Q_j the quotient P_j/S and by i_j the inclusion of S into P_j.

An easy computation shows that the following diagram (with obvious maps) represents the Λ Auslander-Reiten sequences beginning at the P_j's.

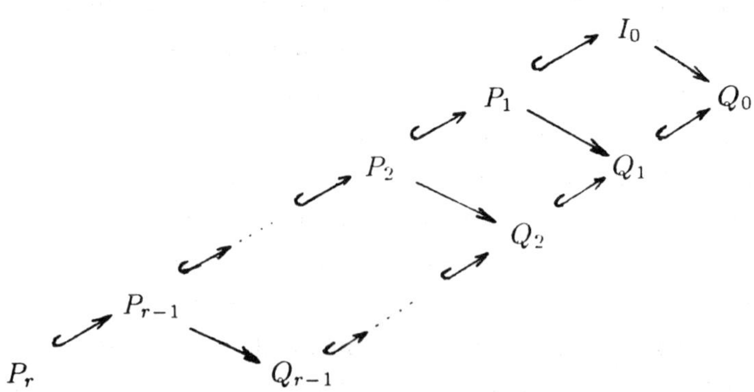

In fact, if $X \xrightarrow{f} Q_j$ is not an isomorphism (X indecomposable) but f is epic, we get a monomorphism from the projective P_j into X, which implies that X is contained in I_0, i.e., X is some $P_{j'}$ with $j' < j$. But this forces X to be equal to P_j. If f is not epic, it factors through Q_{j+1}.

Now we use the above information to compute the $\widetilde{\Lambda}$ Auslander-Reiten sequences beginning at the modules $(0, 0, P_j)$ $(j = 1, \ldots, r)$. As we know, the middle term will be of the form $(P, \binom{g}{h}, P_j \oplus H)$, where P is some k-vector space and where cok $\binom{g}{h} = P_{j-1} \oplus Q_j$ and cok $h = Q_{j-1}$. A good guess for (P, h, H) seems to be (k, i_{j-1}, P_{j-1}), so that we pose the following claim.

Claim. The $\widetilde{\Lambda}$ Auslander-Reiten sequence beginning at $(0, 0, P_j)$ is the following.

$$0 \longrightarrow (0, 0, P_j) \longrightarrow \left(k, \begin{pmatrix} i_j \\ i_{j-1} \end{pmatrix}, P_j \oplus P_{j-1}\right) \longrightarrow (k, i_{j-1}, P_{j-1}) \longrightarrow 0$$

To prove the claim, we show the four properties of Prop. 3.1. To begin, we see that 1) and 4) are obviously true. Let $P_j \xrightarrow{\lambda} Y$ be a morphism with Y indecomposable that is not split epic. If λ is not monic, $\lambda i_j = 0$. If λ is monic it is easy to deduce that Y must be of the form $Y = P_{j'}$, with j' less than j. Then, 2) follows readily. Finally, let us be given morphisms $TX \xrightarrow{x} P_j$, $TX \xrightarrow{y} Y$ such that x does not factor through y. In order to show property 3), we can assume that $X = k$ and that Y is indecomposable. Hence, x may be identified to i_j and we may assume that $y \neq 0$ (otherwise we can take $a = 1$, $b = 0$, $c = 0$). We deduce as before that Y has to be of the form $Y = P_{j'}$, with j' less than j. Then we can take $a = 1$, $b = 0$ and c equal to the inclusion of P_j in $P_{j'}$. The proof is complete. ■

References

[FGR] R.M. Fossum, I.A. Griffith & I. Reiten. *Trivial extensions of abelian categories*. Springer L.N.M. **456**, 1975.

[MA] Marmaridis, N. *Comma categories in representation theory*, Comm. Algebra **11**(1983), 1919-1943.

[McL] *Categories for the working mathematician*. New York, Springer, 1971.

[HM$_1$] H. Merklen. *Auslander-Reiten sequences for triangular matrix algebras*, preprint, 1990.

[HM$_2$] H. Merklen. *Artin algebras which are equivalent to a hereditary algebra module preprojectives*. Springer L.N.M. **1177** (1986), 232-255.

[HM_3] H. Merklen. *Hereditary relatively injective subquivers and equivalence modulo preprojectives*, Comm. Algebra **18**(1990), 3145-3181.

Instituto de Matemática e Estatística
Universidade de São Paulo
Caixa Postal 20570 (Agência Iguatemi)
CEP 01498 - São Paulo-SP - Brasil

Canadian Mathematical Society
Conference Proceedings
Volume 11 (1991)

LATTICES $L(A_n^\nu)$

V.A. PONOMAREV

ABSTRACT. Γ^ν is an equipped graph, if the numbers **0** or **1** are assigned to the ends of each edge. So, we have in Γ^ν four types of edges: $\circ\overset{0\quad 0}{\text{———}}\circ$ $\circ\overset{0\quad 1}{\text{———}}\circ$ $\circ\overset{1\quad 0}{\text{———}}\circ$ $\circ\overset{1\quad 1}{\text{———}}\circ$. We associate with Γ^ν a modular lattice $L(\Gamma^\nu)$, which is defined by generators $\{v_i, w_b\}$, $i \in \Gamma_0$, $b \in \Gamma_1$ and some relations which depends on the equipment ν. We prove that there is one to one correspondence between indecomposable representations of $L(A_n^\nu)$ and positive roots of a Dynkin diagram A_n. We associate to positive rooot $\alpha = \sum_{i \le j \le k} \alpha_j$ of A_n the indecomposable representation τ_α of $L(A_n^\nu)$ /where α_j are the simple roots/. An element h of modular lattice L is perfect one if for every indecomposable representation τ of L we have $\tau(h)=0$ or $\rho(h) = V_\tau$ where $V_\tau \cong K^m$ is the space of representation τ and K is a field). We prove that a lattice D_n^ν of perfect elements of $L(A_n^\nu)$ is distributive one and D_n^ν is isomorphic to the lattice of ideal of of poset $P(A_n^\nu)$ of positive roots. The partial order $\le^\nu$ on $P(A_n^\nu)$ is such that $(\alpha=(i,k) \le^\nu \alpha'=(i',k')) \Leftrightarrow (i \le^\pi i'$ and $k \le^\lambda k')$ where π and λ two order on the set of vertices A_{n0} /so A_{n0}^π and A_{n0}^λ are a chains, and they are defined on equipment ν/.

0. BASIC DEFINITIONS AND RESULTS.

01. EQUIPPED GRAPHS. Let $\Gamma = (\Gamma_0, \Gamma_1, \vartheta)$ be a graph without orientation, where Γ_0 is its set of vertices, Γ_1 is its set of edges, and ϑ is a function that assigns to each edge b a set $\vartheta(b) = \{i,j\}$ of vertices $i,j \in \Gamma_0$. An *equipped* **[1]** graph Γ^ν is a graph $\Gamma = (\Gamma_0, \Gamma_1, \vartheta)$ without

1991 *Mathematics Subject Classification*. Primary 06C05, 06B15; Secondary 05C75.

loops and without orientation, having a collection $\{\nu_b\}$ of functions $\nu_b: \vartheta(b) \longrightarrow \{0,1\}$, where $0,1 \in \mathbb{N}$. We can say in other words, that Γ is an equipped graph, if the numbers **0** or **1** are assigned to the ends of each edge. Hence in Γ^ν we have four types of " equipped " edges: $\overset{0\quad 0}{\circ\!\!-\!\!-\!\!-\!\!\circ}$ $\overset{0\quad 1}{\circ\!\!-\!\!-\!\!-\!\!\circ}$ $\overset{1\quad 0}{\circ\!\!-\!\!-\!\!-\!\!\circ}$ $\overset{1\quad 1}{\circ\!\!-\!\!-\!\!-\!\!\circ}$. In some cases, it is convenient to present these edges as $\circ\!\!\leftarrow\!\!-\!\!\rightarrow\!\!\circ$ $\circ\!\!\leftarrow\!\!-\!\!\leftarrow\!\!\circ$ $\circ\!\!\rightarrow\!\!-\!\!\rightarrow\!\!\circ$ $\circ\!\!\rightarrow\!\!-\!\!\leftarrow\!\!\circ$. Each oriented graph Γ_Λ without loops can be related to an equipped graph Γ^ν in the following way: the edge $i \circ\!\!-\!\!\leftarrow\!\!-\!\!\circ j$ in Γ_Λ transforms into the edge $i\ \overset{0\ 1}{\circ\!\!-\!\!-\!\!\circ} j$ in Γ^ν.

0.2. LATTICES $L(\Gamma^\nu)$. We associate with Γ^ν a modular lattice $L(\Gamma^\nu)$, which is defined by the generators $\{v_i, w_b\}$, $i \in \Gamma_0$, $b \in \Gamma_1$ and the relations[(1)]:

L_1: $v_i(\sum_{j \neq i} v_j) = 0$ for each vertex $i \in \Gamma_0$;

L_2: if $\vartheta(b) = \{i,j\}$ then $w_b \subseteq (v_i + v_j)$ for each edge $b \in \Gamma_1$;

L_{3a}: if $\nu_b(i) = 0$, then $w_b v_i = 0$;

L_{3b}: if $\nu_b(i) = 1$, then $w_b + v_j = v_i + v_j$.

According to the definitions, a lattice $L(A_n^\nu)$ is a lattice with generators $\{v_i, w_j\}$ $i \in \{1,2,\dots,n\}$; $j \in \{1,\dots,(n-1)\}$ and with the relations: 1° $v_i(\sum_{j \neq i} v_j) = 0$ for all i; 2° $w_j \subseteq v_j + v_{j+1}$, 3° $w_j v_s = 0$, if $\nu_j(s) = 0$, where $s \in \{j,(j+1)\}$, 4° $w_j + v_t = v_s + v_t$, where $\{s,t\} = \{j,(j+1)\}$, $\nu_j(s) = 1$.

Example 1°. A_4^ν: $\underset{1}{\circ}\overset{1\quad 1}{\text{———}}\underset{2}{\circ}\overset{1\quad 0}{\text{———}}\underset{3}{\circ}\overset{0\quad 0}{\text{———}}\underset{4}{\circ}$. The lattice $L(A_4)$ has generators $\{v_1,v_2,v_3,v_4,w_1,w_2,w_3\}$ and the relations 1° $v_i(\sum_{j \neq i} v_j)$ for all $i \in \{1,2,3,4\}$, 2° $w_j \subseteq (v_j + v_{j+1})$ for all $j \in \{1,2,3\}$; 3°a) $w_1 + v_2 = w_1 + v_1 = v_1 + v_2$, b) $w_2 + v_3 = v_2 + v_3$, $w_2 v_3 = 0$, c) $w_3 v_3 = w_3 v_4 = 0$. Let us notice on 3b! The main aim of this paper is the investigation of lattices $L(A_n^\nu)$, using the methods of the representation theory.

0.3. Let L be a modular lattice which maximal and minimal elements v and 0. Let V be a finite-dimentional vector space over a commutative field K and let $\mathcal{L}(V)$ be the lattice of all vector subspaces of V. A *representations* ρ of L in V is a lattice morphism $\rho: L \longrightarrow \mathcal{L}(V)$ such that $\rho(0)=0$ and $\rho(v)=V$. Thus ρ associates with any $x, y \in L$ subspaces

---

(1) The operations of meet and join in the modular lattice L are denoted by $\cdot$ and $+$, so that $x \cap y = xy$, $x \cup y = x + y$ and $\bigvee_i x_i = \sum_i x_i$.

$\rho(x)$ and $\rho(y)$ of V in such a way that $\rho(xy) = \rho(x)\rho(y)$ and $\rho(x+y) = \rho(x) + \rho(y)$.

If ρ_1 and ρ_2 are representations of the same lattice L in spaces V_1 and V_2 (over the same field K) then a *morphism of representations* $\psi: \rho_1 \longrightarrow \rho_2$ is a linear map $\psi: V_1 \longrightarrow V_2$ such that $\psi\rho_1(x) \subseteq \rho_2(x)$ for every $x \in L$. It is easy to check that we have thus defined the *category of representation* $\mathcal{R}(L, K)$ of L in finite-dimentional vector spaces over K . In this category the notions of direct sum of representations, and of decomposable and indecomposable representations are defined in the usual way.

0.4. THEOREM. *Let Γ_Λ be an oriented graph without loops and Γ^ν be the corresponding equipped graph. Then the category $\mathcal{L}_K(\Gamma_\Lambda)$ of representation of Γ_Λ is equivalent to the category $\mathcal{R}(L(\Gamma^\nu), K)$ of representation of the modular lattice $L(\Gamma^\nu)$.*

Proof. A representation (V,f) of Γ_Λ (see [21]) consists of a family $\{V_i, f_l\}$ of finite dimentional vector spaces V_i over the field K labelled with the vertices i of Γ_Λ $(i \in \Gamma_0)$ and of linear maps $f_l : V_{s(l)} \longrightarrow V_{t(l)}$ /here $s(l), t(l) \in \Gamma_0$ are the starting and terminal points of the edge $l \in \Gamma_1$/. Let us construct a representation ρ of $L(\Gamma^\nu)$ attached to (V,f). Notice that the edge $s(l) \longrightarrow t(l)$ of Γ_Λ corresponds to the edge $s(l) \overset{1 \quad\; 0}{\text{——}} t(l)$ of Γ^ν. The lattice $L(\Gamma^\nu)$ has generators $\{v_i, w_l\}$ $i \in \Gamma_0$, $l \in \Gamma_1$ and the following relations: 1) $v_i (\sum_{j \neq i} v_j) = 0$ for all $i \in \Gamma_0$ 2) $w_l \subseteq v_{s(l)} + v_{t(l)}$ for each $l \in \Gamma_1$; 3) $w_l v_{t(l)} = 0$; 4) $w_l + v_{t(l)} = v_{s(l)} + v_{t(l)}$. Let us define ρ as the representation in the space $V = \bigoplus_{i \in \Gamma_0} V_i$, such that $\rho(v_i) = \{(0, \ldots, 0, x_i, 0, \ldots, 0) \mid x_i \in V_i\}$ and $\rho(w_l) = \{(0, \ldots, 0, x_{s(l)}, 0, \ldots, 0, f_l(x_{s(l)}), 0, \ldots, 0) \mid x_{s(l)} \in V_{s(l)}, f_l(x_{s(l)}) \in V_{t(l)}\}$.

It is easy to check that $\rho(w_l)\rho(v_{t(l)}) = 0$ and $\rho(w_l) + \rho(v_{t(l)}) = \rho(v_{s(l)}) + \rho(v_{t(l)})$. This means that we have defined a representation ρ of the modular lattice $L(\Gamma^\nu)$. This construction admits inversion, i.e. starting on $\rho \in \mathcal{R}(L(\Gamma^\nu), K)$ we can construct the object $(V,f) \in \mathcal{L}_K(\Gamma_\Lambda)$. It is also easy to verify that there is equivalence between the categories $\mathcal{R}(L(\Gamma^\nu), K)$ and $\mathcal{L}_K(\Gamma_\Lambda)$. ∎

SURVEY OF MAIN RESULTS.

0.5. THEOREM. *There is one-to-one correspondence between the indecomposable representations of a lattice $L(A_n^{\nu})$ and the positive roots of the Dynkin diagram A_n.*

This theorem generalizes the results obtained by Gabriel [2]. It will be proved at the end of the paper. Each positive root α of A_n can be written [3] in the form $\alpha = \sum_{i \le j \le k} \alpha_j$, where $1 \le i \le k \le n$ and α_j is a simple root. We shall denote such a root as $\alpha = \alpha_{i,k}$ and as $\tau_\alpha = \tau_{i,k}$ the indecomposable representation corresponding to it. The representation $\tau_{i,k}$ is a representation in a vector space $T_{i,k} \cong K^{(k-i+1)}$ over a field K. $T_{i,k}$ has a basis $\{e_j\}$ $/i \le j \le k/$ such that

$$\tau_{i,k}(v_j) = \begin{cases} 0, & \text{if } j<i \text{ or } k<j, \\ Ke_j, & \text{if } i \le j \le k. \end{cases} \qquad \tau_{i,k}(w_j) = \begin{cases} 0, & \text{if } j<(i-1), \\ K(e_j + e_{j+1}), & \text{if } i \le j < k, \\ 0, & \text{if } k<j. \end{cases}$$

$$\tau_{i,}(w_{i-1}) = \begin{cases} 0, & \text{if } \nu_{i-1}(i) = 0. \\ Ke_i, & \text{if } \nu_{i-1}(i) = 1 \end{cases} \qquad \tau_{i,}(w_k) = \begin{cases} 0, & \text{if } \nu_k(k) = 0. \\ Ke_k, & \text{if } \nu_k(k) = 1 \end{cases}$$

Let us introduce (as in [4]) an order $\le^{\nu}$ on a set $\mathcal{D}(A_n^{\nu})$ of the isomorphism classes of the indecomposable representation of $L(A_n^{\nu})$ in the following way. We write $\tau_\beta \le^{\nu} \tau_\alpha$ if there is a sequence $\tau_1, \tau_2, \dots, \tau_m \in \mathcal{D}(A_n)$ such that $\tau_1 \cong \tau_\alpha, \tau_m \cong \tau_\beta$ and $\mathrm{Hom}(\tau_j, \tau_{j+1}) \ne 0$ for all $1 \le j < m$. Then $\mathcal{D}(A_n^{\nu})$ becomes a poset. Since $\mathcal{D}(A_n^{\nu})$ is isomorphic to the set $P(A_n^{\nu})$ of all positive roots of A_n we get a partial order on $P(A_n^{\nu})$.

This order can be described independently in the following way. Let us introduce two partial orders $\le^{\pi}$ and $\le^{\lambda}$ on the set of vertices $A_{n,0} \cong \{1,2,\dots,n\}$. The relation $\le^{\pi}$ is defined by the word P_n from a sequence $P_1, P_2, \dots, P_n$. Let $P_1 = 1$ and further by induction $P_i = P_{i-1}i$, if $\nu_{i-1}(i) = 0$ or $P_i = iP_{i-1}$, if $\nu_{i-1}(i) = 1$. If we obtain $P_n = i_1 i_2 \dots i_n$, then we define $i_1 \le^{\pi} i_2 \le^{\pi} \dots \le^{\pi} i_n$. Analogously, $\le^{\lambda}$ is defined by the word $L_n \in \{L_j\}$. Let $L_1 = n$, and further by induction $L_{n+1-i} = L_{n-i}i$, if $\nu_i(i) = 0$ or $L_{n+1-i} = iL_{n-i}$, if $\nu_i(i) = 1$. If $L_n = j_1 j_2 \dots j_n$, then $j_1 \le^{\lambda} j_2 \le^{\lambda} \dots \le^{\lambda} j_n$. We shall denote by $\le$ the trivial order on the set $A_{n,0}$: $1 \le 2 \le \dots \le n$.

0.6.THEOREM. *The set $\mathcal{P}(A_n)$ is ordered in the following way. If $\tau_\alpha = \tau_{i,k}$ and $\tau_{\alpha'} = \tau_{i',k'}$, then $(\tau_\alpha \leq^\nu \tau_{\alpha'}) \Leftrightarrow (i \leq^\pi i'$ and $k \leq^\lambda k')$.*

We call a subset I of a partially ordered set (poset) P an *ideal* if from $x \leq y$ and $y \in I$ follows $x \in I$. An element h a modular lattice L is *perfect* if for each indecomposable representation τ of L we have $\tau(h)=0$ or $\tau(h)=V_\tau$ (where V_τ is the space of the representation τ). The set of all indecomposable representations $\tau \in \mathcal{P}$ such that $\tau(h) = V_\tau$ is called *the characteristic set* of perfect element h. It is easy to prove that the characteristic set of each perfect element h of $L(A_n^\nu)$ is an ideal of the poset $\mathcal{P}(A_n^\nu)$. It is more difficult to prove that each ideal I of $\mathcal{P}(A_n)$ is characteristic one, i.e. there exists a perfect element $h \in L(A_n^\nu)$ such that $I = \{\tau \in \mathcal{P}(A_n^\nu) \mid \tau(h) = V_\tau\}$.

Let us define elements $v_{i_j k} = v_{i_j} v_{j k} \in L(A_n)$ /$i,j,k \in A_{n0}$, $i \leq j \leq k$/ which play the leading role in the further investigation and in the construction of perfect elements.

$$v_{i_i} = \begin{cases} v_i, & \text{if } i=1 \text{ or } \nu_{i-1}(i)=0, \\ v_i w_{i-1}, & \text{if } \nu_{i-1}(i)=1, \end{cases} \qquad v_{kk} = \begin{cases} v_k, & \text{if } k=n \text{ or } \nu_k(k)=0, \\ v_k w_k, & \text{if } \nu_k(k)=1, \end{cases}$$

if $i<j$ then $v_{i_j} = (v_{i_i} + w_i + w_{i+1} + \dots + w_{j-1}) v_j$, and if $j<k$ then $v_{jk} = v_j(w_j + w_{j+1} + \dots + w_{k-1} + v_{kk})$.

0.7.THEOREM. *A lattice $D(A_n^\nu)$ of perfect elements $h \in L(A_n^\nu)$ is distributive one, it is isomorphic to the lattice $\mathcal{J}(\mathcal{P})$ of ideal of the poset $\mathcal{P} = \mathcal{P}(A_n^\nu)$. An arbitrary perfect element $h \in D(A_n^\nu)$ can be written in the form:*

$$h = \sum_{(i,k) \in H} \sum_{\substack{j \\ i \leq j \leq k}} v_{i_j k},$$

where H - ideal of poset $P = \mathcal{P}(A_n^\nu) \cong P(A_n^\nu)$.

We give now some remarks on the further sections. The sectoin 1 contains the description of lattices $L(A_2^\nu)$ and their representations. We discuss some auxiliary propositions about projective intervals in §2.Main property of representations $\tau_{i,k}$ are disscuss in §3. Section 4 includes the basic properties of elemens $v_{i_j k}$ and orders $\leq^\lambda$ and $\leq^\pi$. The main theorem that a representation ρ is decomposed into direct sum of representations $\tau_{i,k}$ will be proved in §5.The perfect elements a_i, b_k

$\in L(A_n)$ are described in §6. The §7 is devoted to remaining problems.

§1. Lattices $L(A_2^{\nu})$ and their representations.

1.1. There are 4 types of different equipment of the graph A_2: $\alpha)\ \overset{0\quad 0}{\circ\!-\!\!-\!\circ}$, $\beta)\ \overset{0\quad 1}{\circ\!-\!\!-\!\circ}$, $\gamma)\ \overset{1\quad 0}{\circ\!-\!\!-\!\circ}$, $\delta)\ \overset{1\quad 1}{\circ\!-\!\!-\!\circ}$. We shall denote corresponding lattices as $L(A_2^{\nu})$, or $L(A_2^{00})$, $L(A_2^{01})$, $L(A_2^{10})$, $L(A_2^{11})$. Each $L(A_2^{\nu})$ has 3 generators v_1, v_2, w the minimum element 0 and two type of relations: 1a) $v_1v_2 = 0$, 1b) $w \subseteq (v_1+v_2)$. /relations 1a, 1b are fulfilled for any equipment/, and relations, which depend on the equipment ν: α) if $\nu=(0,0)$, then $wv_1 = wv_2 = 0$; β) if $\nu=(0,1)$, then $wv_1 = 0$, $w+v_1 = v_1+v_2$; γ) $\nu=(1,0) \Rightarrow wv_2=0$ $w+v_2=v_1+v_2$; δ) $\nu=(1,1) \Rightarrow$ $w+v_1 = w+v_2 = v_1+v_2$.

It is easy to see that the lattices $L(A_2^{01})$ and $L(A_2^{10})$ are isomorphic, and this isomorphism is given by the change $v_1 \longrightarrow v_2$ $v_2 \longrightarrow v_1$. It is known [5] that the free modular lattice with three generators is finite and has 28 different elements. Using this, it is easy to find the structure of the latticies $L(A_2^{\nu})$, besides, it's not difficult to prove this directly. We produce diagrams of these latticies:

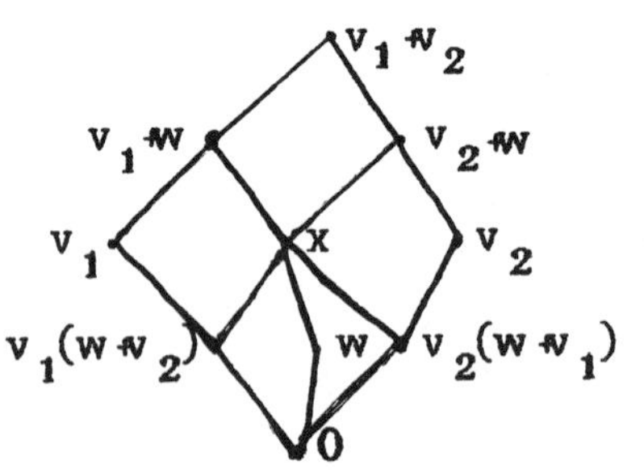

Fig.1.1: $L(A_2^{00})$
where $x=(v_1+w)(v_2+w)=$
$= w+v_1(w+v_2) = w+v_2(w+v_1)$
$= v_1(w+v_2)+v_2(w+v_1)$

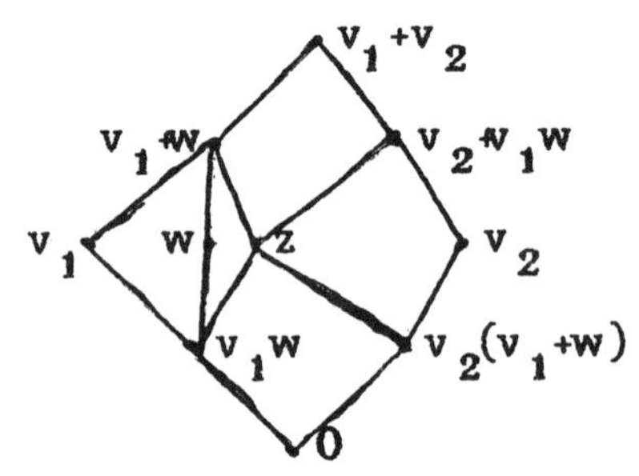

Fig.1.2.: $L(A_2^{10})$
where $z = v_1w + v_2(v_1+w) =$
$=(v_1+w)(v_2+v_1w)$; $v_1+w =$
$= v_1+z = w+z$; $v_1w = v_1z = wz$.

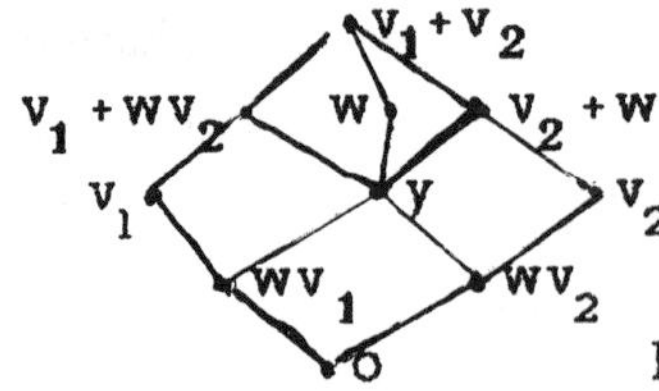

Fig.1.3.: $L(A_2^{11})$

We have the following equality in Fig.1.3: $y = v_1w + v_2w =$
$w(v_1+wv_2) = w(v_1+wv_2) =$
$= (v_1+wv_2)(v_2+wv_1)$.

It follows from these diagrams that the lattices $L(A_2^{00})$ and $L(A_2^{11})$ are dual.

1.2. REPRESENTATIONS OF $L(A_2^\nu)$. If ρ is a representation of $L(A_2^\nu)$ in a vector space V over a field K, we shall denote $\rho(v_i)=V_i$ /i=1,2/ and $\rho(w)=W$. If $A_2^\nu=A_2^{10}$, we have $WV_2 = V_1V_2=0$ and $W+V_2= =V_1+V_2$. This means that each vector $\xi_1 \in V_1$ can be uniquely represented in the form: $\xi_1 = \eta-\xi_2$, where $\eta \in W$ and $\xi_2 \in V_2$. It is easy to check that the correspondence $\xi_1 \longmapsto \xi_2$ is a linear one /i.e. if we have another pair $\xi_1' \longmapsto \xi_2'$, then $(\lambda\xi_1+\mu\xi_1') \longmapsto (\lambda\xi_2+\mu\xi_2')$ for each $\lambda,\mu \in K$/. If we denote $\xi_2 = A\xi_1$ for this correspondence, then the subspace W is isomorphic to the graph of the linear map A / gr.A = $\{(\xi_1, A\xi_1) \mid \xi_1 \in V_1\} \subseteq V_1 \oplus V_2$/. Notice that Ker A = V_1W.

If $A_2^\nu = A_2^{00}$, it is easy to see from Fig.1.1 that three elements $w, v_1(w+v_2)$, and $v_2(w+v_1)$ form an *elementary modular triplet*, i.e. the following equalities are fulfilled: $w(v_1(w+v_2)) = w(v_2(w+v_1)) = (v_1(w+v_2))(v_2(w+v_1)) = 0$ and $w+v_1(w+v_2) = w+v_2(w+v_1) = = v_1(w+v_2)+v_2(w+v_1) = (w+v_1)(w+v_2)$. If ρ is a representation of $L(A_2^{00})$, the same equalities are fulfilled tor the three subspaces $W, V_1(W+V_2)$, and $V_2(W+V_1)$. Hence, an arbitrary vector $\xi_1 \in V_1(W+V_2)$ can be written uniquely in the form $\xi_1 = \eta-\xi_2$, where $\eta \in W$ and $\xi_2 \in V_2(W+V_1)$. The correspondence $\xi_1 \longrightarrow \xi_2$ is the isomorphism I from the $V_1(W+V_2)$ to $V_2(W+V_1)$, / $V_1(W+V_2)$ is a region of Definition: DefI/, and W is equivalent to the graph of I. If ρ is a representation of $L(A_2^{11})$, then the additive relation W defines the isomorphism between the factor spaces V_1/V_1W and V_2/V_2W.

1.3. INDECOMPOSABLE REPRESENTATIONS AND PERFECT ELEMENTS OF $L(A_2^\nu)$. It is well known that the graph /the Dynkin diagram/ A_2^ν has three positive roots: two simple roots α_1 and α_2 and the root $\alpha=\alpha_1+\alpha_2$. According to this, each lattice $L(A_2)$ has three different indecomposable representations $\tau_{11}, \tau_{22}, \tau_{12}$. They have the following properties: 1) both τ_{11} and τ_{22} are representations in one-dimentional spaces $T_{11} \cong K\xi_1$ and $T_{22} \cong K\xi_2$, respectively. Namely $\tau_{11}(v_1)=K\xi_1$, $\tau_{11}(v_2)=0$ and $\tau_{22}(v_1)=0$, $\tau_{22}(v_2)=K\xi_2$. 2) τ_{12} is the representation in a two-dimentional space $T_{12} \cong K\xi_1 \oplus K\xi_2$, such that $\tau_{12}(v_1)=K\xi_1$, $\tau_{12}(v_2)=K\xi_2$, $\tau_{12}(w)= K(\xi_1+\xi_2)$.

The properties 1 and 2 are independent of the equipment ν.

3) If $\nu(1)=0$, then $\tau_{11}(w)=0$, if $\nu(1)=1$, then $\tau_{11}(w)=K\xi_1$.

4) If $\nu(2)=0$, then $\tau_{22}(w)=0$; if $\nu(2)=1$, then $\tau_{22}(w)=K\xi_2$.

We produce now a partial order on a set $\mathcal{P}(A_2^{\nu})$ of the isomorphism classes of all inde- composable representations of $L(A_2^{\nu})$.

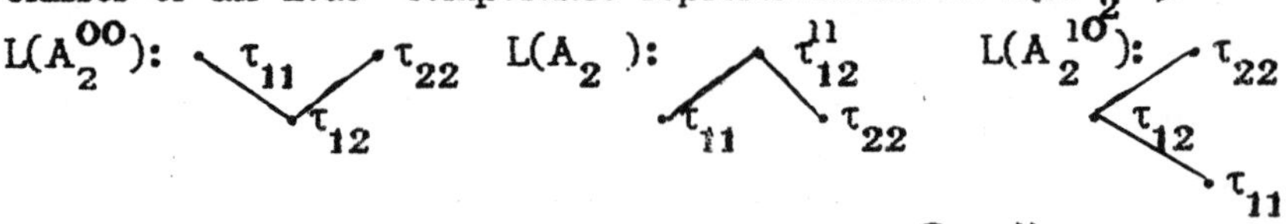

We shall prove later on, that each poset $\mathcal{P}(A_2^{\nu})$ is isomorphic to a poset of V-irreducible perfect elements[1]. Such elements in $L(A_2^{\nu})$ are the following ones:

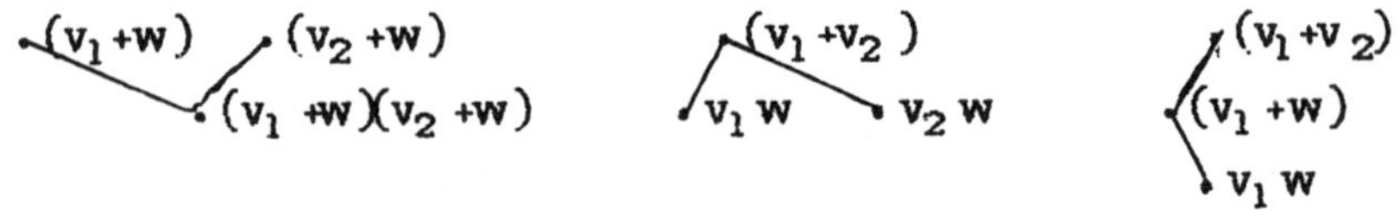

$L(A_2^{00})$ $L(A_2^{10})$ $L(A_2^{10})$

Notice that only lattices $L(A_2^{\nu})$ are finite, all another $L(A_n)$ /$n\geq 3$/ contain infinitely many elements, and it is convenient to investigate them by general methods.

1.4. We discuss here some additional properties of representations of a modular triplet (v,v',w). If ρ is a triplet (v,v',w) in a vector space E over a field K, we denote $V=\rho(V)$, $V'=\rho(v')$, $W=\rho(w)$ and $V_0=VW$, $V_0'=V'W$, $V_1=V(W+V')$, $V_1'=V'(W+V)$. It is easy to see, that $V_0\subseteq V_1\subseteq V$ and $V_0'\subseteq V_1'\subseteq V'$.

1.4 LEMMA. *Let (v,v',w) be a modular triplet with 0 and e and relations: $vv'=0$, $v+v'=e$. Let ρ be a representation of this tiplet in a vector space $E=\rho(e)$. Suppose that U_1, U_2 and $U_1' U_2'$ are subspaces of E with the following properties: a) $V_1=V_0\oplus U_1$, b) $V=V_1\oplus U_2$; c) $V_1'=V_0'\oplus U_1'$; d) $V'=V_1'\oplus U_2'$. Then ρ is decomposed in the direct sum on subspaces $V_0, V_0', U_2, U_2', U_{11}$ where $U_{11}=U_1+U_1'$ and $\rho_{11}=\rho|_{U_{11}}$ is isomorphic to a representation of the simple modular triplet.*

[1] A perfect element h is V-irreducible, if $h=\sum_{i=1} p_i$ is a decomposition in a sum of perfect elements p_i, then it exists i, such that $h\cong p_i$, and all another $p_j=0$ /$j\neq i$/.

PROOF. Notice, that $V_1+V_1'=V(W+V')+V'(W+V)=(V+W)(V(W+V')+V')=(V+W)(V'+W)(V+V')=(V+W)(V'+W)E=(V+W)(V'+W)$. Since $U_1\subseteq V_1=V(W+V')\subseteq V$ we get $0=U_1V_0=U_1VW=U_1W$. We obtain analogously $U_1'W=0$.

Let us denote $W_1=W(U_1+U_1')$. We shall show that (U_1,U_1',W_1) is a simple modular triplet, i.e. $U_1U_1'=U_1W_1=U_1'W_1=0$ and $U_1+U_1'=U_1+W_1=U_1'+W_1$.

In fact, $U_1U_1'\subseteq V_1V_1'=0$, and $U_1W_1=U_1W(U_1+U_1')=U_1W=U_1VW==0$, and similarly $U_1'W_1=0$. We obtain $U_1+W_1=U_1+W(U_1+U_1')=(U_1+W)(U_1+U_1')$. Notice, that $U_1+W=U_1+WV+W=V(W+V')+W=(V+W)(V'+W)=V_1+V_1'$. Analogously $U_1'+W=V_1+V_1'$. Using this, we get $U_1+W_1=(U_1+W)(U_1+U_1')=(V_1+V_1')(U_1+U_1')=U_1+U_1'$ and similarly $U_1'+W_1=U_1+U_1$

We shall show that each element of the lattice L with generators V,V',W is decomposed on subspaces V_0,V_0',U_2,U_2',U_{11}. It is sufficient to prove this for the generators. Let us denote by B a lattice generated by elements $V_0,V_0',U_1,U_1',U_2,U_2'$. It is easy to see that B is the Boolean lattice. We have, according to definition, $V=V_1\oplus U_2=V_0\oplus U_1\oplus U_2$, i.e. $V\in B$. Hence $V=V(V_0+U_{11}+U_2+V_0'+U_2')=VV_0+VU_{11}+VU_2+VV_0'+VU_2'=V_0+(V_0+U_1+U_2)(U_1+U_1')+U_2=V_0+U_1+U_2=VV_0+VU_{11}+VU_2$. We have obtained that element V is decomposed on the elements V_0,V_0',U_{11},U_2,U_2'. We can prove the same property of element V'. We claim that $W=WU_{11}+WV_0+WV_0'=W_1+WV+WV'$. In fact $W_1+WV+WV'=W(U_1+U_1')+WV+WV'=W(U_1+U_1'+WV+WV')=W(U_1+V_0+U_1'+V_0')==W(V_1+V_1')=W(V(W+V')+V'(W+V))=W(W+V)(W+V')=W$. ∎

1.4.2. COROLLARY. *Let the conditions of lemma 1.4.1. are true and let $U_{0i},U_{1j},U_{2k},U_{0s}',U_{2t}'$ is a family of subspaces with the following properties: a) $V_0=\bigoplus_i U_{0i}$, $i\in I_0$, b) $U_1=\bigoplus_j U_{1j}$, $j\in I_1$, c) $U_2=\bigoplus_k U_{2k}$, $k\in I_2$, d) $U_0'=\bigoplus_s U_{0s}'$, $s\in I_0'$, e) $U_2'=\bigoplus_t U_{2t}'$, $t\in I_2'$. We can choose the subspaces $U_{1j}'\subseteq U_1'$ which are concordant with U_{1j}, i.e. the relation W give the isomorphism between U_{1j} and U_{1j}'. Then 1) $U_1'=\bigoplus_j U_{1j}'$, 2) the representation ρ is decomposed on the family of subspaces: $U_{0i},U_{2k},U_{0s}',U_t',T_j$, where $T_j=U_{1j}\oplus U_{1j}'$.*

We omit a simple proof of this. Notice only that subspace U_{1j}' isomorphic to $U_{1j}\subseteq U_1$ can be constructed as $U_{1j}'=U'(W+U_{1j})$.

§2. SOME RESULTS ABOUT PROJECTIVE INTERVALS.

This section contains some auxiliary results obout projective intervals.

2.1. LEMMA. *Let $x \subseteq y$ and w some elements of modular lattice M.*

a) Intervals (x, y) and $(x+w, y+w)$ are perspective if $xw = yw$.

b) Intervals (x, y) and (xw, yw) are perspective if $x+w = y+w$.

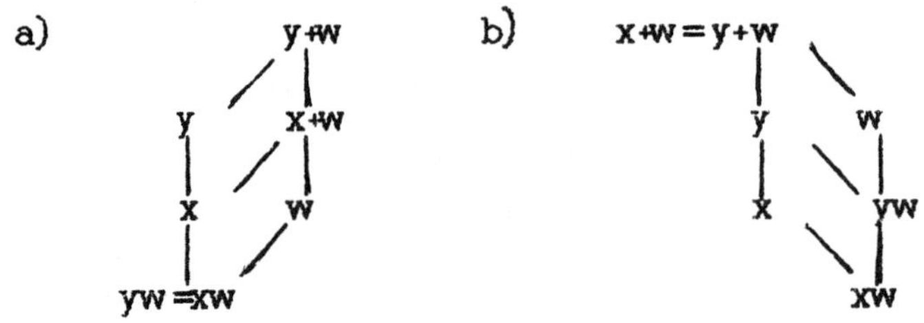

Proof. a) Since $x \subseteq y$ and $xw = yw$, we get $(x+w)+y = x+y+w = y+w$ and $y(x+w) = = x + yw = x+xw = x$. b) Using $x \subseteq y$ and $x+w = y+w$, we get $x(yw) = xyw = xw$ and $x+yw = y(x+w) = y(y+w) = y$. ∎

2.2. PROPOSITION. *Let M be a modular lattice with an unit e and generators v, v', w, x, x', y, y' and relations:*

1^0 $vv' = 0$, $v+v' = e$; 2^0 $x \subseteq y \subseteq v$, $X' \subseteq y' \subseteq v'$; 3^0 $xw = yw$, $x'w = y'w$; 4^0 $x + vw = v(w+x)$, $y+vw = v(w+y)$, $x'+v'w = v'(w+x)$, $y'+v'w = v'(w+y)$; Then intervals (x, y) and (x', y') are projective.

PROOF. It is easy to show that the following diagram is correct:

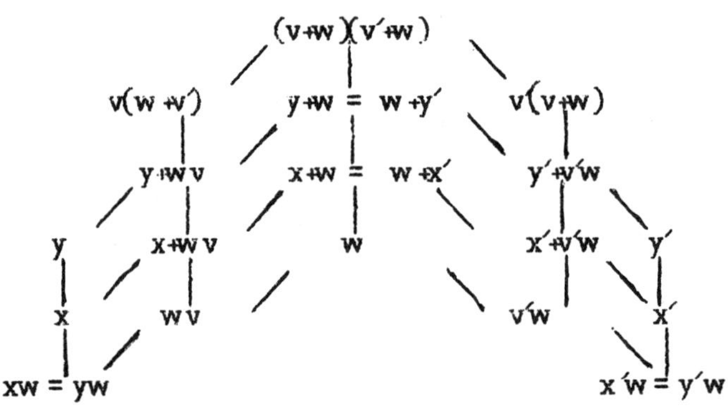

Fig. 2.1.

We get for example $w+x' = w+(x'+v'w) = w+v(x+w) = (x+w)(w+v') = w+x(w+v') = w+xv(w+v')$. According to 4^0 $x \subseteq v(w+x) \subseteq v(w+v')$. Hence $xv(w+v') = x$ and we get $w+x' = w+x$. Similarly $w+y' = w+y$.

We obtain xwv=xw= yw =ywv, hence using Lemma 2.1 intervals (x,y) and (x+wv,y+wv) are perspective. Since xw $\subsetneq$ vw, we have w(x+wv) = wx+wv = wv and analogously w(y+wv)= wy+wv = wv. Hence intervals (x+wv, y+wv) and (x+w,y+w) are also perspective.

We get in a similar way that intervals (x', y') (x'+v'w, y'+v'w) and (x'+w, y'+w) are perspective.

Therefore we obtain that intervals (x,y) and (x', y') are proective.∎

2.3.PROPOSITION. *Let ρ be a representation of the lattice L from proposition 2.2 in a vector space E. Assume $\rho(x) = X$, $\rho(y) = Y$, $\rho(v) = V$ and so on. Suppose that Z is such a subspace, that $ZX = 0$, $Z+X = Y$. Then we can choose a subspace Z' with the following properties: $Z'X' = 0$, $Z'+X'=Y'$, and such that additive relation W define an isomorphism between the subspaces Z and Z'.*

PROOF. Using the conditions of Proposition 2.2 we get: 1^0 VV'=0, V+V'=E; 2^0X$\subsetneq$Y$\subsetneq$ V, X'$\subsetneq$Y'$\subsetneq$V', ^{0}XW=YW, X'W=Y'W; 4^0X+VW = V(W+X'),Y+VW=V(W+Y'), X'+V'W=V'(W+X),Y'+V'W=V'(W+Y).

The latest proof is devided into the several steps. We claim at first that the subspace Z has two following properties:

2.3.1 ZW=0, Z(X+VW) = 0.

As far as Z$\subsetneq$Y, ZX = 0 and XW = YW we get ZW=ZYW=ZXW=0 and analogously Z(X+VW)= ZY(X+VW)= Z(X+YVW)=Z(X+YW)= Z(X+XW)=ZX=0.

2.3.2. Let us choose some basis $\xi_1, \xi_2, \ldots, \xi_s$ in the subspace Z$\subsetneq$Y. We show how to chose such vectors $\xi_1', \xi_2', \ldots, \xi_s'$ in Y' that $(\xi_i+\xi_i')$ $\in$ W. Let us $\xi\in Z\subsetneq Y$ some a non-zero vector. Since ZX=0, we have $(K\xi)VW=0$. Since $K\xi\subsetneq Y \subsetneq Y+WV= V(W+Y')$, the vector ξ can be written in the form $\xi=\omega-\xi'$, where $\omega\in W$ and $\xi'\in Y'$. We shall prove that 1) $\xi'\neq 0$, 2)$\xi'\notin X'$, 3)$\xi'\notin W$. If we suppose $\xi'=0$, or $\xi'\in W$, then we get $\xi=\omega\in W$ or $\xi=(\omega-\xi')\in W$. It contradicts $(K\xi)W=0$. If we suppose $\xi'\in X'$, then using $V(W+X')=X+VW$ and $\xi=\omega-\xi'$ we obtain $\xi\in (X+VW)$. It contradicts $(K\xi)(X+VW)\subsetneq ZY(X+VW)= 0$.

Denote by Z' a subspace stretched on the vectors ξ_i'. We claim that Z' has the following properties: 1) dimZ' = dimZ, 2)Z'X'= 0, 3)Z'+X'= Y'. It means that Z' is the required subspace.

2.3.3. We show, at first, that ξ'_i are linear independent. We suppose on the contrary that there is a linear combination $\eta' = \sum \alpha_i \xi'_i = 0$. Denote $\eta = \sum_i \alpha_i \xi_i$. Then $\eta + \eta' = \eta \in W$. i.e. $\eta \in ZW$. So, if there are $\alpha_i \neq 0$, then $ZW \neq 0$. It is a contradiction, hence $ZW = 0$. Thus the vectors ξ'_i are linear independent, and $\dim Z' = \dim Z$.

2.3.4. We claim, that $Z'X' = 0$. Suppose, on the contrary, that there is a non-zero vector $\zeta' = \sum_i \beta_i \xi'_i \in Z'X'$. We set $\vartheta = \sum_i \beta_i(\xi_i + \xi'_i) = \sum_i \beta_i \xi_i + \zeta'$ and denote $\sum_i \beta_i \xi_i = \zeta$. So $\zeta = \vartheta - \zeta'$. We suppose that $K\zeta' \subseteq X'$, thus $K\zeta \subseteq Z\,V(W+X') = Z(X+WV)$. We have shown in 2.3.1. that $Z(X+WV) = 0$. Hence $\zeta = 0$. It is possible only when all $\beta_i = 0$, i.e. $\zeta' = 0$. We get the contradiction.

2.3.5. LEMMA. *Let v, v', w be such a modular triplet that $vv' = 0$, $w \subseteq (v+v')$; and let $x_1, x_2 \subseteq v$ be such elements, that a triplet (x_1, x_2, y), where $y = v(w+v')$ is distributive one. Then $v'(w+x_1+x_2) = v'(w+x_1) + v'(w+x_2)$.*

PROOF. Since $(x_1 + x_2) \subseteq v$ the distributivity of the triplet $x_1, x_2, v(w+v')$ is equivalent to the distributivity of the triplet $x_1, x_2, (w+v')$. Therefore $v'(w+x_1+x_2) = = v'(w+v')(w+x_1+x_2) = v'(w+(w+v')(x_1+x_2)) =$
$= v'(w+x_1(w+v')+x_2(w+v')) = = v'((w+x_1)(w+v')+(w+x_2)(w+v')) =$
$= v'(w+v'(w+x_1)+v'(w+x_2)) = v'w+v'(w+x_1)+v'(w+x_2) = v'(w+x_1)+v'(w+x_2)$.

2.3.6. We claim, that $Z'+X' = Y'$. As $Z \subseteq Y \subseteq (Y+WV) = V(W+Y') \subseteq V(W+V')$, the triplet $(X, Z, V(W+V'))$ is distributive one and we can use Lemma 2.3.5. Therefore $V'(W+Y) = V'(W+X+Z) = V'(W+X)+V'(W+Z) = V'W+X'+V'(W+Z)$. We have, according to construction of Z', that $V'(W+Z) = V'W+Z'$. Hence $V'W+Y' = V'(W+Y) = V'W+X'+Z'$. It follows from Fig.2.1 that $Y' = Y'(V'W+Y') = Y'(V'W+X'+Z') = Y'V'W+X'+Z' = Y'W+X'+Z' = X'W+X'+Z' = X'+Z'$.∎

§3. MAIN PROPERTIES OF REPRESENTATIONS $\tau_{i,k}$.

3.1. A PARTIAL ORDER ON A SET $\mathcal{D}(A_n^\nu)$. We define now two functions ν^-, ν^+ from A_{n0} to $\{0,1\}$, as $\nu^-(1) = 0$ and $\nu^-(i) = \nu_{i-1}(i)$ if $i > 1$, $\nu^+(n) = 0$ and $\nu^+(i) = \nu_i(i)$ if $i < n$.

3.2. LEMMA. *Let $\tau_{i',k'}$ be the representation of (0--0) type, i.e. $\nu^-(i') = 0$ and $\nu^+(k') = 0$, and $\tau_{i,k}$ is the representation of (0—1) type, i.e. $\nu^-(i) = 0$ and $\nu^+(k) = 1$ Suppose that $i \leq i'$ and k is such that*

$i \leq k' \leq k$, then $Hom_K(\tau_{1,k}, \tau_{1',k'}) = K$.

PROOF. Let $\{e_j\}$ /$i \leq j \leq k$/ be the basis of representation $\tau_{1,k}$ and $\{e'_j\}$ /$i' \leq j \leq k'$/ be the basis of $\tau_{1',k'}$/see §0/. We assume $\mu \in Hom(\tau_{1',k'}, \tau_{1,k})$, /$\mu \neq 0$/. According to definition $\mu(\tau_{1',k'}(v_s)) \subseteq \tau_{1,k}(v_s)$ and $\mu(\tau_{1',k'}(w_t)) \subseteq \tau_{1,k}(w_t))$ for all generators v_s and w_t. It means that $\mu(e'_i) = \xi_i e_i$ and $\mu(e'_{i+1}) = \xi_{i+1}(e_{i+1})$, where $\xi_i, \xi_{i+1} \in K$. Since $\mu: T'_{1',k'} \longrightarrow T_{1,k}$ is a linear map, then $\mu(e'_i + e'_{i+1}) = \xi_i e_i + \xi_{i+1} e_{i+1}$. We have assumed $K(e'_i + e'_{i+1}) = \tau_{1',k'}(w_i)$ and $K(e_i + e_{i+1}) = \tau_{1,k}(w_i)$. We must have $\mu(\tau_{1',k'}(w_i)) \subseteq \tau_{1,k}(w_i))$. It can be realize if and only if $\xi_i e_i + \xi_{i+1} e_{i+1} = \xi(e_i + e_{i+1})$, or if $\xi_i + \xi_{i+1} = \xi$. if we continue these arguments we get that for all s such that $i' \leq s \leq k$ $\mu(e'_s) = \xi e_s$, whith the same $\xi \in K$.

If we consider another vectors e_t /$k < t \leq k'$/, we get $\mu(e'_t) = 0$ and $\mu(e_k + e_{k+1}) = \mu(e_k) + \mu(e_{k+1}) = \xi e_k + 0 = \xi e_k$. It means that $\mu(\tau_{1',k'}(w_k)) \subseteq \tau_{1,k}(w_k)$, since according with supposition $\nu_k(k) = 1$ and hence $\tau_{1,k}(w_k) = \tau_{1,k}(v_k) = Ke_k$. The representation $\tau_{1',k'}$ is such that $\tau_{1',k'}(v_p) = 0$ for $p < i'$. Therefore for such p $\mu(\tau_{1',k'}(v_p)) = 0 \subseteq \tau_{1,k}(v_p)$.

It follows easily from all these arguments that an arbitrary non-zero morphism $\mu^\wedge \in Hom(\tau_{1',k'}, \tau_{1,k})$ is proportional to μ, i.e. $\mu^\wedge = \varkappa\mu, \varkappa \in K$. ∎

Some modification of Lemma 3.2 shows validity of proposition.

3.3. PROPOSITION. *Let $\tau_{1',k'}$ and $\tau_{1,k}$ are representation of $L(A_n^\nu)$. Then $Hom_K(\tau_{1',k'}, \tau_{1,k}) = K$ if $(i \leq^\pi i'$ and $k \leq^\lambda k$). Otherwise $Hom(\tau_{1',k'}, \tau_{1,k}) = 0$.* ∎

3.4. THEOREM. *All representation $\tau_{1,k}$ are indecomposable.*

PROOF. Notice that all representation $\tau_{1,1}$ are indecomposabie, since each of them is a representation in one-dimentional vector space $T_{1,1} \cong K$. Let us suppose that $\tau_{1,k}$ /$i<k$/ is decomposable. It means that there are non-zero subspaces Y_1, Y_2 such that each subspace $X = \tau_{1,k}(x)$, where $x \in L(A_n^\nu)$, can be represented in the form: $X = XY_1 + XY_2$. It must be true in particular for the generators v_i and w_j of $L(A_n^\nu)$. If $i \leq j \leq k$ then according to the definition the subspace $\tau_{1,k}(v_j) = Ke_j = V_j$ is one-dimentional, and the equality $V_j = V_jY_1 + V_jY_2$ can be true if and only if either $V_j \in Y_1$ or $V_j \in Y_2$. It means that there is such s /$i \leq s < k$/ that $V_s \in Y_p$ and $V_{s+1} \in Y_q$ ($\{p,q\} = \{1,2\}$). Then $\tau_{1,k}(w_s) = K(e_s + e_{s+1}) = W_s$ is obviously that $W_sY_1 = W_sY_2 = 0$. It contradicts to the equality $W_s = W_sY_1 + W_sY_2$, which must be fulfilled if $\tau_{1,k}$ is decomposable. ∎

§4. ELEMENTS v_{i_jk} AND PARTIAL ORDERS $\leq^{\pi}$ AND $\leq^{\lambda}$.

4.1. Let us define operations φ_i^- ($i > 1$) and φ_i^+($i < n$) in a lattice $L(A^{\nu}_n)$. We set $\varphi_i^- x_i = v_{i-1}(w_{i-1} + x_i)$ if $i>1$, and $\varphi_i^+ x_i = v_{i+1}(w_i + x_i)$ if $i<n$ for each $x_i \in v_i$.

We give now the new definitions of the elements v_{i_j} and v_{jk}. We shall show later on that these definitions are equivalent with definitions given in Theorem 0.5.

$$v_{i_i} = \begin{cases} v_i, \text{ if } i=1, \text{ or } \nu_{i-1}(i) = 0, \\ w_{i-1} v_i, \text{ if } \nu_{i-1}(i) = 1; \end{cases} \qquad v_{kk} = \begin{cases} v_k, \text{ if } i=n \text{ or } \nu_k(k) = 0, \\ w_k v_k, \text{ if } \nu_k(k) = 1. \end{cases}$$

We set $v_{j-1k} = \varphi_j^- v_{jk}$ if $j \leq k$, and $v_{i_{j+1}} = \varphi_j^+ v_{i_j}$ if $i \leq j$. Elements v_{i_j} $(i<j)$ and v_{jk} $(j<k)$ can be written in a short form, if we use the lemma:

4.2. LEMMA. *If* $x_i \subseteq v_i$, $x_j \subseteq v_j$ *and* $1 \leq i < j \leq n$, *then*

4.2.1. $$\varphi^-_{i+1} \varphi^-_{i+2} \cdots \varphi^-_{j-1} \varphi^-_j x_j = v_i(w_i + w_{i+1} + \cdots + w_{j-1} + x_j).$$

4.2.2. $$\varphi^+_{j-1} \varphi^+_{j-2} \cdots \varphi^+_{i+1} \varphi^+_i x_i = v_j(x_i + w_i + w_{i+1} + \cdots + w_{j-1}).$$

We shall prove 4.2.1 inductively on difference $j-i$. If $(j-i)=1$, then 4.2.1 is true by the definition. Assume that for $1<i$ $\varphi^-_{i+1}\varphi^-_{i+2}\cdots\varphi^-_j x_j = v_i(w_i + w_{i+1} + \cdots + w_{j-1} + x_j)$. Then $\varphi^-_i \varphi^-_{i+1}\varphi^-_{i+2}\cdots\varphi^-_j x_j = v_{i-1}(w_{i-1} + v_i(w_i + w_{i+1} + \cdots + w_{j-1} + x_j))$. Let us denote temporarily $\sigma = w_i + w_{i+1} + \cdots + w_{j-1} + x_j$. Notice that $v_{i-1}(v_i + \sigma) \subseteq v_{i-1}(v_i + v_{i+1} + \cdots + v_{j-1} + v_j) = 0$. Hence a triplet (v_{i-1}, v_i, σ) is distributive one and $(v_{i-1} + v_i)\sigma = v_{i-1}\sigma + v_i\sigma = 0 + v_i\sigma = v_i\sigma$. Therefore, using $w_{i-1} \subseteq (v_{i-1} + v_i)$, we obtain $\varphi^-_i \varphi^-_{i+1}\varphi^-_{i+2}\cdots\varphi^-_j(x_j) = v_{i-1}(w_{i-1} + v_i\sigma) = v_{i-1}(w_{i-1} + (v_i + v_{i-1})\sigma) = v_{i-1}(v_{i-1} + v_i)(w_{i-1} + \sigma) = v_{i-1}(w_{i-1} + \sigma) = v_{i-1}(w_{i-1} + w_i + \cdots + w_{j-1} + x_j))$. We prove 4.2.1. The formula 4.2.2 can be proved analogously. ∎

We obtain from this lemma

4.3. COROLLARY. *Elements* v_{i_j} / $i<j$/ *and* v_{jk} / $j<k$/ *can be written in the form*

$$v_{i_j} = (v_{i_1} + w_i + w_{i+1} \cdots + w_{j-1})v_j,$$
$$v_{jk} = v_j(w_j + w_{j+1} \cdots + w_{k-1} + v_{kk}).$$

4.4. LEMMA. *The operations* φ_i^- *and* φ_i^+ *are concordant with the order relation: i.e. if* $x_i \subseteq x_i'$, *then* $\varphi_i^- x_i \subseteq \varphi_i^- x_i'$ *and* $\varphi_i^+ x_i \subseteq \varphi_i^+ x_i'$.

A proof follows obviously from the definitions.

4.5.PROPOSITION. *Let j be a fixed element of $A_{n0} = \{1,2,\dots,n\}$. Then*
(i) The elements $\{v_{jk}\}$ $(j \le k)$ form a chain: $(v_{jk} \subseteq v_{jk'}) \Leftrightarrow (k \le^{\lambda} k')$.
(ii) The elements $\{v_{i_j}\}$ $(i \le j)$ form a chain: $(v_{i_j} \subseteq v_{i'_j}) \Leftrightarrow (i \le^{\pi} i')$.

We shall prove 4.5.i by induction on the difference $m = n+1-j$ which is equal to the number of elements in the chain $\{v_{jk}\}$ $(1 \le j \le k \le n)$. If $j = n$, i.e. $m = 1$, then this chain contains one element v_{nn} and 4.5.i is obviously true. Assume, that 4.5.i has been proved for some $m = (n+1-j) > 1$. We shall prove it for $m+1$. According to the supposition, the elements $\{v_{jk}\}$ /$j \le k$/ form a chain, i.e. by lemma 4.4 the elements $v_{j-1k} = \{\varphi^{-}_{j} v_{jk}\}$ /$j \le k$/ also form a chain. We show that v_{j-1j-1} also belong to this chain. 1) if $\nu_{j-1}(j-1) = 0$, than we have by definition $v_{j-1j-1} = v_{j-1}$, it means that v_{j-1j-1} is greater than the greatest element of $\{v_{j-1k}\}$ /$j \le k$/. 2) if $\nu_{j-1}(j-1) = 1$, than we have $v_{j-1j-1} = v_{j-1} w_{j-1}$. Since $v_{j-1} w_{j-1} \subseteq v_{j-1}(w_{j-1} + w_j + \dots + w_{k-1} + v_{kk}) = v_{j-1k}$, it means that v_{j-1j-1} is less than the least element in $\{v_{j-1k}\}$ / $j \le k$/. If we set $(v_{j-1k} \subseteq v_{j-1k'}) \Leftrightarrow (k \le^{\#} k')$, then it is easy to check see, that new order $\le^{\#}$ coicides with $\le^{\lambda}$ i.e. $(v_{j-1k} \subseteq v_{j-1k'}) \Leftrightarrow (k \le^{\lambda} k')$.

4.6. Each v_j contains two chains: $\{v_{i_j}\}$ /$1 \le i \le j$/ and v_{jk} /$j \le k \le n$/. If $j = 1$ or $j = n$ one of these chains is degenerate one, i.e. it contains only one element $v_{1_1} = v_1$ or $v_{n_n} = v_n$ correspondingly.

Let us denote by D_j a lattice generated by two chains $\{v_{i_j}\}$ and $\{v_{jk}\}$ It is known [5], that such a lattice is distributive one and it can be described as the lattice of ideals of the poset of indecomposable in a sum elements. It can be proved, that such elements in D_j are /see 3.3/

4.6.1 $v_{i_jk} = (v_{i_i} + w_i + w_{i+1} + \dots + w_{j-1}) v_j (w_j + w_{j+1} + \dots + w_{k-1} + v_{kk})$, where $1 \le i \le j \le k \le n$.

Let j be fixed and let $v_{i^*_j}$ and v_{jk^*} be the maximal elements in the chains $\{v_{i_j}\}$ and $\{v_{jk}\}$, then $v_{i^*_j} = v_{jk^*} = v_j = v_{i^*{}_jk^*}$. Hence $v_{i_jk^*} = v_{i_j} v_{jk^*} = v_{i_j}$ and $v_{i^*{}_jk} = v_{jk}$. We have defined two functions ν^{-}, ν^{+}: $A_{n0} \longrightarrow \{0,1\}$, as $\nu^{-}(1) = 0$ and $\nu^{-}(i) = \nu_{i-1}(i)$ if $i > 1$, $\nu^{+}(n) = 0$ and $\nu^{+}(i) = \nu_{i}(i)$ if $i < n$.

4.7.PROPOSITION. *Let j be a fixed index, then the maximum element $v_{i^*_j}$ in the chain $\{v_{i_j}\}$ /$i \le j$/ can be defined in the following way:*

4.7.PROPOSITION. *Let j be a fixed index, then the maximum element $v_{i^*_j}$ in the chain $\{v_{i_j}\}$ /$i \leq j$/ can be defined in the following way:*
$(v_{i^*_j} = v_j$ *and* $i^* = j) \Leftrightarrow (\nu^-(j) = 0$,
$(v_{i^*_j} = v_j$ *and* $i^* < j) \Leftrightarrow (\nu^-(i^*) = 0$ *and* $\nu^-(i) = 1$ *for all such* i
that $i^* < i \leq j)$.

PROOF. If $\nu^-(j)=0$ then $v_{j_j} = v_j$ according to definition. If $\nu^-(i^*) = 0$ and $i^* < i \leq j$, we have $v_{i^*_{i^*}} = v_{i^*}$ and $v_{i^*_j} = (v_{i^*} + w_{i^*} + \ldots + w_{j-1})v_j$. Since $\nu^-(i) = 1$ for all such i that $1 < i^* < i \leq j$, we have $w_s + v_s = v_s + v_{s+1}$ for all s such that $i^* \leq s < j$. Therefore $v_{i^*_j} =$
$= (v_{i^*} + w_{i^*} + w_{i^*+1} + \ldots + w_{j-1})v_j = (v_{i^*} + v_{i^*+1} + w_{i^*+1} + \ldots + w_{j-1})v_j =$
$(v_{i^*} + v_{i^*+1} + v_{i^*+2} + \ldots + v_{j-1} + w_{j-1})v_j = (v_{i^*} + v_{i^*+1} + v_{i^*+2} + \ldots + v_{j-1} + v_j)v_j = v_j$. ∎

4.7.1. $(v_{i^*_j} = v_j) \Leftrightarrow i^* = \max_i\{i \mid i \leq j \text{ and } \nu^-(i) = 0\}$.[1)]

4.8.PROPOSITION. *Let j be a fixed index, then the maximum element $v_{j_{k^*}}$ in the chain $\{v_{j_k}\}$ /$j \leq k$/ can be defined in the following way:*
$(v_j = v_{j_{k^*}}$ *and* $j = k^*) \Leftrightarrow (\nu^+(j) = 0)$
$(v_j = v_{j_{k^*}}$ *and* $j < k^*) \Leftrightarrow (\nu^+(k^*) = 0$ *and* $\nu^+(t) = 1$ *for all such* t
that $j \leq t < k^*)$. ∎

4.8.1. $(v_{j_{k^*}} = v_j) \Leftrightarrow k^* = \min_k\{k \mid j \leq k \text{ and } \nu^+(k) = 0\}$.[2)] ∎

4.9.PROPOSITION. *Let j /$1 \leq j \leq n$/ be a fixed index, then elements $v_{i_j k}$ /$1 \leq i \leq j \leq k$/ are ordered in the following way:*
$$(v_{i_j k} \subseteq v_{i'_j k'}) \Leftrightarrow (i \leq^{\pi} i' \text{ and } k \leq^{\lambda} k')$$

PROOF. Acording to definition $v_{i_j k} = v_{i_j} v_{j_k}$. Hence, using 4.5 we get: if $i \leq^{\pi} i'$ and $k \leq^{\lambda} k'$, then $v_{i_j k} v_{i'_j k'} = v_{i_j} v_{j_k} v_{i'_j} v_{j_{k'}} = v_{i_j} v_{i'_j} v_{j_k} v_{j_{k'}} = v_{i_j} v_{j_k} = v_{i_j k}$. ∎

Each positive root α of the Dynkin diagram A_n can be written in the form $\alpha = \sum_{i \leq j \leq k} \alpha_j$. We shall call the pair (i,k) /$1 \leq i \leq k \leq n$/ the *index* of such the root α. We introduce the partial order $\leq^{\nu}$ on the set $P(A_n)$ of indices (i,k) /$1 \leq i \leq k \leq n$/ in the following way: $((i,k) \leq (i',k')) \Leftrightarrow (i \leq^{\pi} i'$ and $k \leq^{\lambda} k')$. We shall prove later on that this order is isomorphic to the partial order on the set $\mathcal{P}(A_n^{\nu})$ of isomorphism classes of all indecomposablle representation $\tau_{\alpha} = \tau_{i,k}$.

4.10.EXAMPLE. We give here the construction of orders $\leq^{\pi}$ and $\leq^{\lambda}$ through a sequences $\{P_j\}$ and $\{L_j\}$ for the graph A_4^{ν} /Example 1°/:

1,2) Here max and min are in the sence of trivial ordering: $1 < 2 < \ldots < n$.

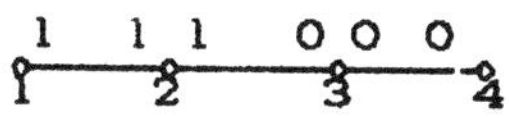

A sequence $\{P_j\}$ is defined by the meanings $\nu_{i-1}(i)$ $/1<i\leq n/$ (i.e by 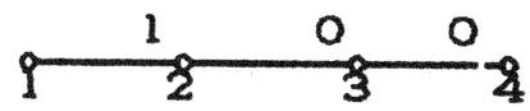

) $P_1=1$, $P_2=21$, $P_3=213$, $P_4=2134$. Hence we obtain the order $\leq^\pi$: $2\leq^\pi 1\leq^\pi 3\leq^\pi 4$. A sequence $\{L_j\}$ is defined by the meanigs $\nu_1(i)$ $/1\leq j<n/$ (i. . by 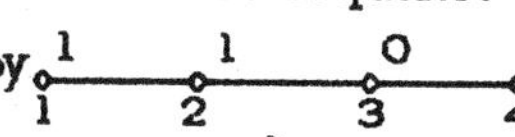

) $L_1=4$, $L_2=43$, $L_3=243$, $L_4=1243$. Hence we obtain the order $\leq^\lambda$: $1\leq^\lambda 2\leq^\lambda 4\leq^\lambda 3$. The partial order $\leq^\nu$ on the set P_4 of the indices (i,k) $/1\leq i\leq k\leq 4/$ is represented on the following diagram:

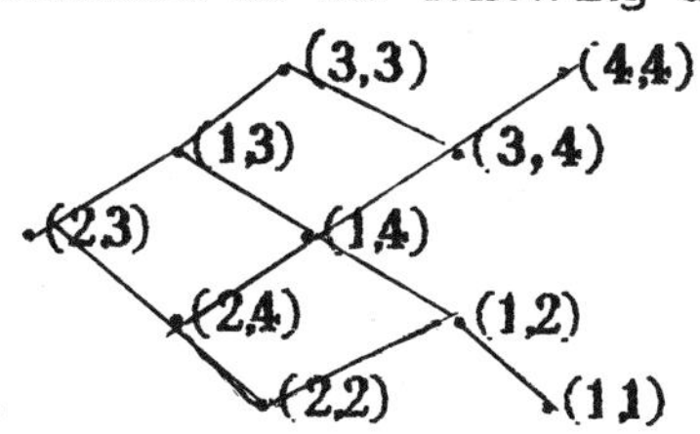

Fig.4.10.1.

Posets of elements $v_{i_j k}$ $/i\leq j\leq k/$ are represented in diagrams:

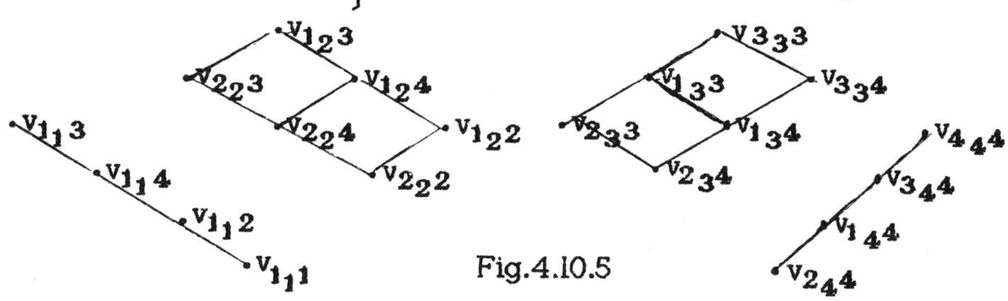

Fig.4.10.5

We collect now elements $v_{i_j k}$ together and " project" them on the set P_4 of the indices $(i,k)/1\leq i\leq k\leq 4/$. A projection operation p act on indices in the following way: $p(i_j k)\longmapsto(i,k)$. The corresponding diagram is depicted in the next page. The projection p will be used later on by a construction of perfect elements.

4.11.PROPOSITION. *The operations* φ_j^- *and* φ_j^+ *act on an arbitrary element* $v_{i_j k}$ $/i\leq j\leq k/$ *in the following way.*

a) if $i=j$ *and* $\nu_{j-1}(j)=0$, *then* $v_{j_j k}=v_{j k}$ *and* $\varphi_j^- v_{j_j k}=\varphi_j^- v_{j k}=v_{j-1 k}$;
if $i=j$ *and* $\nu_{j-1}(j)=1$, *then* $v_{j_j k}=w_{j-1}\, v_{j k}$ *and* $\varphi_j^- v_{j_j k}=v_{j-1}w_{j-1}$;
if $i<j\leq k$, *then* $\varphi_j^- v_{i_j k}=v_{i_{j-1} k}+v_{j-1}w_{j-1}$;
b) if $j=k$ *and* $\nu_j(j)=0$. *then* $v_{i_j j}=v_{i_j}$ *and* $\varphi_j^+ v_{i_j j}=\varphi_j^+ v_{i_j}=v_{i_{j+1}}$;
if $j=k$ *and* $\nu_j(j)=1$, *then* $v_{i_j j}=v_{i_j}w_j$ *and* $\varphi_j^+ v_{i_j j}=w_j v_{j\,j+1}$;
if $i\leq j<k$, *then* $\varphi_j^+ v_{i_j k}=v_{i_{j+1} k}+w_j v_{j\,j+1}$.

PROOF.a) We shall find at first $\varphi_j^- v_{j_j k}$. It can be two cases:

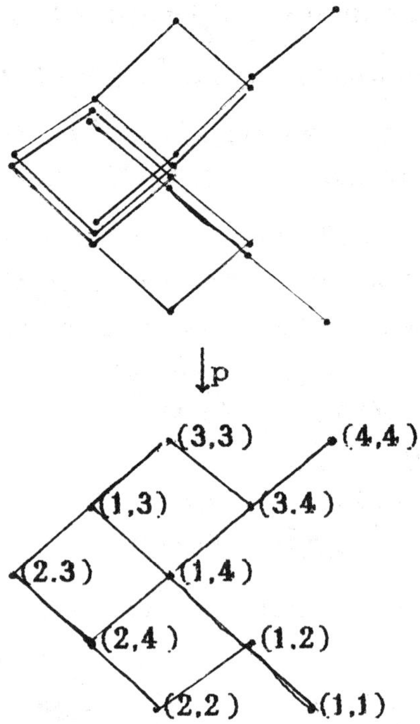

Fig.4.10.3.

1a) If $\nu_{j-1}(j)=0$, then $v_{j_jk} = v_{jk}$. Hence, according to definition 4.1 and colorary 4.3 $\varphi_j^- v_{j_jk} = \varphi_j^- v_{jk} = v_{j-1k} = v_{j-1}(w_{j-1}+w_j+\dots+w_{k-1}+v_k)$.

2a) $\nu_{j-1}(j) = 1$, then $v_{j_jk} = w_{j-1}v_{jk}$ and hence, using that $w_{j-1}v_{jk} \subseteq w_{j-1}$ we obtain $\varphi_j^- v_{i_jk} = v_{j-1}(w_{j-1}+w_{j-1}v_{jk}) = v_{j-1}w_{j-1}$.

3a) We shall find now $\varphi_j^- v_{i_jk}$ in a case $i < j \le k$. We get $\varphi_j^- v_{i_jk} = v_{j-1}(w_{j-1}+v_{i_jk}) = v_{j-1}(w_{j-1}+(v_{i_i}+w_i+\dots+w_{j-1})v_j(w_j+\dots+w_{k-1}+v_{kk}))=$

$= v_{j-1}(w_{j-1}+(v_{i_{j-1}}+w_{j-1})v_j(w_j+\dots+w_{k-1}+v_{kk}))=$

$= v_{j-1}(v_{i_{j-1}}+w_{j-1})v_{j-1}(w_{j-1}+v_j(w_j+\dots+w_{k-1}+v_{kk}))=$

$= (v_{i_{j-1}}+w_{j-1})v_{j-1}(w_{j-1}+w_j+\dots+w_{k-1}+v_{kk})$.

If we use now that $v_{i_{j-1}} \subseteq v_{j-1}$ and $w_{j-1}v_{j-1} \subseteq (w_{j-1}+w_j+\dots+w_{k-1}+v_{kk})$ we obtain $\varphi_j^- v_{i_jk} =$

$= (v_{i_{j-1}}+w_{j-1}v_{j-1})(w_{j-1}+w_j+\dots+w_{k-1}+v_{kk})=$

$= v_{i_{j-1}}(w_{j-1}+w_j+\dots+w_{k-1}+v_{kk})+w_{j-1}v_{j-1} = v_{i_{j-1}k}+w_{j-1}v_{j-1}$.

We can prove analogously the proposition 4.1 for operation φ_j^+. ▪

REMARK. The elements $v_{j-1}w_{j-1}$ and w_jv_{j+1} in **4.1** can be written in a canonical form v_{i_jk}.

According to definition $v_{j-1}w_{j-1}=0$, if $\nu_{j-1}(j-1)=0$; and $v_{j-1}w_{j-1}=$

$v_{j-1 j-1}$ if $\nu_{j-1}(j-1)=1$. We have also if $\nu_j(j+1)=0$, then $w_j v_{j+1}=0$ and if $\nu_j(j+1)=1$ then $w_j v_{j+1}=v_{j+1 j+1}$. We can write $v_{j-1 j-1}=v_{i' j-1 j-1}$ / $i'\leq j-1$/, and $v_{j+1 j+1}=v_{j+1 j+1 k'}$ / $j+1\leq k'$/, where i' and k' are the maximum elements/see 4.7 and 4.8 / in the chains $\{v_{i_{j-1}}\}$ and $\{v_{j+1 k}\}$ (where j is fixed, $i\leq j-1$ and $j+1\leq k$ correspondingly).

§5. THEOREM ABOUT A DECOMPOSITION OF REPRESENTATION.

5.1. Let us denote by $[j^-]^\pi$ a subset $\{1,2,\ldots,j\}\subseteq A_{n,0}$, which is ordered by a relation $\leq^\pi$, and by $[j^+]^\lambda$ a subset $\{j,j+1,\ldots,n\}\subseteq A_{n,0}$, which is ordered by a relation $\leq^\lambda$. It is follows easily from the definition, that the poset $[j^-]^\pi$ is a chain and what is more $[j^-]^\pi$ is the *complete subchain* of the chain $A^\pi_{n,0}$ /i.e. if $i\in A_{n,0}$ and $i\notin[j^-]^\pi$, then either $i\leq^\pi s$ or $s\leq^\pi i$ for every $s\in[j^-]^\pi$/, and also the poset $[j^+]^\lambda$ is a chain and $[j^+]^\lambda$ is the complete subchain of the chain $A^\lambda_{n,0}$. We shall say that i *cover* $i^\bullet$ /on $\leq^\pi$/ if $i^\bullet <^\pi x\leq^\pi i$ imply $x=i$/ and shall denote this as $i^\bullet <\cdot^\pi i$. Let us denote by X_{ijk} such a subspace in V_{ijk}, that

$$X_{ijk}=\begin{cases} V_{i^\bullet jk}+V_{ijk^\bullet}, & \text{where } i^\bullet\leq j \text{ and } i^\bullet <\cdot^\pi i,\ j\leq k^\bullet \text{ and } k^\bullet <\cdot^\lambda k,\\ V_{i^\bullet jk}, & \text{where } i^\bullet\leq j \text{ and } i^\bullet <\cdot^\pi i, \text{ and if } k=\min[j^+]^\lambda,^{1)}\\ V_{ijk^\bullet}, & \text{where } j\leq k^\bullet \text{ and } k^\bullet <\cdot^\lambda k, \text{ and if } i=\min[j^-]^\pi,\\ 0, & \text{if } i=\min[j^-]^\pi \text{ and if } k=\min[j^+]^\lambda.\end{cases}$$

Let us denote by R_{ijk} such a subspace in V_{ijk} $(i\leq j\leq k)$ that:

5.1.1. $$X_{ijk}R_{ijk}=0,\ X_{ijk}+R_{ijk}=V_{ijk}.$$

It is easy to show that every V_{ijk} can be represented in the form:

$$V_{ijk}=\bigoplus_{(i',k')}{}' R_{i'jk'}.$$

($\bigoplus'$ means that sum is taken over such (i',k'), that $(i',k')\leq^\lambda(i,k)$).

Let us suppose that indices i and k are fixed and $i<k$, we say that subspaces R_{ijk} are chosen in a *concordant* way if subspaces R_{ijk} and R_{ij+1k} /$(j+1)\leq k$/ are isomorphic with respect to the relation W_j for every j such that $i\leq j<k$. The proof of the following theorem is very technical. It can be omited by the first reading.

5.2. THEOREM. *Subspaces R_{ijk} can be chosen in a concordant way.*

PROOF. We start a choice of subspaces R_{ijk} from $R_{11k}\subseteq V_1$. If k_0 is the minimum of chain $[1^+]^\lambda=A^\lambda_{n,0}$, we choose $R_{11k_0}=V_{11k_0}$. Otherwise we choose R_{11k} such that $V_{11k^\bullet}R_{11k}=0$ and $V_{11k^\bullet}+R_{11k}$

---

1) $k=\min[j^+]^\lambda$ means that k is minimum of the chain $[j^+]^\lambda$.

Otherwise we choose R_{1_1k} such that $V_{1_1k^*} \cdot R_{1_1k} = 0$ and $V_{1_1k^*} + R_{1_1k} = V_{1_1k}$, where $k^* <\cdot^\lambda k$. The further construction of R_{i_jk} will be given by means of mapping R_{i_jk} from V_j to V_{j+1} using relation W_j. Let us suppose that subspaces R_{i_jk} have been chosen in all $V_1, \ldots, V_j$ /$1 \leq j$/ in a concordant way. The following arguments depends on a type of W_j.

5.2.1. W_j is $\overset{0}{\circ}\!\!-\!\!\overset{0}{\circ}$ type, i.e. $\nu_j(j) = \nu_j(j+1) = 0 \Leftrightarrow W_jV_j = W_jV_{j+1} = 0$. Such a relation give a isomorphism between the subspaces $V_j(W_j + V_{j+1})$ and $V_{j+1}(W_j + V_j)$ /see Fig. 1.1 /. According to Propositions 4.7 and 4.8 and since $\nu_j(j) = \nu^+(j) = 0$ we can write $V_j = V_{jj} = V_{i^*_j} = V_{i^*_j j}$ where $i^* = \max[j^-]^\pi$ and similarly since $\nu_j(j+1) = \nu^-(j+1) = 0$ we can write $V_{j+1} = V_{j+1_{j+1}} = V_{j+1_{k^\wedge}} = V_{j+1_{j+1}k^\wedge}$, where $k^\wedge = \max[(j+1)^+]^\lambda$.

Hence $V_{j+1}(W_j + V_j) = V_{j+1}(W_j + V_{i^*_j}) = \varphi^+_j(V_{i^*_j}) = V_{i^*_{j+1}} = V_{i^*_{j+1}k^\wedge}$, analogously, $V_j(W_j + V_{j+1}) = \varphi^-_{j+1}(V_{j+1}) = \varphi^-_{j+1}(V_{j+1k^\wedge}) = V_{jk^\wedge} = V_{i^*_jk^\wedge}$.

We have supposed that $V_{i^*_jk^\wedge}$ can be represented in the form:

$$V_{i^*_jk^\wedge} = \bigoplus_{(i,k)}' R_{i_jk}. \tag{5.2.2}$$

(notice, here $\oplus'$ means that the sum is taken over (i, k) such that $((i, k) \leq^\nu (i^*, k^\wedge)) \Leftrightarrow (i \leq^\pi i^*$ and $k \leq^\lambda k^\wedge)$ with the same fixed j).

Since $j = \max[j^+]^\lambda$ we have $k \leq^\lambda k^\wedge <\cdot^\lambda j$ for all k such that $j < k$. /notice we use two different order relattion on $A_{n,0}$: trivial order $\leq$ and $\leq^\lambda$, so that $k <^\lambda j$ and $j < k$/. Let us denote $\varphi^+_j(R_{i_jk}) = R'_{i_{j+1}k}$ if $j < k$. As far as subspaces $V_j(W_j + V_{j+1}) = V_{i^*_jk^\wedge}$ and $V_{j+1}(W_j + V_j) = V_{i^*_{j+1}k^\wedge}$ are isomorphic under the action of relation W_j /in other words, under the action of the mappings φ^+_j and φ^-_{j+1} /, we obtain $V_{i^*_{j+1}k^\wedge} = \varphi^+_j(V_{i^*_jk^\wedge}) = \varphi^+_j(\bigoplus_{(i,k)}' R_{i_jk}) = \bigoplus_{(i,k)}' (\varphi^+_j R_{i_jk}) = \bigoplus_{(i,k)}' R'_{i_{j+1}k}$.

We receive similarly that for each $V_{i'_jk'} \subseteq V_{i^*_jk^\wedge}$

$$\varphi^+_j(V_{i'_jk'}) = V_{i'_{j+1}k'} = \bigoplus_{(i,k)}' R'_{i_{j+1}k}. \tag{5.2.3}$$

The mapping φ^+_j sets isomorphism between such $V_{i'_jk'}$ and $V_{i'_{j+1}k'}$. This means that the equalities 5..1.1 are fulfilled for the subspaces $R'_{i_{j+1}k}$, and hence they can be designate as $R_{i_{j+1}k}$.

Remembe we have $i^* <\cdot^\pi (j+1)$ and $V_{j+1_{j+1}k^\wedge} = V_{j+1_{j+1}} = V_{j+1}$. Hence it is sufficient to choose subspaces $R_{j+1_{j+1}k}$ in $V_{j+1_{j+1}k} = V_{j+1k}$. Let us suppose that $\min[(j+1)^+]^\lambda = m$, in other words V_{j+1m} is minimum in the chain $\{V_{j+1k}\}$. We can depict all such subspaces of the type $V_{j+1_{j+1}k}$ and $V_{i^*_{j+1}k}$ in the following diagram (5.4.2). After that we choose subspaces $R_{j+1_{j+1}k}=R_{\gamma}$ as usually: $X_\gamma R_\gamma=0$, $X_\gamma+R_\gamma=V_\gamma$. If $k=m$ then $X_{j+1_{j+1}m}=V_{i^*{}_{j+1}m}$ and if $m<^\gamma k$ then $X_{j+1_{j+1}k}=V_{i^*{}_{j+1}k}+X_{j+1_{j+1}k}$ and $k'<\cdot^\lambda k$.

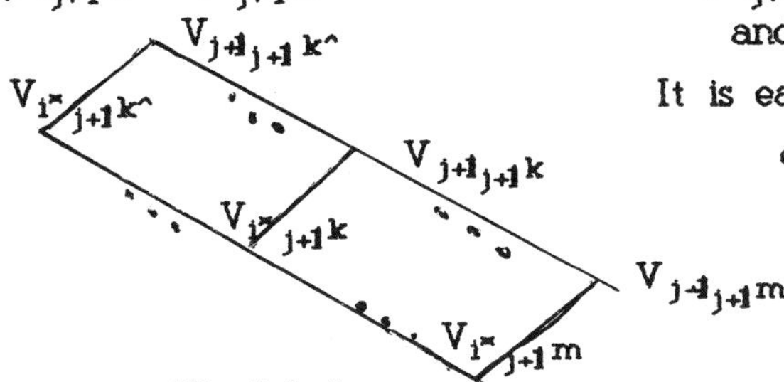

Fig. 5.2.4.

It is easy to prove now, that each subspace $V_{j+1_{j+1}k}$ may be represented in in the form:

(5.2.5) $$V_{j+1_{j+1}k}=\bigoplus_{(i',k')}{}' R_{j'_{j+1}k'}.$$

where the sum is taken over such pair (i',k'), that $i'\leq(j+1)\leq k'$ and $(j',k')\leq^\nu(i,k)$. The equality (5.2.5) is fulfilled in particular for $V_{j+1_{j+1}k^\wedge}=V_{j+1}$. It means that each subspace $V_{i_{j+1}k}\subseteq V_{j+1}$ can be represented in the form:

$$V_{i_{j+1}k}=\bigoplus_{(i',k')}{}' R_{i'_{j+1}k'}.$$

5.2.6. The relation W_j is (1,0) type, i.e. $\nu_j(j)=1$ and $\nu_j(j+1)=0 \Leftrightarrow W_j+V_{j+1}=V_j+V_{j+1}$ and $V_jW_{j+1}=0$. In another words W_j is isomorphic to the graph of a linear map $\omega_j: V_j\longrightarrow V_{j+1}$. Notice that $\mathrm{Ker}\,\omega_j=W_jV_j$ and $\mathrm{Im}\,\omega_j=\varphi_j^+(V_j)=V_{j+1}(W_j+V_j)$. If we write the subspaces V_j and V_{j+1} in the canonical form we get $V_j=V_{i^*{}_jk^*}$, and $V_{j+1}=V_{i^\wedge{}_{j+1}k^\wedge}$, where $i^*=\max[j^-]^\pi$, $k^*=\max[j^+]^\lambda$, $i^\wedge=\max[(j+1)^-]^\pi$ and $k^\wedge=\max[(j+1)^+]^\lambda$. Since $\nu_j(j+1)=0$ we get $i^\wedge=\max[(j+1)^-]^\pi$, i.e. $V_{j+1}=V_{j+1_{j+1}k^\wedge}$ and $k^*<\cdot^\pi(j+1)$. Since $\nu_j(j)=1$ than according to definition $W_jV_j=V_{jj}$ it means $j=\min[j^+]^\lambda$ and hence $i^*=i^\wedge$.

We suppose on induction that each subspace $V_{i_jk}\subseteq V_j=V_{i^*{}_jk^*}$ can be written in the form:

$$V_{i_jk}=\bigoplus_{(i',k')}{}' R_{i'_jk'}.$$

Since $W_jV_{j+1}=0$, we obtain using Proposition 4.11, that for $j<k$ $\varphi_j^+(V_{i_jk})=V_{i_{j+1}k}+W_jV_{j+1}=V_{i_{j+1}k}$. If k is such that $j<k$ we can denote $\varphi_j^+(R_{i_jk})=R'_{i_{j+1}k}$. If we have $j=k$, then $V_{i_jj}\subseteq V_{jj}=W_jV_j$, and hence $\varphi_j^+(V_{i_jj})=V_{j+1}(W_j+V_{i_jj})=V_{j+1}W_j=0$. The subspace $V_j=V_{i^*{}_jk^*}$ can be written in the form $V_{i^*{}_jk^*}=(\bigoplus_{(i,k)}R_{i_jk})\oplus(\bigoplus_i R_{i_jj})=$

$=(\bigoplus_{(i,k)}R_{i_jk})\oplus V_jW_j=U\oplus V_jW_j$, where $U=(\bigoplus_{(i,k)}R_{i_jk})$ and sum is taken over such pairs (i,k) that $j<k$ and $(i,k)\leq^{\nu}(i^*,k^*)$. If we use lemma 1.4.1 we obtain that the relation W_j set a isomorphism between the subspaces U and $\varphi_j^+(V_j)=\varphi_j^+(V_{i^*{}_jk^*})=V_{i^*{}_{j+1}k^*}$, and that $\varphi_j^+(V_{i^*{}_jk^*})=\varphi_j^+(U)=\varphi_j^+(\bigoplus_{(i,k)} R_{i_jk})=\bigoplus'_{(i,k)} R'_{i_{j+1}k}=V_{i^*{}_{j+1}k^*}$.

Using a similar argument we obtain, that each subspace $V_{i_jk}\subseteq V_j$, has the following property: if $j<k$, then $\varphi_j^+(V_{i_jk})=V_{i_{j+1}k}=\bigoplus'_{(i',k')} R'_{i'_{j+1}k'}$. Hence we can designate $R'_{i_{j+1}k}=R_{i_{j+1}k}$.

Another subspaces $R_{j+1_{j+1}k}$ which doesn't belong to $V_{i^*{}_{j+1}k^*}=\varphi_j^+(V_j)$, can be chosen in the same way as we have done before /when W_j was (0,0) type/.

5.2.7. The relation W_j is (0,1) type, i.e. $\nu_j(j)=0$ and $\nu_j(j+1)=1 \Leftrightarrow V_jW_j=0$ and $W_j+V_j=V_j+V_{j+1}$. It means that W_j is isomorphic to the graph of a linear map $\omega_j^-:V_{j+1}\longrightarrow V_j$. Since $\nu_j(j+1)=1$ we can write $\mathrm{Ker}\,\omega_j^-=W_jV_{j+1}=V_{j+1_{j+1}}$. If we express the subspace V_{j+1} in the canonical form we get $V_{j+1}=V_{i^*{}_{j+1}k^*}$, where $i^*=\max[(j+1)^-]^{\pi}$ and $k^*=\max[(j+1)^+]^{\lambda}$. As far as $\nu_j(j+1)=1$, we obtain that $(j+1)=\min[(j+1)^-]^{\pi}$ and hence $i^*\neq(j+1)$.

Using $\nu_j(j)=\nu^+(j)=0$, we get $V_j=V_{i^{\vee}{}_jj}$, where $i^{\vee}=\max[i^-]^{\pi}$. It is easy to verify that $i^{\vee}=i^*$. Therefore $V_j=V_{i^*{}_jj}$. Such W_j give the isomorphism between the subspace $V_j(W_j+V_{j+1})=\varphi_{j+1}^-(V_{j+1})=\varphi_{j+1}^-(V_{i^*{}_{j+1}k^*})=V_{i^*{}_jk^*}$ and factorspace $V_{j+1}/\mathrm{Ker}\,\omega_j^-=V_{j+1}/W_jV_{j+1}=V_{i^*{}_{j+1}k^*}/V_{j+1_{j+1}k^*}$. We have supposed that each subspace $V_{i_jk}\subseteq V_{i^*{}_jj}=V_j$ can be written in the form

(5.2.8) $$V_{i_jk}=\bigoplus'_{(i',k')} R_{i'_jk'}$$

The equality 5.2.8 can be used in particular for $\mathrm{Im}(\omega_j)=V_{i^*{}_jk^*}$

We start now to construct subspaces $R_{i_{j+1}k}$ in $V_{j+1}=V_{i^*{}_{j+1}k^*}$. We begin this construction from $R_{j+1_{j+1}k}\subseteq V_{j+1_{j+1}k^*}=\mathrm{Ker}\,\omega_j^-$. Notice that this subspace $V_{j+1_{j+1}k^*}$ is the maximum in the chain $\{V_{j+1_{j+1}k}\}$ where $k\in[(j+1)^+]^{\lambda}$. We choose subspaces $R_{j+1_{j+1}k}=R_{\gamma}$ as usual, $X_{\gamma}R_{\gamma}=0$ and $X_{\gamma}+R_{\gamma}=V_{\gamma}$. Let us denote temporarily $m=\min[(j+1)^-]^{\lambda}$. As far as $(j+1)=\min[(j+1)^-]^{\pi}$, we get $X_{j+1_{j+1}m}=0$ and hence $R_{j+1_{j+1}m}=V_{j+1_{j+1}m}$, and if $m<^{\lambda}k$ then $X_{j+1_{j+1}k}=V_{j+1_{j+1}k'}$, where $k'<\cdot^{\lambda}k$. It is easy to see now that every subspace $V_{j+1_{j+1}k}$ in $\mathrm{Ker}\,\omega_j^-=V_{j+1_{j+1}k^*}$ can be expressed in the form $V_{j+1_{j+1}k}=\bigoplus R_{j+1_{j+1}k'}$, here the sum is taken over all such k', that $k'\leq^{\lambda}k$ and $(j+1)\leq k'$.

We choose now subspaces $R_{i_{j+1}k}$ with $i<(j+1)$. We have chosen $R_{i_jk} = R_\gamma$ in such a way, that $X_\gamma R_\gamma = 0, X_\gamma + R_\gamma = V_\gamma$. The spaces R_{i_jj} doesn't belong to $\mathrm{Im}\omega_j^- = V_{i^*_jk^*}$ and they aren't subject of further discussion. We shall show later on that a space $R_{i,j} = \sum R_{i_qj}$ split off as a direct summand, here the sum is taken over all R_{i_qj} with the same fixed i,j and such q, that $i \le q \le j$.

It is easy to check that intervals (X_{i_jk}, V_{i_jk}) and $(X_{i_{j+1}k}, V_{i_{j+1}k})$ satisfy the conditions of the Proposition 2.2, and hence these intervals are projective. Therefore we can use the Proposition 2.3 and hence choose the subspace $R_{i_{j+1}k} = R_\gamma$ as $X_\gamma R_\gamma = 0, X_\gamma + R_\gamma = V_\gamma$, and so that W_j give the isomorphism between R_{i_jk} and $R_{i_{j+1}k}$. It means that R_{i_jk} and $R_{i_{j+1}k}$ are chosen in a "concordant" way.

5.2.9. The relation W_j is (1,1) type, i.e. $\nu_j(j) = \nu_j(j+1) = 1 \Leftrightarrow W_j + V_{j+1} = W_j + V_j = V_j + V_{j+1}$. Such W_j give the isomorphism between factorspaces V_j/V_jW_j and $V_{j+1}/V_{j+1}W_j$. The spases V_j, V_{j+1} can be written in canonical form. We obtain $V_j = V_{i^*_jk^*}$, where $i^* = \max[j^-]^\pi$ and $k^* = \max[j^+]^\lambda$. We also get $V_{j+1} = V_{i^\vee_{j+1}k^\vee}$, where $i^\vee = \max[(j+1)^-]^\pi$ and $k^\vee = \max[(j+1)^+]^\lambda$. Using $\nu^+(j) = \nu^-(j+1) = 1$, we find that $i^* = i^\vee$ and $k^* = k^\vee$, i.e. $V_{j+1} = V_{i^*_{j+1}k^*}$. We also get $W_jV_j = V_{jj} = V_{i^*_jj}$ and $W_jV_{j+1} = V_{j+1_{j+1}} = = V_{j+1_{j+1}k^*}$.

We have supposed that each $V_{i_jk} \subseteq V_{i^*_jk^*} = V_j$ can be written as

$$V_{i_jk} = \bigoplus_{(i'k')}{}' R_{i'_jk'} \tag{5.2.10}$$

The expression 5.2.10 can be used in particular for $\mathrm{Ker}(\omega_j) = V_jW_j = V_{i^*_jj}$. The subspaces $R_{i_{j+1}k}$ in $V_{j+1} = V_{i^*_{j+1}k^*}$ are chosen in the same manner, as for W_j of (0,1) type. We choose at first $R_{j+1_{j+1}k}$ in $\mathrm{Indef}\, W_j = W_jV_{j+1} = V_{j+1_{j+1}k^*}$. Notice that $V_{j+1_{j+1}k^*}$ contain a chain $\{V_{j+1_{j+1}k}\}$. If $V_{j+1_{j+1}m}$ is the minimum of this chain, then we choose $R_{j+1_{j+1}m} = V_{j+1_{j+1}m}$. Otherwise we select $R_{j+1_{j+1}k} = R_\gamma$ so, that $V_{\gamma'}R_\gamma = 0$, $V_{\gamma'} + R_\gamma = V_\gamma$, where $\gamma' = (j+1_{j+1}k')$ and $k' \lessdot^\lambda k$. It is easy to see that each $V_{j+1_{j+1}k} \subseteq \mathrm{Indef}\,\omega_j = V_{j+1_{j+1}k^*}$ can be expressed as $V_{j+1_{j+1}k} = \bigoplus_{k'} R_{j+1_{j+1}k'}$, where the sum is taken over all such k', that $k' \le^\lambda k$ and $(j+1) \le k'$.

We select now $R_{i_{j+1}k}$ with $i<(j+1)$. We have chosen $R_{i_jk} = R_\gamma$ in such a way, that $R_\gamma X_\gamma = 0$ and $R_\gamma + X_\gamma = V_\gamma$. /Remind that R_{i_jj}

belongs to $\mathrm{Ker}\,\omega_j = V_j W_j = V_{i^*j}$ and therefore $\varphi_j^+(R_{i_j j}) = V_{j+1}(W_j + R_{i_j j}) \subseteq V_{j+1}(W_j + V_j W_j) = V_{j+1} W_j$. We shall show later on. that $R_{i,j} = \sum R_{i_q j}$. /where the sum is taken over all such $R_{i_q j}$, that $i \leq q \leq j$ and i. j are fixed/ split off as a direct summand/.

We can now verify that intervals $(X_{i_j k}, V_{i_j k})$ and $(X_{i_{j+1} k}, V_{i_{j+1} k})$ satisfy the conditions of the Propsition 2.2,and hence these intervals are projective. Therefore we can use the Proposition 2.3 and select a subspace $R_{i_{j+1} k}$ in a "concordant" way with $R_{i_j k}$. ▪

5.3.THEOREM. *Let $R_{i_j k}$ be chosen in a concordant way, and $R_{ik} = \bigoplus_j R_{i_j k}$, where the sum is taken over all such j, that $i \leq j \leq k$, with the same fixed i and k. Then a representation ρ is decomposed into a direct sum $\rho = \bigoplus_{(i,k)} \rho_{ik}$, where $\rho_{ik} = \rho / R_{ik}$, and every ρ_{ik} is isomorphic to the direct sum $\rho_{ik} \cong \underbrace{\tau_{ik} \oplus \ldots \oplus \tau_{ik}}_{m_{i,k}}$ of the indecomposable representation τ_{ik}, where $m_{ik} = \dim R_{i_j k}$.*

PROOF. It is sufficient to show that for every generator $x \in L(A_n^\nu)$ he corresponding subspace $X = \rho(x)$ can be represented in the form $X = \sum_{(i,k)} XR_{ik}$. Each subspace $V_j = \rho(v_j)$ can be written as $V_j = \bigoplus_{(i,k)} R_{i_j k}$, where the sum is taken over all such (i,k) that $i \leq j \leq k$ and j is fixed. Hence we must show that $V_j = \sum_{(i,k)} V_j R_{ik}$. Notice that a lattice B_R generated by all the spaces $R_{i_j k}$ is distributive one (and what is more it is a Boolien lattice).Using this fact we obtain that $R_{i_j k} R_{i'_{j'} k'}$ is different from zero if and only if $i = i', j = j'$ and $k = k'$ Therefore

$$V_j R_{ik} = \Big(\sum_{(i',k')} R_{i'_j k'}\Big)\Big(\sum_{j'} R_{i_{j'} k}\Big) = \begin{cases} R_{i_j k}, & \text{if } i \leq j \leq k, \\ 0, & \text{otherwise.} \end{cases}$$

Hence $\sum_{(i,k)} V_j R_{ik} = \sum_{(i,k)} R_{i_j k} = V_j$.

We wont to prove now that $\rho(w_j) = W_j = \sum_{(i,k)} W_j R_{ik}$. According to the definition $W_j \subseteq (V_j + V_{j+1})$. It means that $W_j R_{ik} = W_j (V_j + V_{j+1}) R_{ik}$. If we use the distributivity of the lattice B_R we get as above:

$$(5.3.1) \qquad (V_j + V_{j+1}) R_{ik} = \begin{cases} R_{i_j j}, & \text{if } k = j, \\ R_{i_j k} + R_{i_{j+1} k}, & \text{if } i \leq j < k, \\ R_{j+1_{j+1} k}, & \text{if } i = j+1, \\ 0, & \text{otherwise.} \end{cases}$$

Let us fix j and after that denote in 5.3.1 $R_{i_j j} = U_i$, $(R_{i_j k} + R_{i_{j+1} k}) = U_{ik}$, and $R_{j+1_{j+1} k} = U_k$. We wont to prove,that W_j is decomposed on the family of subspaces: $\{U_i, 1\leq i\leq j; U_{ik}, i\leq j<k; U_k, (j+1)\leq k\leq n\}$. It is not difficult to prove, that conditions of Lemma 1.4.1 and Corollary 1.4.2. are fulfilled. Therefore W_j is decomposed on this family. It remains to show that a representation $\rho_{ik}=\rho|_{R_{ik}}$ is decomposed into a direct sum of representation τ_{ik}. If $i=k$, then this statement is obvious. Next we consider cases $i<k$. Let $\{\xi_{i,q}\}/q=1,2,\dots,m_{ik}/$ be a basis in R_{i_ik}. We have chosen the subspaces R_{i_jk} in a concordant way, i.e. the relation W_j gives the isomorphism between R_{i_jk} and $R_{i_{j+1}k}$. Hence there are vectors $\xi_{i+1,q}\in V_{i_{i+1}q}$, such that $(\xi_{i,q}+\xi_{i+1,q})\in W_i$, and $\{\xi_{i+1,q}\}$ is a basis in $V_{i_{i+1}k}$. If we continue this process be obtain a basis $\{\xi_{j,q}\}$ in each subspace R_{i_jk} $(i\leq j\leq k)$. Let us denote by $T_{ik,q}$ the subspace stratched on the vectors $\xi_{i,q},\xi_{i+1,q},\dots,\xi_{k,q}$ and by $\tau_{ik,q}=\rho|_{T_{ik,q}}$. It is easy to show that each representation $\tau_{ik,q}$ is isomorphic to the indecomposable representation $\tau_{i,k}$. Therefore $\rho_{ik}=\underline{\tau_{ik}\oplus\dots\oplus\tau_{ik}}_{m_{i,k}}$, where $m_{ik}=\dim V_{i_jk}$. ∎

Theorem 5.3. and result of the section 3 show that each indecomposable representations τ is isomorphic to one of $\tau_{i,k}$ and hence the following theorms are true:

5.4. THEOREM. *The set $\mathcal{D}(A_n^\nu)$ of isomorphism classes of all indecomposable representations of $L(A_n^\nu)$ is ordered in the following way. If $\tau_\alpha=\tau_{i,k}$ and $\tau_{\alpha'}=\tau_{i',k'}$, then $\tau_\alpha\leq^\nu\tau_{\alpha'}) \Leftrightarrow (i\leq^\pi i'$ and $k\leq^\lambda k')$.* ∎

5.5. THEOREM. *There is one-to-one correspondence between the indecomposable representations of a lattice $L(A_n^\nu)$ and the positive roots of the Dynkin diagram A_n. If α is a positive root and $\alpha=\alpha_i+\dots+\alpha_k$, where α_j is a simple root, then the indecomposable represantation $\tau_{i,k}$ corresponds to such the root α.* ∎

§6. CONSTRUCTION OF PERFECT ELEMENTS.

6.1. Let us denote by a_i and b_k the following elements of $L(A_n^\nu)$:

$$a_i=\begin{cases}\sum_{j=1}^{i} v_j+\sum_{j=i+1}^{n} v_{i_j}, & \text{if } i=1 \text{ or } \nu_{i-1}(i)=0,\\ \sum_{j=i}^{n} v_{i_j}, & \text{if } \nu_{i-1}(i)=1,\end{cases}$$

$$b_k=\begin{cases}\sum_{j=1}^{k-1} v_{jk}+\sum_{j=k}^{n} v_j, & \text{if } k=n \text{ or } \nu_k(k)=0,\\ \sum_{j=1}^{k} v_{jk}, & \text{if } \nu_k(k)=1.\end{cases}$$

6.2.PROPOSITION. *The elements $\{a_i\}$ and $\{b_k\}$ forms a chains:*

6.2.1. $(a_i \subseteq a_{i'}) \Leftrightarrow (i \leq^{\pi} i')$,

6.2.2. $(b_k \subseteq b_{k'}) \Leftrightarrow (k \leq^{\lambda} k')$.

A proof follows easily from property 4.5 of elements v_{i_j} and v_{jk}. ∎

REMARK. The maximums between $\{a_i\}$ and $\{b_k\}$ are equal $a_{i^*} = b_{k^*} = e = v_1 + v_2 + \ldots + v_n$. Their index can be found in the following way:

$$(a_{i^*} = e) \Leftrightarrow (i^* = \max A^{\pi}_{n0}) \text{ and } (b_{k^*} = e) \Leftrightarrow (k^* = \max A^{\lambda}_{n0})$$

6.3.THEOREM. *The elements a_i and b_k are perfect.*

Before we start to prove this theorem, let us remind that we say that H / $H \subseteq \mathcal{P}(A^{\nu}_n) \cong P(A^{\nu}_n)$/ is a *characteristic* subset of a perfect element h, if $(\tau \in H) \Leftrightarrow (\tau(h) = V_{\tau}$, where V_{τ} is the spase of representation τ). /Here $P(A^{\nu}_n)$ is the set of all positive roots of the Dynkin diagram A_n/.

We shall show that the characteristic subset $h(a_i)$, corresponding to the element a_i, is the following one: $(\tau_{i',k} \in h(a_i)) \Leftrightarrow (i' \leq^{\pi} i$ and k is arbitrary, i.e. $i' \leq k$).

Let ρ be a representation of $L(A^{\nu}_n)$. According to the definition

$$\rho(a_i) = \begin{cases} \sum_{j=1}^{i} V_j + \sum_{j=i+1}^{n} V_{i_j}, \text{if } i=1 \text{ or } \nu_{i-1}(i)=0. \\ \sum_{j=1}^{n} V_{i_j}, \text{ if } \nu_{i-1}(i)=1. \end{cases}$$

where $V_j = \rho(v_j)$ and $V_{i_j} = \rho(v_{i_j})$. /Notice that $V_{1_1} = V_1$ and $V_{i_i} = W_{i-1} V_i$ if $\nu_{i-1}(i) = 1$/. We wont to show that $\rho(a_i)$ contains all indecomposable representation $\tau_{i'k}$, such that $i' \leq^{\pi} i$, and doesn't contains $\tau_{i'',k}$, such that $i <^{\pi} i''$. Let $R_{i',k}$ and $R_{i'',k}$ be the spaces of representations $\rho_{i',k}$ and $\rho_{i'',k}$, which are multiple of $\tau_{i',k}$ and $\tau_{i'',k}$. than it is sufficient to show, that

(6.3.1) $$R_{i',k} \subseteq \rho(a_i), \quad \text{if } i' \leq^{\pi} i,$$

(6.3.2) $$\rho(a_i) R_{i'',k} = 0, \text{ if } i <^{\pi} i''.$$

We shall proove these properties, using induction on $\leq^{\pi}$.

6.3.3. We start to prove (6.3.1) for the minimum element a_{i_m} in a chain $\{a_j\}$ / remind that $(a_{i'} \subseteq a_i) \Leftrightarrow (i' \leq^{\pi} i)$/. It is easy to check that (a_{i_m} is the minimum) $\Leftrightarrow (i_m = \min A^{\pi}_{n,0}) \Leftrightarrow (i_m = \max_1 \{i \mid \nu^1(i) = 1\})$, here $\max_1$ in a sence of the trivial ordering on $A_{n,0}$ and where $\nu^1(i)$

$$= \begin{cases} \nu_{i-1}(i), \text{ if } i \neq 1 \\ 1, \text{if } i = 1. \end{cases}$$

We calculate $R_{i.k}=\sum_{j=i}^{k} R_{i_jk} \subseteq \sum_{j=i}^{k} V_{i_jk} \subseteq \sum_{j=1}^{n} V_{i_j} \subseteq \rho(a_i)$.

6.3.4. Let us suppose that we have proved 6.3.1 for several sequential members of the chain $\{a_j\}$ starting from a minimum element a_{i_m} and ending with an element $a_{i'}$. Suppose that $(a_{i'} \subseteq\cdot a_i)/\Leftrightarrow (i' <\cdot^{\pi} i)$. We assume on induction that $(R_{i^\wedge,k} \subseteq \rho(a_{i'}))$ for all $i^\wedge \leq^{\pi} i'$. Since $\rho(a_{i'}) \subseteq \rho(a_i)$, it is sufficient to show that $R_{i,k} \subseteq \rho(a_i)$ for any k /$i \leq k$/. We get $R_{i.k}=\sum_{j=i}^{k} R_{i_jk} \subseteq \sum_{j=i}^{k} V_{i_jk} \subseteq \sum_{j=i}^{n} V_{i_j} \subseteq \rho(a_i)$.

6.3.5. We shall prove 6.3.2 inductively, discending from the maximum element $a_{i^*}=e$ in the chain $\{a_j\}$. Then $\rho(a_{i^*})=E=\bigoplus_j V_j$, and 6.3.2 is obviously true since each $\rho_{i,k} \in E$.

Let us suppose that 6.3.2. has been proved for some $\rho(a_i)$. We shall prove 6.3.2. for $\rho(a_{i'})$, where $a_{i'} \subset\cdot a_i$ /$i' <\cdot^{\pi} i$/. Since $\rho(a_{i'}) \subseteq \rho(a_i)$, we get $\rho(a_i)R_{i'',k}=0$, for all i'' such that $i <^{\pi} i''$. Therefore we must prove that $\rho(a_{i'})R_{i,k}=0$, for all admissible k /i.e. $i \leq k$/. All subspaces R_{i_jk} form a Boolean lattice B_R, and $R_{i,k}, \rho(a_i) \in B_R$. Hence for the proof $\rho(a_{i'})R_{i,k}=0$, it is sufficient to show that $\rho(a_{i'})V_jR_{ik}=0$ for all j. Notice that $V_jR_{i,k}=0$, if $j<i$ or $k<j$; otherwise, if $i \leq j \leq k$, $V_jR_{i,k}=R_{i_jk}$. So we must prove that $\rho(a_{i'})R_{i_jk}=0$ for all such j that $i \leq j \leq k$. It can be three possibility: a) $\nu_{i'-1}(i')=0$, b) $i'=1$, c) $\nu_{i'-1}(i')=1$. It follows easily from from the definition of the ordering $\leq^{\pi}$, that in cases a) and b) we have $i'<i$ /in trivial ordering/. Hence we obtain in these cases that if $i'<i \leq j$ then $\rho(a_{i'})V_j=V_{i'_j}$. Therefore $\rho(a_{i'})V_jR_{i,k}=V_{i'_j}R_{i_jk}=V_{i'_j}V_{i_jk}R_{i_jk}=V_{i'_j}V_{i_j}V_{jk}R_{i_jk}=V_{i'_j}V_{jk}R_{i_jk}=V_{i'_jk}R_{i_jk}$. Using the definition of R_{i_jk} and $i' <\cdot^{\pi} i$ we get $V_{i'_jk}R_{i_jk}=0$ and hence $\rho(a_{i'})V_jR_{i,k}=0$. If we have $\nu_{i'-1}(i')=1$, then $\rho(a_{i'})=\sum_{j=i'}^{n} V_{i'_j}$. It can be in this case either $i'<i$ or $i<i'$. If $i'<i$, then we get in the same way as we have done above, that for j /$i'<i \leq j \leq k$/ $\rho(a_{i'})V_jR_{i,k}=0$ and hence $\rho(a_{i'})R_{i,k}=0$. We assume now that $i<i'$. If j is such that $i \leq j < i'$, then $\rho(a_{i'})V_j=0$ and hence $\rho(a_{i'})V_jR_{i,k}=0$. If $i' \leq j$ then $\rho(a_{i'})V_jR_{i,k}=V_{i'_j}R_{i_jk}=V_{i'_j}V_{i_jk}R_{i_jk}=V_{i'_jk}R_{i_jk}=0$. It follows from both these cases that $\rho(a_{i'})R_{i,k}=0$.

We obtain from 6.3.1 and 6.3.2 that ρ is decomposed in a direct sum:

$$\rho = \rho_{a_1} \oplus \left(\bigoplus_{(i'',k)} \rho_{i'',k} \right),$$

where the sum $\bigoplus_{(i'',k)} \rho_{i'',k}$ is taken over such (i'',k) that $i \leq^{\pi} i''$ and $i'' \leq k$, and where $\rho_{a_1} = \rho|_{\rho(a_1)} = \bigoplus_{(i',k)} \rho_{i',k}$, and this sum is taken over such (i',k) that $i' \leq^{\pi} i$, and k is arbitrary admissible /i.e. $i' \leq k$/. Notice that both $\rho_{i',k}$ and $\rho_{i'',k}$ are multiplies of the indecomposable $\tau_{i',k}$ and $\tau_{i'',k}$ respectively.

This property of the space $\rho(a_1)$ means that element $a_1 \in L(A_n^{\nu})$ is perfect one. A characteristic subset $H_{a_1} \subseteq \mathcal{P}(A_n^{\nu}) \cong P(A_n^{\nu})$ of a_1 consists of such indecomposable $\tau_{i',k}$ that $i' \leq^{\pi} i$, and k is arbitrary admissible, /i.e. $i' \leq k$/.

We can prove analogously that each element b_k is also perfect. The characteristic subset H_{b_k} of b_k consists of such indecomposable $\tau_{i,k'}$ that $k' \leq^{\lambda} k$, and i is arbitrary admissible /i.e. $i \leq k'$/. ▪

6.4. THEOREM. *A lattice $D(A_n^{\nu})$ of perfect elements $h \in L(A_n^{\nu})$ is isomorphic to the lattice of ideal $\mathfrak{I}(\mathcal{P})$ of poset $\mathcal{P} = \mathcal{P}(A_n^{\nu})$. An arbitrary perfect element $h \in D(A_n^{\nu})$ can be written in the form:*

$$h = a_{i_1} b_{k_1} + a_{i_2} b_{k_2} + \ldots + a_{i_m} b_{k_m},$$

where all the pairs (i_q, k_q) are admissible (i.e. $i_q \leq k_q$), and indices $\{i_q\}$ and $\{k_q\}$ are ordered in the following way: $i_1 <^{\pi} i_2 <^{\pi} \ldots <^{\pi} i_m$ and $k_1 >^{\lambda} k_2 >^{\lambda} \ldots >^{\lambda} k_m$.

PROOF. Let $H \subseteq P(A_n^{\nu})$ an arbitrary ideal of the poset $P(A_n^{\nu})$ of positive roots /$P(A_n^{\nu}) \cong \mathcal{P}(A_n^{\nu})$/. H can be represented as a join $H = \cup H_\alpha$ of $\vee$- irrduceble ideal H_α /$H_\alpha = \{\beta \mid \beta \leq^{\nu} \alpha, \beta \in P(A_n^{\nu})\}$. /Such the ideal H_α is principal one/. We shall write $H_\alpha = H_{i,k}$, where (i,k) /$1 \leq i \leq k \leq n$/ is the index of the root α. Hence $H_\alpha = H_{i,k} = \{(i',k') \mid i' \leq^{\pi} i \text{ and } k' \leq^{\lambda} k\}$. It is not difficult to verify that the meet ab and the join $a+b$ of perfect elements a and b are also perfect. We can check, starting from the definition, that the characterstic subset of perfect element $a_i b_k$ /$i \leq k$/ coincides with the subset $H_{i,k}$. Therefore an arbitrary ideal $H = \cup_\alpha H_\alpha = \cup_{(i,k)} H_{i,k} \subseteq P(A_n^{\nu})$ is the chracteristic subset of the perfect element $h = \sum_{(i,k)} a_i b_k$. It is easy to check that: 1) $a_i(b_{k_1} + b_{k_2} + \ldots + b_{k_s}) = a_i b_{k'}$, where $b_{k'}$ is the maximum of the elements $b_{k_1}, b_{k_2}, \ldots, b_{k_s}$, 2) $a_{i_1} b_k + a_{i_2} b_k + \ldots + a_{i_t} b_k = a_{i'} b_k$, where $a_{i'}$ is the maximum of $\{a_{i_q}\}$, 3) if $i' \leq^{\pi} i$ and $k' \leq^{\lambda} k$ then $a_{i'} b_{k'} + a_i b_k = a_i b_k$. Using these equalities, we can reduce $h = \sum_{(i,k)} a_i b_k$ to the canonical form $h = a_{i_1} b_{k_1} + a_{i_2} b_{k_2} + \ldots + a_{i_m} b_{k_m}$, where $(a_{i_1} \subset a_{i_2} \subset \ldots \subset a_{i_m}) \Leftrightarrow (i_1 \leq^{\pi} i_2 \leq^{\pi}$

$\cdots \le 1^\pi$) and $(b_{k_1} \supset b_{k_2} \supset \dots \supset b_{k_m}) \Leftrightarrow (k_1 \overset{\lambda}{\ge} k_2 \overset{\lambda}{\ge} \cdots \overset{\lambda}{\ge} k_m)$. ∎

§7. REMAINING PROBLEMS.

7.1. We shall discuss here some remaining problem. The whole structure of the latticies $L(A_n^\nu)$ $(n \ge 3)$ is not yet known. We say that the elements $x,y \in L(A_n^\nu)$ are *linearly equivalent* and write $x \sim y$ if for any representation ρ (over an arbitrary field K) $\rho(x) = \rho(y)$. We shall denote by $\underline{L(A_n^\nu)}$ the lattice $L(A_n^\nu)$ which is factored out on the linear equivalence.

Let us consider the representation $\tau_{\mathcal{P}} = \bigoplus_{(i,k)} \tau_{i,k}$, where (i,k) runs through whole set $\mathcal{P} = P(A_n^\nu) \cong \mathcal{P}(A_n^\nu)$. Let $V_{\mathcal{P}}$ be the space of this representation over the field $\mathbb{Q}$.

If we use the definition of the representations $\tau_{i,k}$, we can provide $V_{\mathcal{P}}$ with a basis $\{e_{i_jk}\}$ where $1 \le i \le j \le k \le n$. We have $\tau_{\mathcal{P}}(v_j) = V_j = \bigoplus_{(i,k)} \mathbb{Q} e_{i_jk}$ where the sum is taken over all (i,k) such that $i \le j \le k$. We omit here expession for $\tau_{\mathcal{P}}(W_j)$ which depends on equipment ν. It is easy to show that $\dim V_j = j(n+1-j)$ for each j, and $\dim V = (1/6)n(n+1)(n+2)$. If we set $T_{i,k} = \sum_j \mathbb{Q} e_{i_jk}$, where the sum is taken over all such j that $i \le j \le k$, then $\tau_{\mathcal{P}}|_{T_{i,k}} \cong \tau_{i,k}$. Let us denote by i' and k' such elements of A_{n0} that $i' \overset{\pi}{\lessdot} i, k' \overset{\lambda}{\lessdot} k$ if such a indices are exist.

We define the linear maps $\psi_i : T_{i,k} \longrightarrow T_{i',k}$ and $\psi'_k : T_{i,k} \longrightarrow T_{i,k'}$ (if there are such i' and k') as follows: we set $\psi_i(e_{i_jk}) = e_{i'_jk}$ if $i' \le j$ and $\psi_i(e_{i_jk}) = 0$ otherwise, also $\psi'_k(e_{i_jk}) = e_{i_jk'}$ if $j \le k'$ and $\psi'_k(e_{i_jk}) = 0$ otherwise. It is not dificult to verify that ψ_i and ψ_k can be extended until morphisms of representations: $\psi_i : \tau_{i,k} \longrightarrow \tau_{i',k}$ and $\psi'_k : \tau_{i,k} \longrightarrow \tau_{i,k'}$. We can also prove that the representations $\tau_{i,k}$ are *irreduceble* i.e. $\tau_{i,k}(L(A^\nu)) \cong \mathcal{L}(\mathbb{Q}^{k-i+1})$ if $(k-i) \ge 2$ /where $\mathbb{Q}$ is the field of rational numbers, and $\mathcal{L}$ is the lattice of all subspaces in $\mathbb{Q}^{k-i+1}$/.

We call a family of subspaces $U_{i,k}$ / $U_{i,k} \subseteq T_{i,k}$ / *admissible* if 1) $U_{i,i}$ is equal to 0 or $T_{i,i} \cong \mathbb{Q} e_{i_ii}$, 2) $U_{i,i+1}$ is equal to one of the five subspaces: $0, \mathbb{Q} e_{i_ii+1}, \mathbb{Q} e_{i_{i+1}i+1}, \mathbb{Q}(e_{i_ii+1} + e_{i_{i+1}i+1}), T_{i,i+1}$; 3) $\psi'_i U_{i,k} \subseteq U_{i',k}$ for all i,k if i' exists, and $\psi'_k U_{i,k} \subseteq U_{i,k'}$ for all i,k if k' exists.

Conjecture. The lattice $\underline{L(A_n^\nu)}$ is isomorphic to the lattice of all admissible subspaces in $V_{\mathcal{P}} = \bigoplus_{(i,k)} T_{i,k}$.

BIBLIOGRAPHY

[1] I.M.Gel'fand,V.A.Ponomarev: Gabriel's theorem is also trrue for representations of equipped graphs by relations. // Funkc. Anal. i Priloj.— 1981. V. 15, Vyp. 2.— P. 71-72.

[2] Gabriel P. Unzerlegbare Darstellungen, Manuscripta Math. 6 (1972), 71-103.

[3] Bourbaki N. Groupes et algebres de Lie. Chap. IV—VI.

[4] Gel'fand I.M.,Lidskiy B.V.,Ponomarev V.A. A preprojective reduction of a free modular lattice D^r.— Doklady Akademii Nauk USSR.— 1987 V293(3) p. 524-528.

[5] Birkhoff G. Lattice theory.- New York: Amer Math. Soc.,1948.

[6] Grätzer G. General lattice theory. — Berlin: Akademie-Verlag,1978.

Institite of Biophysics
Academy of Science
142292, t.Putschino
USSR

Canadian Mathematical Society
Conference Proceedings
Volume 11 (1991)

THE STRUCTURE OF THE PRINCIPAL p–BLOCK WITH CYCLIC NORMAL DEFECT GROUP

Klaus W. Roggenkamp[1]

ABSTRACT. In this note we shall give an explicit description of the principal p–block of $\mathbb{Z}_p^\wedge G$ for a finite group G with a normal cyclic Sylow p–subgroup. Here $\mathbb{Z}_p^\wedge$ stands for the ring of p–adic integers.

Let R be a complete Dedekind domain of characteristic zero with residue field $\mathbb{F}_p$ of characteristic $p > 0$ and field of fractions K, and let G be a finite group with a normal cyclic Sylow p–subgroup $P_n = < c_n : c_n^{p^n} = 1 >$. If $O_{p'}(G)$ is the largest normal subgroup of G of order prime to p, then $G/O_{p'}(G) = P_n \cdot S$ is a semi–direct product, where $S = < s : s^q = 1 >$ with $q \mid (p-1)$, and action ${}^s q = q^\sigma$ induced by viewing $P_n \cdot S$ as a subgroup of the affine group $\mathbb{F}_{p^n} \cdot \mathbb{F}_{p^n}^\times$.

The <u>principal block of</u> RG is the the group ring $RG/O_{p'}(G)$, and so we shall <u>assume</u> that $O_{p'}(G) = 1$.

For a normal subgroup N of G we denote by $I_R(N) \cdot G$ the induced augmentation ideal, i.e. the kernel of the natural map $RG \longrightarrow RG/N$.

The group ring RP_n is then a pullback – $P = P_n$ –

$$\text{(1a)}\qquad \begin{array}{ccccccccc} 0 & \longrightarrow & I_R(P) & \longrightarrow & RP & \longrightarrow & R & \longrightarrow & 0 \\ & & \downarrow id & & \downarrow \alpha & & \downarrow \beta & & \\ 0 & \longrightarrow & I_R(P) & \longrightarrow & \Lambda_n & \longrightarrow & R/p^n \cdot R & \longrightarrow & 0 \end{array},$$

and the group ring of RG is a pullback

[1]1980 <u>Mathematics Subject Classification</u> (1985 <u>Revision</u>). 16A20
This research was partially supported by the Deutsche Forschungsgemeinschaft.

$$
\text{(1b)}\qquad
\begin{array}{ccccccccc}
0 & \longrightarrow & I_R(P)\cdot G & \longrightarrow & RG & \longrightarrow & RG/P & \longrightarrow & 0 \\
 & & \downarrow \mathrm{id} & & \downarrow \alpha' & & \downarrow \beta' & & \\
0 & \longrightarrow & I_R(P)\cdot G & \longrightarrow & \Gamma & \longrightarrow & R/p^n\cdot R(G/P) & \longrightarrow & 0
\end{array}
\quad .
$$

Since R contains all q^{th} roots of unity, we conclude

$$
\text{(2)}\qquad RS = \Pi_{i=0}^{q-1} R_i \ \text{ and } \ R/p^nR(S) = \Pi_{i=0}^{q-1} R_i/p^nR_i \ ,
$$

where R_i is R with S acting in the appropriate way; i.e. $s : 1 \longrightarrow \sigma^i$, via the Teichmüller Character [La].

We put

$$
\text{(3)}\qquad \Omega_n = \Lambda_n^{\ S} \ , \text{ the } S\text{–fixed points in } \Lambda_n
$$

under the S–conjugation action; Ω_n can then be identified with $\mathrm{End}_{RS}(\Lambda_n)$. We put

$$
\omega_n = \Pi_{i=0}^{q-1} ({}^{s^i}c_n - 1)\ , \text{ viewed in } \Lambda_n \ .
$$

THEOREM 1[2]. *The group ring* RG *is the pullback*

$$
\begin{array}{ccc}
RG & \longrightarrow & RS = \Pi_{i=0}^{q-1} R_i \\
\downarrow & & \downarrow \\
\mathfrak{R} & \longrightarrow & \mathfrak{R}/\varrho\cdot\mathfrak{R} \simeq \Pi_{i=0}^{q-1} R_i/p^n\cdot R_i
\end{array}
\ ,
$$

where

$$
\Gamma \simeq \mathfrak{R} =: \begin{bmatrix}
\Omega & \cdot & \cdots & \cdots & \cdot & \Omega \\
\omega\cdot\Omega & \Omega & & & & \vdots \\
\vdots & \omega\cdot\Omega & \ddots & & & \vdots \\
\vdots & & \ddots & \ddots & & \vdots \\
\vdots & & & \vdots & \Omega & \vdots \\
\omega\cdot\Omega & \cdot & \cdots & \cdots & \omega\cdot\Omega & \Omega
\end{bmatrix}_{q\times q} \ , \textit{ and}
$$

$$
\varrho \simeq \begin{bmatrix}
0 & 1 & 0 & \cdots & \cdots & 0 \\
0 & 0 & 1 & 0 & \cdots & 0 \\
\vdots & 0 & \ddots & \ddots & & \vdots \\
\vdots & & \ddots & \ddots & \ddots & \vdots \\
0 & & & \vdots & \ddots & 1 \\
\omega & 0 & \cdots & \cdots & \cdots & 0
\end{bmatrix}_{q\times q} \ ,
$$

moreover, $\varrho\cdot\mathfrak{R}$ *becomes identified with* $I_R(P_n)\cdot G$. (For simplicity we have written here ω for ω_n and Ω for Ω_n .

[2]The structure theorem for general blocks with cyclic normal defect groups is considerably more complicated and will be published elsewhere.

REMARKS. 1.) *In case* P_n *has order* p *and* $R = \mathbb{Z}_p$ *is the ring of* p–*adic integers, this gives the results in* [J] *and* [R] .

2.) *This result arose, after the author has talked with Lee Klingler about his description of the group rings of the cyclic group of order* p^2 *as a multiple pullback.*

Before we come to the proof, let us give an inductive description of Λ_n and Ω_n for $R = \mathbb{Z}_p^{\wedge}$, the p–adic completion of $\mathbb{Z}$:

PROPOSITION 1. *For* $n \geq 1$ *we denote by* ζ_n *a primitive* p^n–*th root of unity.*

1.) $\Lambda_1 = R[\zeta_1]$ *and* $\Omega_1 = R[\zeta_1]^S$, *where we view* S *as a subgroup of* $\mathrm{Gal}(R[\zeta_1],R)$.

2.) *For* $n > 1$, Λ_n *is a pullback*

$$\begin{array}{ccc} \Lambda_n & \xrightarrow{\alpha_n} & \Lambda_{n-1} \\ \downarrow \beta_n & & \downarrow \kappa_{n-1} \\ R[\zeta_n] & \xrightarrow{\lambda_n} & \mathbb{F}_p P_{n-1}/f_{n-1}(c_{n-1}^{p^{n-2}}) \end{array} , \tag{4}$$

where

$$f_n(X) = 1 + X^{p^{n-1}} + X^{p^{n-1}\cdot 2} + \cdots + X^{p^{n-1}\cdot(p-1)} \cdot f_{n-1}(c_{n-1}^{p^{n-2}})$$

is the trace of $c_{n-1}^{p^{n-2}}$ *in* $\mathbb{F}_p P_{n-1}$. α_n *and* β_n *are the projections, and*

$$\mathrm{Ker}(\kappa_{n-1}) = p\cdot RP_{n-1} + f_{n-1}(c_{n-1}{}^{p^{n-2}})\cdot RP_{n-1} ,$$

$$\mathrm{Ker}(\lambda_n) = (\zeta_n - 1)^{p^{n-2}\cdot(p-1)}\cdot R[\zeta_n] .$$

3.) Ω_n *is obtained by taking* S–*fixed points, which is an exact functor, so that* Ω_n *is the pullback of*

$$\begin{array}{ccc} \Omega_n & \xrightarrow{\alpha_n} & \Omega_{n-1} \\ \downarrow \beta_n & & \downarrow \kappa\_ \\ R[\zeta_n]^S & \xrightarrow{\lambda_n} & (\mathbb{F}_p P_{n-1}/f_{n-1}(c_{n-1}^{p^{n-1}}))^S \end{array} . \tag{5}$$

PROOF. 1.) and 3.) are obvious — note that $|S|$ is a unit in R and so taking fixed points with respect to S is an exact functor on the relevant modules.
In order to prove 2.) , we note that RP_n may be written as pullback

(6)
$$\begin{array}{ccc} RP_n & \xrightarrow{\gamma_n} & RP_{n-1} \\ \downarrow \delta_n & & \downarrow \nu_{n-1} \\ R[\zeta_n] & \xrightarrow{\mu_n} & \mathbb{F}_p P_{n-1} \end{array} ,$$

where

$$\mathrm{Ker}(\gamma_n) = I_R(<c_n^{p^{n-1}}>)\cdot P_n \simeq \mathrm{Ker}(\mu_n) = (\zeta_n^{p^{n-1}} -1)\cdot R[\zeta_n] ,$$

$$\mathrm{Ker}(\nu_{n-1}) = p\cdot RP_{n-1} \simeq \mathrm{Ker}(\delta_n) = \Sigma_{i=1}^{p} c_n^{p^{n-1}\cdot i}\cdot RP_n .$$

The diagrams (6) and (4) show, that $p\cdot R[\zeta_n] = (\xi_n - 1)^{p^{n-1}\cdot(p-1)} \subset \mathrm{Ker}(\mu_n)$ is contained in the kernel of λ_n. Note, that $f_n(X)$ is the minimum polynomial of ζ_n. Since $f_{n-1}(X)^p \equiv f_n(X) \bmod(p)$, it is easily seen, that $f_{n-1}(\zeta_n) \in \Lambda_n \cap \mathrm{Ker}(\lambda_n)$. However, $f_i(X)$ is the minimum polynomial of ζ_i, and so this is the only relation, which has to be added to $\mathrm{Ker}(\mu_n)$ to obtain $\mathrm{Ker}(\lambda_n)$. Id est,

$$\mathrm{Ker}(\lambda_n) = (\zeta_n - 1)^{p^{n-1}}\cdot R[\zeta_n] + f_{n-1}(\zeta_n)\cdot R[\zeta_n] .$$

However,

(7) $$((X^{p^{n-2}})^p - 1)/(X^{p^{n-2}} - 1) = f_{n-1}(X) ,$$

and so we have modulo p:

(8) $$f_{n-1}(\zeta_n)\cdot R[\zeta_n] = ((\zeta_n - 1)^{p^{n-1}} /(\zeta_n - 1)^{p^{n-2}})\cdot R[\zeta_n] =$$
$$(\zeta_n -)^{p^{n-2}\cdot(p-1)}\cdot R[\zeta_n] .$$

as claimed.

We now have to compute the image of λ_n. Since $\mathrm{Ker}(\lambda_n)/\mathrm{Ker}(\mu_n) = (\zeta_n - 1)^{p^{n-2}}$, we have to find a submodule in the uniserial ring $\mathbb{F}_p P_{n-1}$ of dimension p^{n-2}. However, this is exactly the module

$$f_{n-1}(c_{n-1})\cdot\mathbb{F}_p P_{n-1} = \mathrm{tr}(c_{n-1}^{p^{n-2}}) .$$

This completes the proof of Proposition 1.

For the proof of the theorem we need some preliminary observations:

LEMMA 1. $\Gamma \simeq \begin{bmatrix} \Omega & \cdot & \cdots & \cdots & \Omega \\ \omega\cdot\Omega & \Omega & & & \vdots \\ \vdots & \omega\cdot\Omega & \ddots & & \vdots \\ \vdots & & \ddots & \ddots & \vdots \\ \omega\cdot\Omega & \cdots & \cdots & \omega\cdot\Omega & \Omega \end{bmatrix}_{q\times q}$ *and*

$$\mathrm{I(P)}\cdot\mathrm{G} \simeq \begin{bmatrix} \omega\cdot\Omega & \Omega & \cdots & \cdots & \Omega \\ \omega\cdot\Omega & \omega\cdot\Omega & & & \vdots \\ \vdots & \omega\cdot\Omega & \ddots & & \vdots \\ \vdots & & \ddots & \ddots & \Omega \\ \omega\cdot\Omega & \cdots & \cdots & \omega\cdot\Omega & \omega\cdot\Omega \end{bmatrix}_{q\times q}.$$

(We have written ω for ω_n and Ω for Ω_n.)

Since Λ_n and Ω_n can be written as iterated pullbacks (cf. Proposition 1), this gives an explicit description of Γ.

PROOF. If M is an RG–lattice, then taking fixed point with respect to S is an exact functor on lattices, since $|S|\cdot R = R$, and thus $H^1(S, -) = 0$. Thus applying $H^0(S, -)$ to the diagram (1a), we get the commutative diagram with exact rows $(P = P_n)$

$$(9)\qquad \begin{array}{ccccccccc} 0 & \longrightarrow & I_R(P)^S & \longrightarrow & RP^S & \longrightarrow & R & \longrightarrow & 0 \\ & & \downarrow id & & \downarrow \alpha & & \downarrow \beta & & \\ 0 & \longrightarrow & I_R(P)^S & \longrightarrow & \Lambda_n^S & \longrightarrow & R/p^n\cdot R & \longrightarrow & 0 \end{array}.$$

Since $I_R(P_n) = (c_n - 1)\cdot\Lambda_n$, we conclude

$$(10)\qquad I_R(P_n)^S = \omega_n\cdot\Lambda_n{}^S = \omega_n\cdot\Omega_n\,.$$

CLAIM 1. *The non–isomorphic indecomposable projective left* Γ*–lattices are the modules*

$$(11)\qquad \Lambda_n\,,\ (c_n - 1)\cdot\Lambda_n\,,\ldots,\ (c_n - 1)^{q-1}\cdot\Lambda_n\,,$$

where P_n *acts by left multiplication and* S *acts by conjugation.*

PROOF. 1.) Surely RP_n with S acting by conjugation is the principle indecomposable RG–lattice, but then Λ_n, as projection of RP_n into Γ, is an indecomposable Γ–lattice with $\Lambda_n/\mathrm{rad}_\Gamma(\Lambda_n) = S_o$ the trivial simple RG–module.

2.) *We shall show that* $(c_n - 1)$ *is a unit in* $K\cdot\Gamma$ *and conjugation by* $(c_n - 1)$ *stabilizes* Γ *and thus induces an automorphism on* Γ.

To see this, we note that $\Gamma = RG/(\Sigma_{i=1}^{p^n} c^i)RG$, and observe, that

$$K\cdot I_R(P_n) = K\cdot(c_n - 1)\cdot\Gamma = K\cdot\Gamma .$$

Since conjugation by $(c_n - 1)$ is an automorphism of $K\cdot\Gamma$, we need only to observe that

$$(c_n - 1)\cdot RG = RG\cdot(c_n - 1) .$$

Now the statement follows, since the augmentation ideal $I_R(P_n)\cdot G$ is a two sided characteristic ideal in RG .

3.) Since an automorphism – like conjugation with $(c_n - 1)$ – gives rise to an invertible bimodule, we conclude, that the modules listed in (11) are all indecomposable projective Γ–lattices.

4.) We shall next consider the isomorphism types of the Γ–modules in (11) . Let $S_o, \dots, S_{q-1}$ be the non–isomorphic simple RG–modules $S_i \simeq R_i/\mathrm{rad}(R_i)$ (cf. 2) . These are then also the simple non–isomorphic Γ–modules – note the $\ker(\beta')$ and $I_R(P_n)\cdot G$ lie in the radical of RG (cf. 1b) . We shall fix the numbering of the simple modules in such a way, that

s acts trivially on S_o ,

s acts by $1 \longrightarrow \sigma^i$, on S_i , $1 \le i \le q-1$,

where σ is a fixed primitive q^{th} root of unity in R , such that ${}^s c_n = c_n^{\sigma}$ – with the appropriate identifications – $R/\mathrm{rad}(R) = \mathbb{Z}/p\mathbb{Z}$.

5.) *We shall show next, that*

(12) $$(c_n - 1)^i\cdot\Lambda_n/\mathrm{rad}((c_n - 1)^i\cdot\Lambda_n) \simeq S_i .$$

In order to see this, we note, that we have the relation:

(13) $$(c_n^j - 1) = (1 + c_n + \dots + c_n^{j-1})\cdot(c_n - 1) ,$$

and thus

(14) $$ {}^s(c_n - 1) = (c_n^{\sigma} - 1) = (1 + c_n + \dots + c_n^{\sigma-1})\cdot(c_n - 1)$$
$$\equiv \sigma\cdot(c_n - 1) \ \mathrm{mod}((c_n - 1)^2\cdot\Lambda_n .$$

Similarly one gets

(15) $$ {}^s(c_n - 1)^i \equiv \sigma\cdot(c_n - 1) \ \mathrm{mod}((c_n - 1)^{i+1}\cdot\Lambda_n) .$$

Thus S acts on $(c_n - 1)^i\cdot\Lambda_n/(c_n - 1)^{i+1}\cdot\Lambda_n = R_i/p^n\cdot R_i$ as multiplication with σ^i ,

in particular, this module is uniserial with composition factors all isomorphic to S_i. Consequently, $(c_n - 1)^i \cdot \Lambda_n / \mathrm{rad}((c_n - 1)^{i+1} \cdot \Lambda_n) \simeq S_i$. Moreover, $(c_n - 1)^q \cdot \Lambda_n \simeq \Lambda_n$, since $\sigma^q = 1$. Hence we have constructed a complete set of non isomorphic indecomposable projective Γ–lattices as claimed (cf. (11)).
This proves Claim 1.

We are now in the position to prove Lemma 1: We observe

$$\mathrm{End}_{RG}((c_n - 1)^i \cdot \Lambda_n)) = \Lambda_n{}^S = \Omega_n , \; 0 \le i \le q-1 , \tag{16}$$

$$\mathrm{Hom}_{RG}((c_n - 1)^i \cdot \Lambda_n , (c_n - 1)^j \cdot \Lambda_n) = \Omega_n \text{ for } i > j \text{ and}$$

$$\mathrm{Hom}_{RG}((c_n - 1)^i \cdot \Lambda_n , (c_n - 1)^j \cdot \Lambda_n) = \omega_n \cdot \Omega_n \text{ for } i < j < q .$$

Only the last equation needs some explanation: For $i < j < q$ we have

$$\mathrm{Hom}_{RG}((c_n - 1)^i \cdot \Lambda_n, (c_n - 1)^j \cdot \Lambda_n) =$$
$$\mathrm{Hom}_{RG}(\Lambda_n, (c_n - 1)^{j-i} \cdot \Lambda_n) \subset \mathrm{End}_{RG}(\Lambda_n) = \Omega_n .$$

Since no composition factor of $(c_n - 1) \cdot \Lambda_n / (c_n - 1)^{q-1} \cdot \Lambda_n$ is isomorphic to S_o (cf. (16)), and since $(c_n - 1)^q \cdot \Lambda_n \simeq \Lambda_n$, we conclude, that

$$\mathrm{Hom}_{RG}(\Lambda_n, (c_n - 1)^{j-i} \cdot \Lambda_n) = \mathrm{Hom}_{RG}(\Lambda_n, (c_n - 1)^q \cdot \Lambda_n) .$$

However, since $(c_n - 1) \cdot \Lambda_n = (c_n^{\sigma^i} - 1) \cdot \Lambda_n$, we can write

$(c_n - 1)^q \cdot \Lambda_n = \omega_n \cdot \Lambda_n$ and so we conclude

$$\mathrm{Hom}_{RG}((c_n - 1)^i \cdot \Lambda, (c_n - 1)^j \cdot \Lambda_n) = \omega_n \cdot \Omega_n .$$

Hence

$$\Gamma \simeq \mathfrak{R} =: \begin{bmatrix} \Omega & \cdot & \cdot & \cdots & \cdot & \Omega \\ \omega \cdot \Omega & \Omega & & & & \cdot \\ \vdots & \omega \cdot \Omega & \ddots & & & \vdots \\ \vdots & & \ddots & \ddots & & \vdots \\ \vdots & & & \ddots & \ddots & \cdot \\ \omega \cdot \Omega & \cdot & \cdots & \cdot & \omega \cdot \Omega & \Omega \end{bmatrix}_{q \times q} ,$$

as stated in the Lemma. If one identifies the orders Γ and $\mathfrak{R}$, then the indecomposable projective Γ–lattices $(c_n - 1)^i \cdot \Lambda_n$, $0 \le i \le q-1$, become identified with the $(i+1)^{st}$ column C_{i+1} in $\mathfrak{R}$. We now put

$$\varrho = \begin{bmatrix} 0 & 1 & 0 & \cdot & \cdot & \cdots & \cdot & 0 \\ 0 & 0 & 1 & 0 & \cdot & \cdots & \cdot & 0 \\ \vdots & 0 & \ddots & & \ddots & & & \vdots \\ \vdots & & & \ddots & & \ddots & & \vdots \\ 0 & & & & \ddots & \vdots & \cdot & 1 \\ \omega & 0 & \cdot & \cdot & \cdots & \cdot & \cdot & 0 \end{bmatrix}_{q\times q} ,$$

then $(c_n - 1)^i \cdot \Lambda_n \simeq C_{i+1} = \varrho^i \cdot \mathfrak{R}$; in particular,

$$I(P_n)\cdot G \simeq \quad \varrho\cdot\mathfrak{R} = \begin{bmatrix} \omega\cdot\Omega & \Omega & \cdot & \cdots & \cdot & \Omega \\ \omega\cdot\Omega & \omega\cdot\Omega & \ddots & & & \vdots \\ \vdots & & \omega\cdot\Omega & \ddots & & \vdots \\ \vdots & & & \ddots & \ddots & \Omega \\ \omega\cdot\Omega & \cdot & \cdots & \cdot & \omega\cdot\Omega & \omega\cdot\Omega \end{bmatrix}_{q\times q} ,$$

This completes the proof of Lemma 1.

We now are in the position to prove Theorem 1: We identify Γ with $\mathfrak{R}$ and note that $C_i/\varrho\cdot C_i \simeq R_{i-1}/p^n\cdot R_{i-1}$, and so the group ring RG is the pullback (cf.1b)

$$\begin{array}{ccl} RG & \longrightarrow & RS \quad = \Pi_{i=0}^{q-1} \quad R_i \\ \downarrow & & \qquad\qquad \downarrow \\ \mathfrak{R} & \longrightarrow & \mathfrak{R}/\varrho\cdot\mathfrak{R} = \Pi_{i=0}^{q-1} R_i/p^n\cdot R_i \end{array}$$

This proves Theorem 1.

REFERENCES.

[CR] Curtis, C. W. – I. Reiner, Methods of representation theory, I, II , Wiley–Interscience 1981, 1987

[J] Jacobinski, H. "Hereditary covers and blocks," Lecture at ICRA II, Ottawa, 1979

[La] Lang, S. Cyclotomic fields. Granduate Texts in Mathematics, Springer–Verlag, 1978

[R] Roggenkamp, K. W. Integral Representations and Structure of Finite Group Rings. Les Presses De l'Universite de Montreal, 1980

Klaus W. Roggenkamp
Mathematisches Institut B
Universität Stuttgart
7 Stuttgart, Germany

Canadian Mathematical Society
Conference Proceedings
Volume 11 (1991)

TWO-PEAK POSETS OF FINITE PRINJECTIVE TYPE

DANIEL SIMSON

ABSTRACT. The main aim of this paper is to give a characterization of finite posets J having precisely two maximal elements and having only finitely many isomorphism classes of indecomposable socle projective K-linear representations over a given field K, or equivalently, two-peak posets J having only finitely many indecomposable canonical forms of block matrices (1.5) with coefficients in K with respect to the J-elementary transformations (i) and (ii) defined below. An application to the study of indecomposable lattices over a class of orders is given.

0. INTRODUCTION

Throughout this paper K is a field and we denote by $(J, \preceq)$ a finite poset (i.e. partially ordered set) having exactly two maximal elements $*$ and $+$ with respect to the partial order $\preceq$. The elements $*$ and $+$ are called the peaks of J [9]. We shall write $i \prec j$ if $i \preceq j$ and $i \neq j$. For the sake of simplicity we write J instead of $(J, \preceq)$.

Following Gabriel [3] we define the category $J-\mathrm{sp}$ of J-spaces over the field K as follows. The objects of $J-\mathrm{sp}$ are systems

$$\mathbf{M} = (M_j)_{j\in J}$$

of finite dimensional K-vector spaces M_j such that $M_j \subseteq M_* \oplus M_+$ for all $j \in J$, $\pi_p(M_j) = 0$ for $j \npreceq p \in \{*,+\}$ and $\pi_p(M_i) \subseteq \pi_p(M_j)$ for $i \prec j \prec p \in \{*,+\}$, where $\pi_p : M_* \oplus M_+ \to M_p$ is the natural projection. By a map $f : \mathbf{M} \to \mathbf{M}'$ in $J-\mathrm{sp}$ we shall mean a pair $f = (f_*, f_+)$ of K-linear maps $f_* : M_* \to M'_*$, $f_+ : M_+ \to M'_+$ such that $(f_* \oplus f_+)(M_j) \subseteq M'_j$ for all $j \in J$. It is clear that $J-\mathrm{sp}$ is an additive category with the finite Krull-Schmidt property. Moreover $J-\mathrm{sp}$ is closed under taking kernels and according to [1] $J-\mathrm{sp}$ has Auslander-Reiten sequences as well as source maps and sink maps (see [7]).

The definition above obviously generalizes to the case when J has an arbitrary number of maximal elements. If J has a unique maximal element $*$, then the notion of a J-space coincide with the notion of $(J-\{*\})$-space in the sense of Gabriel [3].

1980 *Mathematics Subject Classification* (1985 *Revision*) Primary 16A64; Secondary 14L30. 15A21.

The paper was written when the author was visiting the University of Bielefeld under a financial support of DFG.

The main result of this paper is Theorem 1.7 containing a characterization of two-peak posets J for which the category $J-\mathrm{sp}$ is of finite type, i.e. has finitely many isomorphism classes of indecomposable objects. Since obviously the category $J-\mathrm{sp}$ contains as a full subcategory the categories $*^{\nabla}-\mathrm{sp}$ and $+^{\nabla}-\mathrm{sp}$, where

$$j^{\nabla}=\{i\in J \mid i\preceq j\}$$

for $j\in J$, then our result is more general than the characterization of one-peak posets of finite type established by Kleiner in [5].

An application of this characterization to the study of lattices over a class of orders is given in Section 2, where the reader is also referred for an example illustaring the use of J-spaces over two-peak posets J.

1. MAIN RESULTS

Throughout this paper we shall consider the two-peak poset J as a quiver with the commutativity relation induced by the ordering $\prec$, and we denote by R the path algebra KJ of J with coefficients in the field K.

Given j in J we denote by e_j the standard primitive idempotent corresponding to j. The right R-modules will be identified with the systems

$$X=(X_i;\ {}_jh_i)_{i,j\in J}$$

where X_i is a finite dimensional K-vector space and ${}_jh_i : X_i \to X_j, i\prec j$, are K-linear maps such that ${}_th_{jj}h_i={}_th_i$ for all $i\prec j\prec i$ in J..

Following [6] and [15] we call the right R-module X *prinjective* if X is finitely generated and the right $K\check{J}$-module Xe is projective, where $\check{J}=J-\{+,\star\}$ and $e=\sum_{j\in\check{J}}e_j$. We denote by $\mathrm{prin}(R)$ and by $\mathrm{mod}_{\mathrm{sp}}(R)$ the categories of finitely generated prinjective right R-modules and socle projective right R-modules [9], respectively. It follows from [6] that the categories $\mathrm{prin}(R)$ and $\mathrm{mod}_{\mathrm{sp}}(R)$ have Auslander-Reiten sequences, source maps and sinks maps. Following [9-11] we define a functor

$$(1.1)\qquad \Theta:\mathrm{mod}(R)\longrightarrow J-\mathrm{sp}$$

by setting $\Theta(X)=(\overline{X}_j)_{j\in J}$, where $\overline{X}_*=X_*$, $\overline{X}_+=X_+$ and

$$\overline{X}_j=\mathrm{Im}[X_j\overset{({}_\star h_j,{}_+h_j)}{\longrightarrow}X_\star\oplus X_+].$$

It is easy to check that the functor Θ induces two functors

$$(1.2)\qquad \mathrm{mod}_{\mathrm{sp}}(R)\ \underset{\sim}{\overset{\Theta'}{\longrightarrow}}\ J-\mathrm{sp}\overset{\Theta_J}{\longleftarrow}\mathrm{prin}(R)$$

such that the functor Θ' is an equivalence of categories, whereas the functor Θ_J is full, dense and $\mathrm{Ker}\Theta_J=[e_jK\check{J};\ j\in J]$, i.e. $\mathrm{Ker}\Theta_J$ consists of all maps in prin(R) having a factorization through a direct sum of copies of the projective $K\check{J}$-modules $e_jK\check{J}$, $j \in \check{J}$. In particular the functor $\boldsymbol{\Theta}_J$ preserves the finite type.

Let us present a useful interpretation of modules in prin(R) in terms of block matrices with coefficients in the field K. For this purpose given a module X in mod(R) we define the coordinate vector $\mathbf{cdn(X)} \in \mathbb{N}^{\mathbf{J}}$ by the following formula (see [6, 9])

$$\begin{aligned}(\mathbf{cdn}(X))(j) &= \dim_K(X_j) && \text{for } j \in \{+,\star\},\\ &= \dim_K(topX)e_j && \text{for } j \in \check{J}.\end{aligned}$$

We consider $\mathbf{cdn}$ as a map $\mathbf{cdn}(X) : J \to \mathbb{N}$. Note that the projective cover $P(X)$ of X has the form

$$P(X) = \bigoplus_{j \in \check{J}} (e_j R)^{(\mathbf{cdn}X)(j)}$$

if X is an indecomposable object in prin(R) or in $J - \mathrm{sp} \cong \mathrm{mod}_{\mathrm{sp}}(R)$, and X is not isomorphic to the simple projective modules e_+R , e_*R.

Given a vector $v \in \mathbb{N}^J$ we consider the affine irreducible K-variety

$$\mathbf{prin}_v^R = \mathrm{Hom}_R(P(v), Q(v))$$

where

$$P(v) = \oplus_{j \in J}(e_jR)^{v(j)}, \; Q(v) = E_R(e_*R)^{v(*)} \oplus E_R(e_+R)^{v(+)}.$$

There is a natural algebraic group action

$$\star : \mathbf{G}(v) \times \mathbf{prin}_v^R \longrightarrow \mathbf{prin}_v^R \tag{1.3}$$

where $\mathbf{G}(v) = \mathrm{Aut}_R(P(v)) \times \mathrm{Aut}_R(Q(v))$. It is easy to see that there is a bijection between the isomorphism classes of modules in prin(R) with $\mathbf{cdn}(X) = v$ and the $\mathbf{G}(v)$-orbits in $\mathbf{prin}_v^R$. The bijection is given by attachig to X the composed map

$$f_X = [P(v) \overset{u}{\hookrightarrow} P(X) \overset{w}{\to} \boldsymbol{\Theta}_J(X) \overset{u'}{\hookrightarrow} E_R(\boldsymbol{\Theta}_J(X))] \tag{1.4}$$

where u and u' are the natural embeddings and w is the natural epimorphism.

Note that given $p \in \{+,\star\}$, the injective envelope $E_R(e_pR)$ of e_pR (considered as a K-linear representation of the poset J) is the constant diagram having the space K over all $j \preceq p$ and the space zero elsewhere. It follows that

$$\begin{aligned}\mathrm{Hom}_R(e_iR, E_R(e_pR)) &\cong K \text{ for all } i \preceq p,\\ &= 0 \text{ for } i \in J - p^{\nabla}.\end{aligned}$$

Therefore if $\check{J} = \{1, \dots, n\}$, then the variety $\mathbf{prin}_v^R$ can be identified with the variety $\mathbf{Mat}_v^J$ of all block matrices (with coefficients in K) of the form

$$(1.5) \qquad A = \underbrace{\begin{array}{|c|c|c|c|} \hline A_{1\star} & A_{2\star} & \cdots & A_{n\star} \\ \hline A_{1+} & A_{2+} & \cdots & A_{n+} \\ \hline \end{array}}_{v(1)\quad v(2)\quad \cdots \quad v(n)} \begin{array}{l} \}v(*) \\ \}v(+) \end{array}$$

where $A_{i\star} = 0$ if $i \in J - \star^\nabla$ and $A_{j+} = 0$ if $j \in J - \star^\nabla$. Moreover the group $\mathbf{G}(v)$ is isomorphic to the group $\mathbf{G}_v^J$ generated by the following elementary transformations on matrices in $\mathbf{Mat}_v^J$

(i) *All elementary transformations on rows inside each horizontal block in* A.

(ii) *All elementary transformations on columns from the* i*-th block to the* j*-th block for all* $i \preceq j$.

It follows that

$$\mathbf{G}_v^J = H_v^J \times \mathrm{Gl}(K, v(\star)) \times \mathrm{Gl}(K, v(+))$$

where H_v^J consists of all matrices of the form

$$(1.6) \qquad h = \begin{pmatrix} h_{11} & h_{12} & \dots & h_{1n} \\ 0 & h_{22} & \dots & h_{2n} \\ \vdots & \vdots & \ddots & \vdots \\ 0 & 0 & \dots & h_{nn} \end{pmatrix} \in \mathrm{Gl}(K,\ v(1) + \dots + v(n))$$

where $h_{ii} \in \mathrm{Gl}(K,\ v(i))$ and $h_{ij} = 0$ if i and j are not comparable in $\check{J}$. Here we suppose that the order in J is such that $i \prec j$ implies that $i < j$ in the natural order. We can always achieve this by a suitable renumbering the elements of $\check{J} = \{1, \dots, n\}$.

Now we are able to prove the main result of this paper.

Theorem 1.7. *Suppose that J is a finite poset having exactly two maximal elements $\star$ and $+$. Let K be an infinite field and let $R = KJ$ be the path algebra of J with coefficiens in K. Then the following conditions are equivalent.*

(a) *The category $J - \mathrm{sp}$ of J-spaces is of finite type.*

(b) *The category $\mathrm{prin}(R)$ of prinjective right R-modules is of finite type.*

(c) *The algebraic K-variety $\mathbf{prin}_v^R$ has finitely many $\mathbf{G}(v)$-orbits with respect to the action (1.3) for every coordinate vector $v \in \mathbb{N}^J$.*

(d) *The rational Tits quadratic form $q_J : \mathbb{Q}^J \to \mathbb{Q}$:*

$$(1.8) \qquad q_J(z) = \sum_{j \in J} z_j^2 + \sum_{\substack{p \prec q \\ q \neq \star, +}} z_p z_q - \sum_{p \prec \star} z_p z_\star - \sum_{q \prec +} z_q z_+$$

is weakly positive, i.e. $q_J(z) > 0$ for all non-zero vectors $z \in \mathbb{N}^J$.

(e) *The poset* J *does not contain as a full peak subposet the following one peak enlargements* $\mathcal{P}_1 = (1,1,1,1)^*$, $\mathcal{P}_2 = (2,2,2)^*$, $\mathcal{P}_3 = (1,3,3)^*$, $\mathcal{P}_4 = (N,4)^*$, $\mathcal{P}_5 = (1,2,5)^*$ *of the critical posets of Kleiner* [5] *and the following 26 two-peak forms* $\mathcal{P}_6, \ldots, \mathcal{P}_{31}$ *(see* [12; Remark 3.21] and [15]).

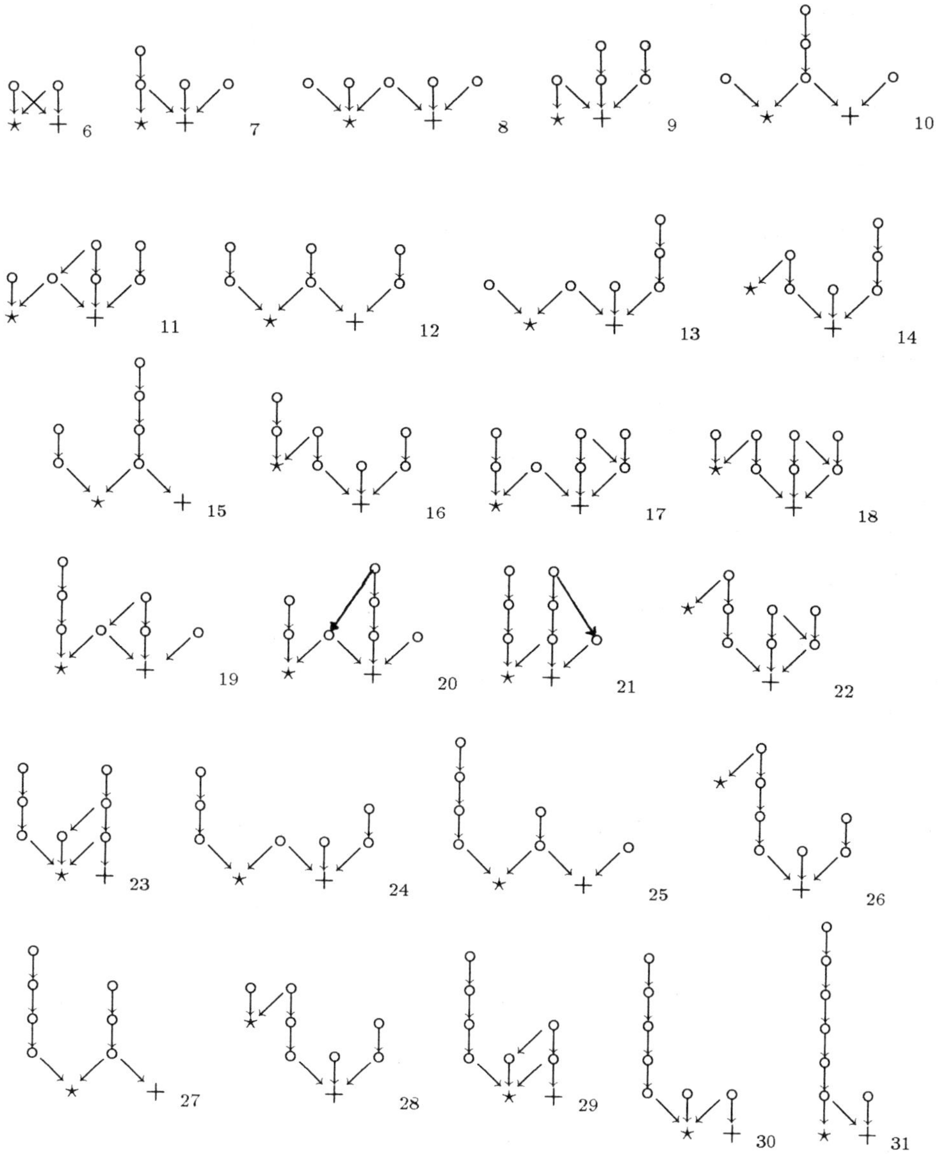

Corollary 1.9. *If the conditions* (a) - (e) *in Theorem* 1.7 *are satisfied then the following statements hold:*

(i) *Every point in J is separating* [2], *the Auslander-Reiten translation quivers* $\Gamma(\text{prin}(R))$ *and* $\Gamma(J-\text{sp})$ *of the categories* prin(R) *and* $J-\text{sp}$ *respectively coincide with their preprojective components. The translation quiver* $\Gamma(J-\text{sp})$ *is obtained from* $\Gamma(\text{prin}(R))$ *by delating the points* $[e_j K\check{J}]$ *with* $j \in \check{J} = J - \{\star, +\}$.

(ii) *The map* $X \mapsto \mathbf{cdn}(X)$ *establishes a bijection between the isomorphism classes of indecomposable modules in* prin(R) *and the positive roots of the quadratic form* q_J, *i.e. vectors* v *in* $\mathbb{N}^J$ *satisfying* $q_J(v) = 1$.

(iii) *If X is an indecomposable object in* prin(R) *or in* $J-\text{sp}$, *then* $\text{End}(X) \cong K$, $\text{Ext}^1_R(X,X) = 0$ *and* $(\mathbf{cdn}X)(j) \leq 6$ *for all* $j \in J$.

(iv) *The poset* $C = \star^\nabla \cap +^\nabla$ *is a chain. There exist a one-peak poset* $\xi_C J$ *containing* C *as a peak subposet and an equivalence of categories*

$$J-\text{sp}/[J_\star-\text{sp}, J_+-\text{sp}] \cong \xi_C-\text{sp}/[C-\text{sp}] \tag{1.10}$$

where $J_\star = J - +^\nabla$ *and* $J_+ = J - \star^\nabla$.

The statement (i) in Corollary 1.9 follows from the proof of Theorem 1.7 given below and from the well-known facts on the Auslander-Reiten translation quivers (see [7]). The final part of (i) follows from the construction of the quiver $\Gamma(\text{prin}(R))$ given in [6; Theorem 3.4]. The statements (ii) and (iii) are a consequence of (i) (see [7]). The proof of (iv) will be given below. □

Proof of Theorem 1.7. The equivalence of (a) and (b) follows from the properties of the functor $\boldsymbol{\Theta}_J$ discussed above.

(b) ⇒ (c) We suppose that $X_1, \ldots, X_r$ is a complete set of indecomposable modules in prin(R) up to isomorphism. Then every X in prin(R) with $\mathbf{cdn}(X) = v$ has a decomposition

$$X = X_1^{t_1} \oplus \cdots \oplus X_r^{t_r}$$

where $t_1, \ldots, t_r > 0$ and the vector $t(X) = (t_1, \ldots; t_r)$ satisfies

$$\| v \| := \sum_{j \in J} v(j) = \| \mathbf{cdn}(X) \| = \sum_{i=1}^{r} t_i \| \mathbf{cdn}(X_i) \| .$$

Then the Krull-Schmidt property together with the observations obove yields

$$t(X) = t(Y) \Leftrightarrow X \cong Y \Leftrightarrow \mathbf{G}(v) \star f_X = \mathbf{G}(v) \star f_Y$$

and therefore (c) follows.

(c) ⇒ (d) It follows from (c) that the affine variety $\mathbf{prin}_v^R$ is a union of finitely many $\mathbf{G}(v)$-orbits $\mathcal{O}_1, \ldots, \mathcal{O}_s$ and therefore $\mathbf{prin}_v^R = \overline{\mathcal{O}}_t$ for some t because the variety $\mathbf{prin}_v^R$ is irreducible. Since the subgroup K^* of $\mathbf{G}(v)$ acts trivially on

$\mathbf{prin}_v^R$ then in view of [6; Proposition 4.2], [10; Theorem 10.1] and the matrix interpretation of $\mathbf{prin}_v^R$ and $\mathbf{G}(v)$ presented above we get the following formula

$$\begin{aligned}\sum_{j\in J} v(j)^2 + \sum_{\substack{p\prec q\\ q\neq\star,+}} v(p)v(q) &= \dim \mathbf{G}(v) \\ &> \dim \mathbf{prin}_v^R = \sum_{p\prec\star} v(p)v(\star) + \sum_{q\prec +} v(q)v(+)\end{aligned}$$

where dim is the variety dimension. Hence we easily conclude that $q_J(z) > 0$ for all non-zero vectors z in $\mathbb{N}^J$ and (d) follows.

In order to prove that (d) implies (e) it is sufficient to show that for every critical poset form L listed in (e) there is a non-zero vector v_L in $\mathbb{N}^L$ such that $q_L(v_L) = 0$. If L is one of the one-peak forms in (e) we take for v_L the vectors $(1,1,1,1;2)$,$(1,1,1,1,1,1;3)$, $(2,1,1,1,1,1,1;4)$, $(2 \to 1 \leftarrow 1 \to 2,1,1,1,1;5)$, $(3,2 \to 2,1,1,1,1,1;6)$, respectively. If L is one of the beginning fourth two-peak forms in (e) we take for v_L the vectors

$$\begin{pmatrix} 1 & 1 \\ 1 & 1 \end{pmatrix}, \begin{pmatrix} 1 & 1 & 2 & 1 & 1 \\ & 2 & & 2 & \end{pmatrix}, \begin{pmatrix} & 1 & 1 \\ 2 & 1 & 1 \\ 1 & 3 & \end{pmatrix}, \begin{pmatrix} 1 & & \\ 1 & 1 & 1 \\ 1 & 2 & \end{pmatrix}$$

respectively. The remaining cases are left to the reader.

(e)$\Rightarrow$(a) First we note that the poset $C = \star^\nabla \cap +^\nabla$ is a chain because otherwise the first two-peak forms of the list in (e) is conained in J. Now it is easy to check that every point in J is separating (see [2] and [14; Proposition 4.9]). It follows from [2] and [6] that $\Gamma(\text{prin}(R))$ has a preprojective component and the assumptions required in the main theorem of Weichert [16] are satisfied. Hence we conclude that J does not contain the critical forms of Weichert [16] and therefore (a) follows. This finishes the proof of the theorem. □

Now we are going to establish a useful relation between the two-peak poset J and the associated one-peak poset $\xi_C J$ required in Corollary 1.9 (iv). We also get an alternative proof of the implication (e)$\Rightarrow$(a) in Theorem 1.7. For this purpose we denote by c the unique maximal element of the chain $C = \star^\nabla \cap +^\nabla$. We set $J_c = J - C$ and we consider the vector space category

$$\mathsf{H}_K^c = \mathrm{Hom}_{KJ_c}(rad(e_c R), J_c - \mathrm{sp}). \tag{1.11}$$

By our assumption the poset J_c is a disjoint union of two one-peak posets $J_* = J - +^\nabla$ and $J_+ = J - \star^\nabla$, which are of finite type according to [5]. It follows that every indecomposable object

$$\overline{y} := \mathrm{Hom}_{KJ_c}(rad(e_c R), y)$$

in the category H_K^c is one-dimensional, and therefore the set $J(\mathsf{H}_K^c)$ consisting of the isomorphism classes of indecomposable objects $\overline{y}$ in H_K^c is a poset with respect to the relation

$$\overline{x} \preceq \overline{y} \Leftrightarrow \mathsf{H}_K^c(\overline{y}, \overline{x}) \neq 0.$$

According to [12; 3.3] we associate to our two-peak poset J the poset disjoint union

$$\xi_C J = C \cup J(\mathsf{H}_K^c) \tag{1.12}$$

with the following new relations

$$\overline{y} \prec c \quad \text{for all } \overline{y} \in J(\mathsf{H}_K^c), \text{ and}$$
$$i \prec \overline{y} \Leftrightarrow \mathrm{Hom}_{KJ_c}(\mathbf{r}(e_i R), y) = 0 \quad \text{for } i \in C - \{c\}, \overline{y} \in J(\mathsf{H}_K^c).$$

Here we denote by $\mathbf{r}(e_i R)$ the restriction of the representation $e_i R$ of J to the subposet J_c. It is easy to see that ξ_J is a finite one-peak poset and it follows from [12; Theorem 3.4] that there is an equivalence of categories (1.10).

A simple combinatorial analysis shows that if J does not contain as full peak subposets the posets listed in (e) then the one-peak poset $\xi_C J$ does not contain one-peak enlargements of the critical poset of Kleiner listed in (e). Then the implication (e)$\Rightarrow$(a) in Theorem 1.7 is a consequence of Corollary 1.9 (iv). Let us consider for example the case when the posets $J_+ - c^\Delta$ and $J_* - c^\Delta$ are chains, where given $j \in J$ we set

$$j^\Delta = \{j \in J | j \succeq c\}.$$

It follows from the Splitting Lemma [10; Section 8] extended naturally to our two-peak situation that the problem reduces to the case when the posets J_* and J_+ are linearly ordered. Since there is a decomposition $\xi_C J = J_* + C + J_+$ and c is a unique maximal element in $\xi_C J$ then in view of the relations $i \prec \overline{y}$ defined above one can easily prove our claim by a straightforward combinatorial analysis. The general case is also reduced by the Splitting Lemma as above to the situation when one of the posetsJ_+, J_* is a chain.

Remark 1.13.(a) A characterization of two-peak posets of tame prinjective type (see [6]) remains an open problem. However, under the assumption that the poset $C = \star^\nabla \cap +^\nabla$ is linearly ordered and the category $J - \mathrm{sp}$ is of tame type, it is easy to check that all objects of the vector space category (1.11) are one-dimensional, $\xi_C J$ is a poset with respect to the ordering defined above and the equivalence (1.10) remains valid. Therefore, in this case, the category $J - \mathrm{sp}$ is of tame type if and only if the poset $\xi_C J$ does not contain as a full subposet the following critical posets $(1,1,1,1,1)$, $(1,1,1,2)$, $(2,2,3)$, $(1,3,4)$, $(N,5)$, $(1,2,6)$ of Nazarova (see [7]). The case when C is not a chain is briefly discussed in [12; Remark 5.2 (d)].

(b) It would be interesting to find a list of all sincere two-peak posets J of finite prinjective type, i.e. posets J having an indecomposable J-space $\mathbf{M}$ such that $(\mathbf{cdnM})(\mathbf{j}) \neq \mathbf{0}$ for all j in J.

(c) For the sake of simplicity of our presentation we restrict our consideration in this paper to the two-peak posets only. However by the arguments applied in the proof of Theorem 1.7 one can extend our main characterization from two-peak posets to posets with an arbitrary number of maximal elements.

(d) A more general problems than the above ones are studied in [13, 14] in a connection with a class of non-schurian vector space categories described by two-separated stratified posets. □

2. AN APPLICATION

In this section we suppose that D is a complete discrete valuation domain with a maximal ideal p such that $D/p = K$. Let

$$(2.1) \qquad A = \begin{bmatrix} D & D_{12} & \dots & D_{1n} \\ p & D & \dots & D_{2n} \\ \vdots & \vdots & \ddots & \vdots \\ p & p & \dots & D \end{bmatrix} \subseteq \mathbf{M}_n(D)$$

be a $(0,1)$-suborder of $\mathbf{M}_n(D)$, i.e. D_{ij} is either D or p, and $D_{ij} = p$ for all $i > j$ (see [17]). We note that the set $I(A) = \{1, \dots, n\}$ is a poset with respect to the following relation

$$i \preceq j \Leftrightarrow D_{ij} = D.$$

Now we suppose that $A_1 \subseteq \mathbf{M}_{n_1}(D)$ and $A_2 \subseteq \mathbf{M}_{n_2}(D)$ are $(0,1)$-orders such that the posets $I(A_1)$ and $I(A_2)$ contain a common intervally closed subposet $I = \{1, \dots, r\}$, where $r \leq n_1$ and $r \leq n_2$. We denote by $I(A_1)^* \coprod_I I(A_2)^+$ the two-peak poset obtained from the disjoint union $I(A_1)^* \cup I(A_2)^+$ by making the identification of the elements in I , where given a poset L we set $L^* = L \cup \{*\}$ with $j \prec \star$ for all $j \in L$.

Let us consider the pull-back D-order

$$(2.2) \qquad A_1 \times_I A_2 = \{(a, a') \in A_1 \times A_2 | a_{ij} \equiv a'_{ij} (\text{ modulo p}) \text{ for all } i, j \in I\}.$$

We denote by $\mathrm{latt}(A_1 \times_I A_2)$ the category of right lattices over the order $A_1 \times_I A_2$, i.e the category of finitely generated right $A_1 \times_I A_2$-modules which are free as D-modules. A D-order B is said to be of finite lattice type if the category $\mathrm{latt}(B)$ is of finite type.

Proposition 2.3. *Under the assumption and the notation above the set $J = I(A_1)^* \coprod_I I(A_2)^+$ is a two-peak poset with the peaks $+$ and $\star$, and the following statements hold.*

(a) *There is a representation equivalence*

$$(2.4) \qquad T : \mathrm{latt}(A_1 \times_I A_2) \longrightarrow J - \mathrm{sp}/[e_* KJ, e_+ KJ].$$

(b) *The order $A_1 \times_I A_2$ is of finite lattice type if and only if the poset I is a chain and the poset J does not contain as a full peak subposet the critical forms* *listed in Theorem* 1.7(*e*).

Proof. First we note that the order $A = A_1 \times_I A_2$ is a suborder of the hereditary order $B = \mathbf{M}_{n_1}(D) \times \mathbf{M}_{n_2}(D)$ and the two-sided ideal $\boldsymbol{\pi} = \mathbf{M}_{n_1}(p) \times \mathbf{M}_{n_2}(p) \subseteq \mathrm{rad(A)}$ is a B-ideal. It is easy to see that the ring

$$R = \begin{bmatrix} A/\boldsymbol{\pi} & B/\boldsymbol{\pi} \\ 0 & B/\boldsymbol{\pi} \end{bmatrix}$$

is a finite dimensional two-peak K-algebra which is Morita equivalent to the path algebra KJ of the poset J. Then according to [4, 8] the composed functor

$$\mathrm{latt}(A) \xrightarrow{\mathsf{F}} \mathrm{mod}_{\mathrm{sp}}(R) \cong \mathrm{mod}_{\mathrm{sp}}(KJ) \cong J - \mathrm{sp}$$

with $\mathsf{F}(X) = (X/X\boldsymbol{\pi}, XB/X\boldsymbol{\pi}, u)$, induces the representation equivalence T required in (a). Here u is the embedding induced by the natural monomorphism $X \hookrightarrow XB$ (see [8]). Since (b) follows from (a) together with Proposition 1.7, then the proposition is proved. □

Example 2.5. Let I be the poset $1 \to 2 \to 3$ and consider two 5×5 D-orders

$$A_1 = \begin{pmatrix} D & D & D & D & p \\ p & D & D & D & p \\ p & p & D & p & p \\ p & p & p & D & p \\ p & p & p & p & D \end{pmatrix}, \quad A_2 = \begin{pmatrix} D & D & D & p & D \\ p & D & D & p & D \\ p & p & D & p & p \\ p & p & p & D & D \\ p & p & p & p & D \end{pmatrix}$$

Then the posets $I(A_1)^*$, $I(A_2)^+$, $J = I(A_1)^* \coprod_I I(A_2)^+$ and $\xi_C J$ have the following forms

$I(A_1)^*$:

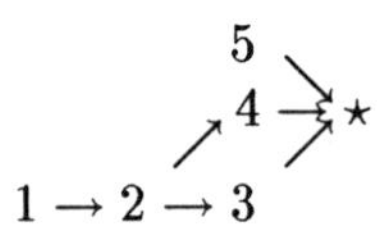

$I(A_2)^+$:

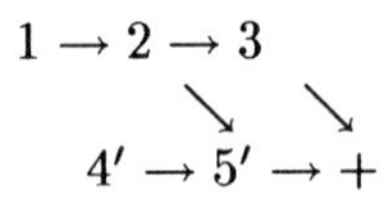

J :

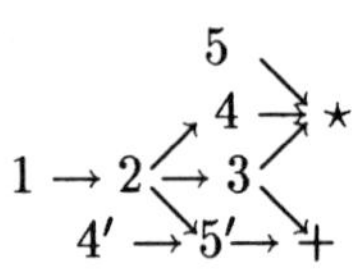

$\xi_C J$:

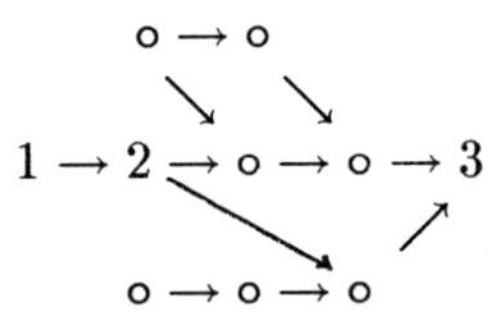

where $C = I$. The category $J - \mathrm{sp}$ is of infinite type because the subposet

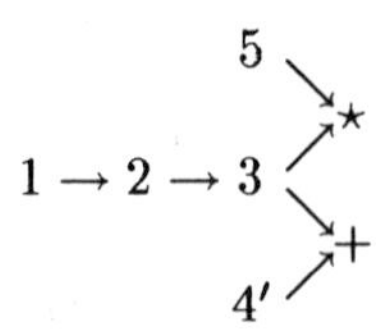

of J appears on the list in Theorem 1.7 (e). Since C is a chain and the posets $J - \{*\}$, $J - \{+\}$ are of finite type, then there exits an equivalence (1.10). It follows from the criterion of Nazarova (see [7]) that the category $\xi_C J - \mathrm{sp}$ is of tame type. Then applying (1.10) and (2.4) we can get a complete classification of the indecomposable lattices over $A_1 \times_I A_2$. □

The author was informed by Ju.A.Drozd that a result similar to our Theorem 1.7 was recently obtained by N.Golovashchuk in Kiev.

References

[1] M. Auslander and S. O. Smalø, *Almost split sequences in subcategories*, J. Algebra 69 (1981), 426 - 454.

[2] K. Bongartz, *A criterion for finite representation type*, Math. Ann. 269 (1984), 1 - 12.

[3] P.Gabriel, *Indecomposable representations* I, Manuscripta Math. 6(1972),71-103.

[4] E. L. Green and I. Reiner, *Integral representations and diagrams*, Michigan Math. J. 25 (1978), 53 - 84.

[5] M. M. Kleiner, *Partially ordered sets of finite type*, *in* Zap. Naychn. Sem. Leningrad. Otdel. Mat. Inst. Steklov. (LOMI)28 (1972), 32 - 41 [In Russian].

[6] J. A. de la Peña and D. Simson, *Prinjective modules, reflection functors, quadratic forms and Auslander-Reiten sequences*, Trans. Amer. Math. Soc. 1991 (to appear).

[7] C. M. Ringel, *Tame Algebras and Integral Quadratic Forms*, Lecture Notes in Math. Vol. 1099, Springer-Verlag, Berlin-Heidelberg-New York-Tokyo, 1984.

[8] C. M. Ringel and K. W. Roggenkamp, *Diagrammatic methods in the representation theory of orders*, J. Algebra 60 (1979), 11 - 42.

[9] D. Simson, *Socle reductions and socle projective modules*, J. Algebra 103 (1986), 18 - 68.

[10] D. Simson, *On Representations of Partially Ordered Sets*, Paderborn, July 1986, 188 p.

[11] D. Simson, *Moduled categories and adjusted modules over traced rings*, Dissertationes Math. 269(1990), 1-67.

[12] D. Simson, *Peak reductions and waist reflection functors*, Fund. Math. 137(1991), (to appear).

[13] D. Simson, *On the representation type of stratified posets*, C. R.Acad. Sci. Paris 311(1990), 5-10.

[14] D. Simson, *Right peak algebras of two-separate stratified posets, their Galois coverings and socle projective modules*, (to appear).

[15] D. Vossieck, *Représentations de bifoncteurs et interprétation en termes de modules*, C. R. Acad. Sci. Paris, 307 (1988), 713 - 716.

[16] Th. Weichert, *Darstellungstheorie von Algebren mit projektivem Sockel*, Doctoral Thesis, Universität Stuttgart, 1989.

[17] A.G.Zavadskij and V.V.Kirichenko, *Torsion-free modules over prime rings,in* Zap. Naychn. Sem. Leningrad. Otdel. Mat. Inst. Steklov. (LOMI), 57(1976), 100-116 [In Russian].

INSTITUTE OF MATHEMATICS
NICHOLAS COPERNICUS UNIVERSITY
UL.CHOPINA 12/18, 87-100 TORUŃ, POLAND

Canadian Mathematical Society
Conference Proceedings
Volume 11 (1991)

AN ALGORITHM FOR POSETS WITH AN EQUIVALENCE RELATION

A.G.Zavadskij

ABSTRACT. An algorithm of differentiation for posets with an equivalence relation is constructed. It consists of five main operations (and several additional), and in each case a functor of differentiation is described. The algorithm can be applied to any tame matrix problem determined by poset with an equivalence relation.

1. INTRODUCTION. Many classification problems of linear algebra arise in the modern representation theory. A natural way to deal with these problems is to construct some algorithms being able to reduce any problem from a given class to another one belonging to the same class but having smaller dimension.

In the categorical language such a reduction means a transition to new categories which differ a little from the initial ones. And algorithms are none others than functors acting in the corresponding categories and inducing equivalence between them modulo some ideals.

The algorithmic approach was found effective especially in the representation theory of tame posets where the basic points were established with help of differentiation procedures [2,3,5]. Note that the algorithms [2] and [5] act in the class of linear matrix problems of non-gelfand type, and the algorithm [3] in the class of flat (non-linear) gelfand matrix problems.(the term "linear problem" means a problem on classification of objects and morphisms of the subspace or factorspace category of a given vector space category).

In the present paper a new algorithm of differentiation is constructed. It is specially accomodated to the class of linear

1980 Mathematics Subject Classification (1985 Revision).16A64

matrix problems introduced in [4] and determined by posets with an equivalence relation. The algorithm can be applied to any tame problem of this class (including gelfand problems). In particular, it embraces problems which are equivalent to those of considered in [3].

The algorithm consists of five main operations I,...,V. The differentiation I is the old one with respect to a pair of points for ordinary posets [5], and roughly speaking the rest of operations have some common idea with the differentiation I. Nevertheless they don't resemble the differentiation I even outwardly and need more subtle considerations. The algorithm includes also additional, rather trivial, differentiations $\alpha, \bar{\alpha}, \beta, \gamma$.

In each case we define a functor of differentiation but don't study its deep categorical properties. The main aim of this article is to describe the induced by functors operations $\downarrow$ and $\uparrow$ making real differentiation and integration of objects of categories of representations (clearly functors can be also adapted to a description of morphisms if necessary).

Let us give now a general formulation of the main result about the algorithm. Assume $\mathfrak{M}$ is a poset with an equivalence relation and $\mathfrak{M}'$ the corresponding derivative poset with respect to one of the mentioned differentiations I,...,V or $\alpha, \bar{\alpha}, \beta, \gamma$. In the categories of representations of the posets $\mathfrak{M}$ and $\mathfrak{M}'$ we choose some full subcategories R and R' which differ a little from the initial ones and define a differentiation functor $' : R \longrightarrow R'$ inducing the operations $\downarrow$ and $\uparrow$. Among objects of the category R we choose also so called reduced objects not containing some trivial direct summands. Then the following result (established for each case separatly) takes place.

THEOREM 1. The correspondence $U \longmapsto U^{\downarrow}$ determines a natural bijection of the set of all reduced (pairwise nonisomorphic) objects of the category R onto the set of all (pairwise nonisomorphic) objects of the category R'. Moreover, the inverse bijection is determined by the operation $\uparrow$, the indecomposables go into indecomposables and, of course, the inequality $\dim U^{\downarrow} \leqslant \dim U$ holds.

At the end of the paper we formulate some consequences being used in applications.

In the separate paper by V.M.Bondarenko and the author (submitted to these proceedings) the obtained algorithm is used for the description of posets with an equivalence relation of tame type and of finite growth. Some combinatorial properties of the algorithm are established there too.

2. THE CATEGORY OF REPRESENTATIONS. By poset with an equivalence relation we mean a pair $(\mathfrak{M}, \Theta)$ where $\mathfrak{M}$ is a finite poset and Θ is an arbitrary equivalence relation on the set of all points of $\mathfrak{M}$ (we assume that Θ is the set of all equivalence classes). Generally speaking, the both relations - order and equivalence - don't depend from each other.

Denote by $\langle x \rangle$ the equivalence class of a point $x \in \mathfrak{M}$ and by $r(x)$ the number of points in this class. A point x will be called small (big) if $r(x)=1$ $(r(x) \geqslant 2)$. Small and big points will be pictured correspondently by o and • . We write $x \sim y$ if $x \neq y$ and $\langle x \rangle = \langle y \rangle$.

Let k be an arbitrary field. Representations of posets with an equivalence relation (also called weakly completed posets) were introduced by L.A.Nazarova and A.V.Roiter [4] in the matrix language. According to their definition a representation of a poset $(\mathfrak{M}, \Theta)$ over k is any matrix M over k which is devided into vertical stripes M_x numbered by elements $x \in \mathfrak{M}$ with the only one condition: if $x \sim y$ then the numbers of columns of M_x and M_y are equal.

Two representations are called isomorphic if one of them may be turned into the other with help of the following transformations:

1) elementary transformations of rows of the whole matrix M;
2) simultaneous transformations of columns of the matrices M_x and M_y if $x \sim y$;
3) additions of columns of M_x to columns of M_y if $x < y$.

It will be convenient for us to use in this paper the vector space interpretation of representations of posets with an equivalence relation which is analogous to that one given by P.Gabriel [1] for ordinary posets.

Let U be a finite dimensional vector space over k and $K \in \Theta$ some equivalence class. Denote by $U^{|K|}$ a direct sum of $|K|$ copies of the space U (assuming the summands to be numbered by elements of the class K).

If $x, y \in \mathfrak{M}$, denote by $\pi_x : U^{r(x)} \to U$ a natural projection onto the summand with a number x and by $i_x : U \to U^{r(x)}$

a natural injection of the summand with the number x. Let $\varepsilon_{xy}: U^{r(x)} \to U^{r(y)}$ be the composition $\varepsilon_{xy} = \pi_x i_y$ (we write compositions naturally: from the left to the right) inducing an isomorphism of the component with the number x onto the component with the number y.

For a given subspace $E \subset U$ and a subset $X \subset K$ denote by E^X a direct sum $E^X = \bigoplus_{x \in X} E i_x$ contained in $U^{|K|}$ (assume $U^{\emptyset} = 0$).

Suppose that a subspace $U_K \subset U^{|K|}$ corresponds to every class $K \in \Theta$. A collection $(U, U_K)_{K \in \Theta}$ will be called a representation of a poset with an equivalence relation $(\mathfrak{M}, \Theta)$ over k if the following condition holds:

$$U_{\langle x \rangle} \varepsilon_{xy} \subset U_{\langle y \rangle} \quad \text{if} \quad x < y$$

(note that generally speaking $U_{\langle x \rangle} \varepsilon_{xx} \not\subset U_{\langle x \rangle}$). Sometimes we shall write simply U instead of $(U, U_K)_{K \in \Theta}$.

Any k-linear map $\varphi: U \to V$ induces, for every $K \in \Theta$, a natural map $\varphi^{|K|}: U^{|K|} \to V^{|K|}$ being a direct sum of $|K|$ copies of the map φ. A map $\varphi: U \to V$ will be called a morphism from a representation $(U, U_K)_{K \in \Theta}$ into a representation $(V, V_K)_{K \in \Theta}$ if $U_K \varphi^{|K|} \subset V_K$ for each $K \in \Theta$. Denote by $(\mathfrak{M}, \Theta)$-sp just defined category of representation.

One can easily verify that the classification of objects of the category $(\mathfrak{M}, \Theta)$-sp leads to the described above matrix problem. In the case $|K| = 1$ for any $K \in \Theta$ we obtain the category $\mathfrak{M}$-sp of ordinary representations of a poset $\mathfrak{M}$ in the sense of [1].

For any vector space W denote by W^* the dual space $W^* = \mathrm{Hom}_k(W, k)$. For a given inclusion $X \subset Y$ let $X^{\perp} = \{\alpha \in Y^* \mid X\alpha = 0\}$.

To every representation $(U, U_K)_{K \in \Theta}$ we associate, in a natural way, the dual representation $(U^*, U_K^*)_{K \in \Theta}$ of the category $(\mathfrak{M}^{op}, \Theta)$-sp, where $U_K^* = U_K^{\perp}$ for the inclusion $U_K \subset U^{|K|}$ (points of the posets $\mathfrak{M}$ and $\mathfrak{M}^{op}$ are naturally identified).

Denote $U_x^+ = U_{\langle x \rangle} \pi_x$; $U_x^- = U_{\langle x \rangle} \cap \mathrm{Im}\, i_x$. Since $U_x^- \subset U_x^+ \subset U$ one can associate with a representation $(U, U_K)_{K \in \Theta}$ the ordinary representations $(U, U_x^-)_{x \in \mathfrak{M}}$ and $(U, U_x^+)_{x \in \mathfrak{M}}$ of the category $\mathfrak{M}$-sp. They play a dual role because $(U_x^+)^{\perp} = (U^*)_x^-$ for the inclusion $U_x^+ \subset U$. Since $U_x^- = U_x^+$ for a small point x then for such points we shall omit the marks $\pm$ and write simply U_x.

Let $\underline{U}_K = \sum U_x^+ i_y = \sum U_{\langle x \rangle} \varepsilon_{xy}$ where $x < y$ and $y \in K$. The dimension of a representation U is the vector $\underline{\dim}\, U = (h, h_K)_{K \in \Theta}$, where h=dim U and $h_K=(U_K/\underline{U}_K)$.

For a subset $A\subset\mathfrak{M}$ let $U_A^+=\sum_{x\in A}U_x^+$ and $U_A^-=\bigcap_{x\in A}U_x^-$ ($U_\emptyset^+=0$ and $U_\emptyset^-=U$).

Sometimes we shall consider infinite posets with an equivalence relation ($\mathfrak{M},\Theta$) and denote by ($\mathfrak{M},\Theta$)-sp the category of all their representations having finite supports. Here by a representation having finite support we mean such a representation U that there exists a finite subset with an equivalence relation ($\mathfrak{M}_1,\Theta_1$), where $\mathfrak{M}_1\subset\mathfrak{M}$ and $\Theta_1\subset\Theta$, for which every space U_K, $K\in\Theta\setminus\Theta_1$, is generated by the spaces U_K, $K\in\Theta_1$, in the sense of summation, i.e. for any $K\in\Theta\setminus\Theta_1$ the following condition holds: $U_K=\underline{U}_K=\sum U_x^+ i_y$ where $x\in\mathfrak{M}_1$; $y\in K$ and $x<y$ (note that the dual representation may not satisfy this condition).

Given a point $a\in\mathfrak{M}$, we denote $a^\Delta=a^\Delta_{\mathfrak{M}}=\{x\in\mathfrak{M}\mid x\geqslant a\}$, $a^\nabla=a^\nabla_{\mathfrak{M}}=\{x\in\mathfrak{M}\mid x\leqslant a\}$. If $A\subset\mathfrak{M}$ then $A^\Delta=\bigcup_{a\in A}a^\Delta$, $A^\nabla=\bigcup_{a\in A}a^\nabla$. By $\min\mathfrak{M}$ ($\max\mathfrak{M}$) denote the set of all minimal (maximal) points of $\mathfrak{M}$. We'll write $\mathfrak{M}=A_1+\ldots+A_n$ if $\mathfrak{M}=A_1\cup\ldots\cup A_n$ and $A_i\cap A_j=\emptyset$ for $i\neq j$. If $a\in A$ then let $A\setminus a=A\setminus\{a\}$.

By k(A), where k is a fixed field and $A\subset\mathfrak{M}$, denote a representation of the form $k(A)=(U,U_K)_{K\in\Theta}$, where U=k and $U_K=k^{K\cap A^\Delta}$. Of course, $k(A)=k(A^\Delta)$ and $(k(A))^*=k((\mathfrak{M}\setminus A^\Delta)^{op})$. We shall write $k(A_1,\ldots,A_n)=k(A_1\cup\ldots\cup A_n)$, moreover, here instead of one-point sets we shall write simply the corresponding elements, for example, $k(A,b)=k(A\cup\{b\})$.

Throughout the paper we shall assume that every equivalence class $K\in\Theta$ contains no more than two elements (it is sufficient to built an algorithm exactly for this case). Sometimes a two-element class $\langle x\rangle=\{x,x_1\}$ will be written in the form (x,x_1) in order to indicate the corresponding numeration of the summands of the space $U^2=U\oplus U$ containing $U_{\langle x\rangle}$. Denote $\Theta_2=\{K\in\Theta\mid |K|=2\}$.

Now we start with a description of the algorithm. The differentiation acts with respect to some specially chosen pair (or even group) of points of a poset $\mathfrak{M}$ that's why we consider five main cases I,...,V and four additional $\alpha,\bar{\alpha},\beta,\gamma$.

Note that sometimes it is useful to involve the matrix language for considerations concerning a substantiation of the algorithm. We recommend the reader to use matrices in those cases when "abstract" formulas seem bulky - the matrix sense is always simple. As for the author, the idea of each differentiation was caught by matrices, but a strict substantiation needed an invariant language.

3. DIFFERENTIATION I. This is the differentiation with respect to a pair of points described in [5]. For a convenience of the reader we briefly remind this procedure. A pair of small points (a,b) of a poset $\mathfrak{M}$ is called I-suitable (i.e. suitable for Differentiation I) if $\mathfrak{M}=a^{\Delta}+b^{\nabla}+C$, where $C=\{c_1<c_2<\ldots<c_n\}$ is a small-point chain (possibly $C=\emptyset$) and, moreover, the points a,b,c_i are mutually incomparable for any i (the last condition isn't essential). The derivative poset (with respect to the pair (a,b) is a poset $\mathfrak{M}'$ obtained from $\mathfrak{M}$ as follows:

1) the chain $c_1<\ldots<c_n$ is replaced by two new chains $c_1^-<\ldots<c_n^-$ and $c_1^+<\ldots<c_n^+$;
2) any of the points c_i^-,c_i^+ inherits all relations of the order which the point c_i had before with points of a subset $\mathfrak{M}\setminus C$;
3) relations $a<c_i^+$; $c_i^-<b$; $a<b$ and $c_i^-<c_i^+$ are added (and, of course, induced relations).

It is supposed that the two-element classes are not changed, i.e. $\Theta_2'=\Theta_2$.

Denote the categories briefly $R=(\mathfrak{M},\Theta)$-sp, $R'=(\mathfrak{M}',\Theta')$-sp and define the differentiation functor $':R\to R'$ which assigns to any object $(U,U_K)_{K\in\Theta}$ the derivative object $(U',U_K')_{K\in\Theta'}$ according to the following rule:

$$
\begin{aligned}
&U'=U;\\
(\text{I.1})\qquad &U'_{c_i^+}=U_a+U_{c_i};\quad U'_{c_i^-}=U_b\cap U_{c_i};\\
&U_b'=U_b+U_a;\\
&U_K'=U_K \text{ for other classes } K\in\Theta'.
\end{aligned}
$$

We also set $\varphi'=\varphi$ for a morphism φ considered as a linear map.

Note that $(k(a))'=k(a,c_i)'=k(a)$ and as a rule U' contains direct summands of the type $k(a)$; the number of these summands is not less then $\dim(U_a+U_b)/U_b$. Denote by $U^{\downarrow}$ such an object of the category R' (defined uniquely up to isomorphism) for which $U'=U^{\downarrow}\oplus k^m(a)$, where $m=\dim(U_a+U_b)/U_b$. In [5] it was pointed out a direct way of construction of the object $U^{\downarrow}$ which follows immediatly from (I.1) as well. Namely, presenting the space U in the form $U=E\oplus W$, where $E\subset U_a$; $E\oplus U_b=U_a+U_b$ and $W\supset U_b$, we obtain

$U^{\downarrow}=(W,W_K)_{K\in\Theta'}$, where $W_K=U'_K\cap W^{|K|}$ for any $K\in\Theta'$, i.e. $W_{c_i^+}=$ $=(U_a+U_{c_i})\cap W$; $W_{c_i^-}=U_{c_i}\cap W$; $W_b=U_b$ and $W_K=U_K\cap W^{|K|}$ for the rest of K. Obviously, $\dim W_* = \dim U - \dim E$ and $\dim E=m$.

The objects of the category R, not containing direct summands of the type $k(a)$ and $k(a,c_i)$, $i=1,\dots,n$, will be called reduced.

There exists an important (in some sense inverse) operation $\uparrow$ which assigns to any object W of the category R' the primitive object $W^{\uparrow}$ of the category R such that $(W^{\uparrow})'=W\oplus k^m(a)$. For its construction write any space $W_{c_i^+}$ in the form $W_{c_i^+}=\underline{W}_{c_i^+}\oplus H_i\oplus F_i$, where $H_i\cap W_b=0$ and $F_i\subset W_b$. Let $f_1^i,\dots,f_{m_i}^i$ be a basis of F_i. Considering a new space E of dimension $m=\sum_{i=1}^{n} m_i$ with a basis e_j^i $(1\leqslant i\leqslant n;\ 1\leqslant j\leqslant m_i)$ we have $W^{\uparrow}=(U,U_K)_{K\in\Theta}$, where

$$
\begin{aligned}
&U=W\oplus E;\\
(\mathrm{I}.2)\qquad &U_K=W_K\oplus E^{K\cap a^{\Delta}} \quad \text{for } K\neq\langle c_i\rangle;\\
&U_{c_i}=U_{c_{i-1}}+W_{c_i^-}+H_i+\{e_1^i+f_1^i,\dots,e_{m_i}^i+f_{m_i}^i\} \text{(assume } U_{c_0}=0).
\end{aligned}
$$

The truth of Theorem 1 for Differentiation I was proved in [5] (in fact, it easily follows from the difinitions given above).

4. SOME SPECIAL NOTATIONS. Denote by $\langle(\mathfrak{M},\Theta)\text{-sp}\dots\rangle$ a full subcategory of the category $(\mathfrak{M},\Theta)$-sp formed by all those objects which satisfy some fixed relations written instead of dots. We shall often use the standard relations $\alpha(a)$, $\beta(b)$, $\gamma(b_1,\dots,b_n)$ and dual to them, where a,b,b_i are some points of the poset $\mathfrak{M}$.

Namely, we say that an object U satisfies the relation $\alpha(a)$, where a is any point of $\mathfrak{M}$, if $U_a^+\subset U^+_{\mathfrak{M}\setminus a^{\Delta}}$ or, more briefly, $a^+\subset(\mathfrak{M}\setminus a^{\Delta})^+$ (the dual relation $\alpha^*(a)$ is $U^-_{\mathfrak{M}\setminus a^{\nabla}}\subset U_a^-$ or $(\mathfrak{M}\setminus a^{\nabla})^-\subset a^-$).

If b is a big point and $b\sim b_1$ then the relation $\beta(b)$, for an object U, is the following: $b_1^-=(b_1^{\nabla}\setminus b_1)^+$ (it means that for a matrix realization of this representation, corresponding to its dimension $(h,h_K)_{K\in\Theta}$, the matrix M_b is non-singular in columns, i.e. $\operatorname{rank} M_b=h_{\langle b\rangle}$). Clearly, the relation $b_1^+=$ $=(b_1^{\Delta}\setminus b_1)^-$ is nothing else as the dual one $\beta^*(b)$.

In conclusion, if $b_1,\dots,b_n \in \mathfrak{M}$ are any points with a condition $b_i^{\nabla} \setminus b_i = B$ for all i, where $B \subset \mathfrak{M}$ is some subset (note, in such a case the points b_i are automatically incomparable), then the relation $\gamma(b_1,\dots,b_n)$ is the following: $\{b_1,\dots,b_n\}^- = B^+$. The dual one $\gamma^*(a_1,\dots,a_n)$ has the form $\{a_1,\dots,a_n\}^+ = A^-$ where $A = a_i^{\Delta} \setminus a_i$ for all i.

Note that $\beta(b) = \gamma(b_1)$ if $b \sim b_1$.

5. ADDITIONAL DIFFERENTIATION α. We shall say that a poset $\mathfrak{N}$ is obtained from a poset $\mathfrak{M}$ by replacing of a point $q \in \mathfrak{M}$ by some poset Q (or by substitution Q instead of q) if $\mathfrak{N} = (\mathfrak{M} \setminus q) + Q$, where inside of any of the subsets Q and $\mathfrak{M} \setminus q$ the initial order relation is preserved and, moreover, any point from Q inherits all the relations of order which the point q had before (with points of the subset $\mathfrak{M} \setminus q$).

Differentiation α is a simple procedure and acts with respect to any big point a of a poset $(\mathfrak{M}, \Theta)$. Let $a \sim a_1$ and $S = \mathfrak{M} \setminus a^{\Delta}$. The derivative poset $(\mathfrak{M}', \Theta')$, corresponding to Differentiation $\alpha = \alpha_a$, is obtained from a poset $(\mathfrak{M}, \Theta)$ in the following way:

1) the point a_1 is replaced by a two-element chain $a_1 < a_1^+$;
2) the set of two-element classes isn't changed: $\Theta_2' = \Theta_2$ (i.e. $a \sim a_1$ as before and a_1^+ is small).

Consider the categories $R = (\mathfrak{M}, \Theta)\text{-sp}$ and $R' = \langle (\mathfrak{M}', \Theta')\text{-sp} \,|\, \alpha(a) \rangle = \langle (\mathfrak{M}', \Theta')\text{-sp} \,|\, a^+ \subset S^+ \rangle$. The functor $' : R \to R'$ maps an object U into the derivative object U' according to the following rule:

$$(\alpha.1) \qquad \begin{aligned} &U' = U; \\ &U'_{a_1^+} = U^+_{a_1}; \\ &U'_{(a,a_1)} = U_{(a,a_1)} \cap (U^+_S, U); \\ &U'_K = U_K \text{ for others } K \in \Theta'. \end{aligned}$$

For a morphism $\varphi : U \to V$ (considered as a linear map) set $\varphi' = \varphi$. Since $U^+_S \varphi \subset V^+_S$ and $U^+_{a_1} \varphi \subset V^+_{a_1}$, the map φ' really is a morphism in the category R', and the definition is correct.

Denote $A = a^{\Delta}_{\mathfrak{M}} \setminus a$; $A' = a^{\Delta}_{\mathfrak{M}'} \setminus a$. Clearly $(k(a))' = k(A')$. Reduced objects of the category R are objects not containing direct summands of the type k(a).

By $U^{\downarrow}$ denote such an object of the category R' for which $U' = U^{\downarrow} \oplus k^m(A')$, where $m = \dim(U^+_a + U^+_S)/U^+_S$. In other words, if

$U_a^+ + U_S^+ = E \oplus U_S^+$ for some subspace $E \subset U_a^+$ and $U = E \oplus W$ for some $W \supset U_S^+$, then $U^{\downarrow} = (W, W_K)_{K \in \Theta'}$ where $W_K = U'_K \cap W^{|K|}$ for each $K \in \Theta'$.

Conversely, for any $W \in \mathrm{Ob}R'$, define now the primitive object $W^{\uparrow}$ for which $(W^{\uparrow})' = W \oplus k^m(A')$. For this goal we present the space $W_{a_1^+}$ in the form $W_{a_1^+} = \underline{W}_{a_1^+} \oplus F$ for some complement F with a basis $f_1, \dots, f_m$. Considering a new space E with a basis $e_1, \dots, e_m$ we define $W^{\uparrow} = (U, U_K)_{K \in \Theta}$, where

$$U = W \oplus E;$$

$$(\alpha.2) \qquad U_K = W_K \oplus E^{K \cap A} \text{ for } K \neq \langle a \rangle;$$

$$U_{(a,a_1)} = W_{(a,a_1)} \oplus \{(e_1, f_1), \dots, (e_m, f_m)\} \oplus E^{A \cap \{a_1\}}.$$

Analysing the constructions of the operations $\downarrow$ and $\uparrow$, we conclude that $W^{\uparrow}$ is necessary reduced, $U^{\downarrow\uparrow} \simeq U$ for a reduced U, and that Theorem 1 is true for the considered in this item categories R and R'.

6. ADDITIONAL DIFFERENTIATION $\overline{\alpha}$. Later on it will be convenient to consider some special case of Differentiation α and to modify it a little. A pair of incomparable points (a,b) will be called suitable for Differentiation $\overline{\alpha}$ if $\mathfrak{M} = a^{\Delta} + b^{\nabla}$, where a is a big point and b is small. Assume $a \sim a_1$. Then the derivative poset $(\mathfrak{M}', \Theta')$ is obtained from $(\mathfrak{M}, \Theta)$ by replacing the point a_1 by a two-element chain $a_1 < a_1^+$ together with addition of a relation $a < b$ (the two-element classes aren't changed). Here we assume $R = (\mathfrak{M}, \Theta)$-sp, $R' = (\mathfrak{M}', \Theta')$-sp, and the functor $' : R \to R'$ is defined exactly by the formulas $(\alpha.1)$ for $S = b^{\nabla}$. The rest of considerations concerning Differentiation α are transfered without changes to Differentiation $\overline{\alpha}$.

7. ADDITIONAL DIFFERENTIATIONS β AND γ. These are trivial operations being connected with each other: Differentiation β is a special case of Differentiation γ.

Consider the case of Differentiation γ. It acts with respect to a subset of incomparable points $\{b_1, \dots, b_n\}$ of a poset $\mathfrak{M}$ with the condition $b_i^{\nabla} \setminus b_i = B$ for all i, where $B \subset \mathfrak{M}$ and $n \geq 1$ (note that the points b_i may be both small and big). The derivative poset $(\mathfrak{M}', \Theta')$ is obtained from $(\mathfrak{M}, \Theta)$ by addition of a new small point c which is placed "between $\{b_1, \dots, b_n\}$ and B", i.e. $c < b_i$ for any i and $B \subset c^{\nabla}$ (naturally $\Theta'_2 = \Theta_2$).

Denote $R=(\mathfrak{M},\Theta)$-sp, $R'=\langle(\mathfrak{M}',\Theta')\text{-sp}|\gamma(b_1,\dots,b_n)\rangle$. The functor $':R\to R'$ maps an object U into an object U' as follows:

$$(\gamma.1)\qquad \begin{array}{l} U'=U; \\ U'_c=U_{\bar{c}}\{b_1,\dots,b_n\}; \\ U'_K=U_K \text{ for others } K. \end{array}$$

Let again $\varphi'=\varphi$ for a morphism φ. Obviously, this functor is an isomorphism of categories and maps indecomposables into indecomposables. That's why one can assume $U^{\downarrow}=U'$. Theorem 1 is true if all objects of the category R are assumed to be reduced.

Differentiation β is a special case of γ when n=1 and the point b_1 is big. Geometrically (but not for representations) it is dual to Differentiation α and gives an isomorphism of the categories $(\mathfrak{M},\Theta)\text{-sp}\cong\langle(\mathfrak{M}',\Theta')\text{-sp}|\beta(b)\rangle$.

8. DIFFERENTIATION II. We shall introduce a new notation. Let $(x_1,y_1),\dots,(x_n,y_n)$ be arbitrary two-element equivalence classes of a poset $(\mathfrak{M},\Theta)$ and $(U,U_K)_{K\in\Theta}$ some its representation. Denote by

$$[\,U_{(x_1,y_1)}-U_{(x_2,y_2)}-\dots-U_{(x_n,y_n)}\,] \qquad (n\geqslant 1)$$

a subspace in U^2 formed by all such pairs (t_o,t_n) for which there exists a sequence $t_o,t_1,\dots,t_n$ of elements of the space U satisfying the conditions $(t_{i-1},t_i)\in U_{(x_i,y_i)}$ for all $1\leqslant i\leqslant n$. Such defined subspace will be called a coupling of the spaces $U_{(x_1,y_1)},\dots,U_{(x_n,y_n)}$.

For instance, later on we'll consider a coupling

$$[U_{(a_1,a)}-U_{(b,b_1)}]=\{(t_o,t_2)\mid (t_o,t_1)\in U_{(a_1,a)} \text{ and } (t_1,t_2)\in U_{(b,b_1)} \text{ for some } t_1\in U\}.$$

A coupling of the form $[U_{(x,y)}-U_{(x,y)}-\dots-U_{(x,y)}]$, consisting of n copies of the same space $U_{(x,y)}$, will be denoted by $U^{[n]}_{(x,y)}$.

A pair of points (a,b) of a poset $\mathfrak{M}$ will be called II-suitable if $\mathfrak{M}=a^{\Delta}+b^{\nabla}$ where a,b are big incomparable points belonging to different equivalence classes. Assume $a\sim a_1$ and $b\sim b_1$. The derivative poset $(\mathfrak{M}',\Theta')$ with respect to pair (a,b) is obtained from $(\mathfrak{M},\Theta)$ in such a way:

1) the point a_1 is replaced by a two-element chain $a_1<a_1^o$

and the point b_1 by a two-element chain $b_1^o < b_1$;

2) a relation $a < b$ is added;

3) a relation $a_1^o \sim b_1^o$ is added (i.e. $\Theta_2' = \Theta_2 \cup \{(a_1^o, b_1^o)\}$).

Denote $A = a^{\Delta}_{\mathfrak{m}} \setminus a$; $B = b^{\nabla}_{\mathfrak{m}} \setminus b$; $A' = \mathfrak{m}' \setminus b^{\nabla}$; $B' = \mathfrak{m}' \setminus a^{\Delta}$. The functor ' will be defined not on the whole category $(\mathfrak{m}, \Theta)$-sp but on its full subcategory $R = \langle (\mathfrak{m}, \Theta)\text{-sp} \mid \alpha(a); \beta(b) \rangle = \langle (\mathfrak{m}, \Theta)\text{-sp} \mid a^+ \subset b^+;\ b_1^- = (b_1^{\nabla} \setminus b_1)^+ \rangle$ (if one of these relations isn't satisfied, one can apply the additional differentiation α or β). Also denote $R' = \langle (\mathfrak{m}', \Theta')\text{-sp} \mid \alpha(a); \beta(b) \rangle = \langle (\mathfrak{m}', \Theta')\text{-sp} \mid a^+ \subset (B')^+;\ b_1^- = (b_1^o)^+ \rangle$ and define the functor $': R \to R'$ which maps an object (U, U_K) into an object (U', U_K') as follows:

$$U' = U;$$

$$U'_{(a,a_1)} = U_{(a,a_1)} \cap (U_B^+, U);$$

(II.1) $$U'_{(b,b_1)} = U_{(b,b_1)} + (U_a^+, 0);$$

$$U'_{(a_1^o, b_1^o)} = [\,U_{(a_1,a)} - U_{(b,b_1)}\,];$$

$$U_K' = U_K \quad \text{for other classes } K,$$

and, as usually, suppose $\varphi' = \varphi$ for $\varphi \in \mathcal{M}or\, R$.

Clearly, $(k(a,b))' = k(A', b)$. The objects of the category R without direct summands $k(a,b)$ will be called reduced.

By $U^{\downarrow}$ denote an object of the category R' satisfying the condition $U' = U^{\downarrow} \oplus k^m(A', b)$, where $m = \dim(U_a^+ + U_B^+)/U_B^+$. If one takes any subspace $E \subset U_a^+$, for which $U_a^+ + U_B^+ = E \oplus U_B^+$, and any $W \supset U_B^+$, for which $E \oplus W = U$, then $U^{\downarrow} = (W, W_K)_{K \in \Theta'}$, where $W_K = U_K' \cap W^{|K|}$ for all K, i.e. $W_{(b,b_1)} = (U_{(b,b_1)} + (U_a^+, 0)) \cap W^2$; $W_{(a_1^o, b_1^o)} = [\,U_{(a_1,a)} - U_{(b,b_1)}\,] \cap W^2$ and $W_K = U_K \cap W^{|K|}$ for others K.

The inverse building of the primitive object $W^{\uparrow}$ with a property $(W^{\uparrow})' = W \oplus k^m(A', b)$ comes true by the following way. Assume $W_{(a_1^o, b_1^o)} = \underline{W}_{(a_1^o, b_1^o)} \oplus F$ where F is some complement with a basis $(\alpha_1, \beta_1), \ldots, (\alpha_m, \beta_m)$. Considering a new vector space E with a basis $e_1, \ldots, e_m$ set $W^{\uparrow} = (U, U_K)$, where

$$U = W \oplus E;$$

(II.2) $$U_K = W_K \oplus E^{K \cap A} \quad \text{for } K \neq \langle a \rangle, \langle b \rangle;$$

$$U_{(a,a_1)} = W_{(a,a_1)} \oplus \{(e_1, \alpha_1), \ldots, (e_m, \alpha_m)\} \oplus E^{A \cap \{a_1\}};$$

$$U_{(b,b_1)}=W_{(b,b_1)}\oplus\{(e_1,\beta_1),\dots,(e_m,\beta_m)\}\oplus E^{A\cap\{b_1\}}.$$

It isn't difficult to verify that $U'=W\oplus k^m(A',b)$ and that U is a reduced object. Moreover, $U^{\downarrow\uparrow}\cong U$ for any reduced U. It is clear now that Theorem 1 is true for the defined here categories R and R'.

9. DIFFERENTIATION III. A pair of points (a,b) of a poset $(\mathfrak{M},\Theta)$ will be called suitable for Differentiation III if $\mathfrak{M}=a^{\Delta}+b^{\nabla}$ where a,b are big incomparable equivalent points. In this situation the derivative poset $(\mathfrak{M}',\Theta')$ is obtained from $(\mathfrak{M},\Theta)$ as follows:

1) the point a is replaced by an infinite decreasing chain $a_1>a_2>a_3>\dots$ and the point b by an infinite increasing chain $b_1<b_2<b_3<\dots$;
2) a relation $a_1<b_1$ is added (and induced relations);
3) Θ_2' is obtained from Θ_2 by rejection the class $\{a,b\}$ and an addition the classes $\{a_n,b_n\}$, $n\geqslant 1$.

Let $A=a^{\Delta}_{\mathfrak{M}}\setminus a$; $B=b^{\nabla}_{\mathfrak{M}}\setminus b$. Denote

$$R=\langle(\mathfrak{M},\Theta)\text{-sp}\,|\,\alpha(a);\beta(b)\rangle=\langle(\mathfrak{M},\Theta)\text{-sp}\,|\,a^+\subset b^+;\ a^-=(a^{\nabla}\setminus a)^+\rangle,$$

$$R'=\langle(\mathfrak{M}',\Theta')\text{-sp}\,|\,\alpha(a_1);\underset{n\geqslant 1}{\beta(b_n)}\rangle=\langle(\mathfrak{M}',\Theta')\text{-sp}\,|\,a_1^+\subset B^+;\ \underset{n\geqslant 1}{a_n^-=a_{n+1}^+}\rangle$$

and define the functor $'$: $R\to R'$ which assigns to an object (U,U_K) the object (U',U_K') as follows:

$$\text{(III.1)}\qquad \begin{aligned} &U'=U;\\ &U_K'=U_K \quad \text{for } K\neq(a_n,b_n),\ n\geqslant 1;\\ &U'_{(a_n,b_n)}=(U_B^+,U)\cap U^{[n]}_{(a,b)}+(0,U_a^+),\ n\geqslant 1. \end{aligned}$$

For a morphism φ of the category R set $\varphi'=\varphi$.

In contrast to other cases, here the operation $\downarrow$ (defined below) will annul an infinite series of indecomposables of the category R (the series of representations of the pencil $\{a,b\}$), that's why we introduce the corresponding objects $\Phi(\lambda,n)$.

Let $\lambda\in k[X]$ be an irreducible polynomial of a degree $l\geqslant 1$ different from X. Denote by $\Phi(\lambda,n)$ an indecomposable object of the category R given by such a collection of matrices $\{M_x\}_{x\in\mathfrak{M}}$ that $M_x=\emptyset$ for $x\neq a,b$; $M_b=E$ is a unit matrix of order ln and M is an indecomposable matrix in the Frobenius normal form (of the same order) with the characteristic polynomial λ^n. Clearly, $\phi(X-1,1)=k(a,b)$.

Note that the formulas (III.1) really define an object of the category $(\mathfrak{m}',\Theta')$-sp. Namely, since $U_B^+ \subset U_b^-$ we have $(U')^+_{b_n} \subset (U')^-_{b_{n+1}}$ and $(U')^-_{a_n}=(U^{[n]}_{(a,b)} \cap (U_B^+,U_a^+))\pi_a=(U')^+_{a_{n+1}}$, so U' is a representation of the poset $(\mathfrak{m}',\Theta')$. Moreover, U' is an object of the category $(\mathfrak{m}',\Theta')$-sp, i.e. a representation with a finite support (see the first item). In order to explain it in a most simple way cosider a matrix realisation of a representation U which can be written in the form

(III.2)

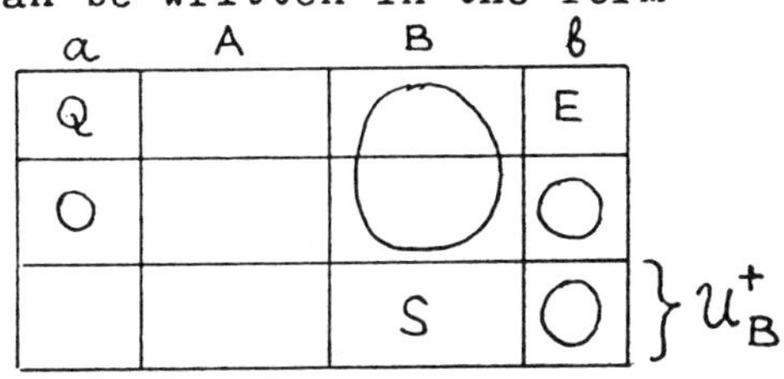

where the striped matrix S is non-singular in rows. Keeping the shape of the unit matrix E invariable, the matrix Q has to be reduced by similarity transformations, and one can suppose that Q is block-diagonal matrix with Frobenius blocks along its diagonal and zeros elswhere. Moreover, every Frobenius block must be singular because in the contrary case U has a direct summand $\phi(\lambda,n)$ with $\lambda \neq X$. Hence one can suppose that exactly Jordan blocks with eigenvalue 0 occupy the diagonal of Q. Now it isn't difficult to verify that $U^{[n]}_{(a,b)}=0$ modulo (U_H^+,U_B^+), where $H=\overset{\triangledown}{a}\setminus a$, for sufficiently big n (bigger than the maximal order of the mentioned Jordan blocks). Note that columns of the matrix a_n are formed by those columns of the matrix a which correspond to zero columns of the Jordan blocks of order n; and columns of b_n by those of b which are connected (via the unit matrix E) with zero rows of the Jordan blocks of order n.

It is easy to see that $(\phi(\lambda,n))'=k^{1n}(A,b_1)$. The reduced objects of the category R are objects without direct summands $\phi(\lambda,n)$ for all possible λ,n.

For $U \in \mathrm{Ob}\, R$ denote by $U^{\downarrow}$ such a direct summand of U' for which $U'=U^{\downarrow}\oplus k^m(A,b_1)$, where $m=\dim(U_a^+ + U_B^+)/U_B^+$. It can be built analogously to the case of Differentiation II. Namely, taking any $E \subset U_a^+$ and $W \supset U_B^+$, for which $U=E\oplus W$ and $U_a+U_B^+=E\oplus U_B^+$, we have $U^{\downarrow}=W$ and $W_K=U'_K \cap W^{|K|}$ for $K \in \Theta'$, i.e. $W_{(a_n,b_n)}=(U^{[n]}_{(a,b)}+(0,U^+))\cap W^2$

for $n\geqslant 1$ and $W_K=U_K\cap W^{|K|}$ for $K\neq(a_n,b_n)$.

The primitive object $W^{\uparrow}$ of the category R is defined for any object W of the category R' in the following way. Let it be $W_{(a_n,b_n)}=\underline{W}_{(a_n,b_n)}\oplus F_n$ for some complement F_n of a dimention p_n (note that $p_n=0$ for almost all n according to the definition of the category $(\mathfrak{M}',\Theta')$-sp for infinite posets). Suppose $(e^n_{01},\dots,e^n_{n1}),\dots,(e^n_{0p_n},\dots,e^n_{np_n})$ be a basis of F_n. For every $n\geqslant 1$ consider a new vector space E_n of dimention $(n-1)p_n$ with a basis e^n_{ij} where $1\leqslant i\leqslant n-1$; $1\leqslant j\leqslant p_n$. Denote $E=\bigoplus_n E_n$, $H=a^{\triangledown}_{\mathfrak{M}}\setminus a$ and set $W^{\uparrow}=(U,U_K)_{K\in\Theta}$,where

$$
\begin{aligned}
&U=W\oplus E;\\
\text{(III.3)}\qquad &U_K=W_K\oplus E^{K\cap A} \text{ for } K\neq(a,b);\\
&U_{(a,b)}=(W^+_H,W^+_B)\oplus T, \text{ where}
\end{aligned}
$$

$$T=\bigoplus_n\bigoplus_{i=1}\left\{(e^n_{i-1,i},e^n_{i1}),\dots,(e^n_{i-1,p_n},e_{ip_n})\right.$$

Comparing this constraction with the operation $\downarrow$ (it's useful to enlist the matrix interpretation (III.2)) we conclude that $W^{\uparrow}$ is a reduced object, $(W^{\uparrow})'=W\oplus k^m(A,b_1)$, where $m=\dim E=\sum_n(n-1)p_n$, and also $U^{\downarrow\uparrow}\cong U$ for a reduced $U\in \mathrm{Ob}\, R$. Hence Theorem 1 is true for Differentiation III.

Note that (for $U=W^{\uparrow}$) the number p_n is equal to the number of Jordan blocks of order n along diagonal of Q in (III.2), and that a restoring of the space $U_{(a,b)}$ means in the matrix language a restoring of all Jordan blocks of Q (the order of Q is equal to $\sum_n np_n$).

10. DIFFERENTIATION IV. A triple of incomparable points (a,b_1,b_2) of a poset $(\mathfrak{M},\Theta)$ will be called IV-suitable if:

a) $\mathfrak{M}=a^{\Delta}+\{b_1,b_2\}^{\triangledown}$ where a is big and b_1,b_2 are small;

b) $b_1^{\triangledown}\setminus b_1=b_2^{\triangledown}\setminus b_2=B$.

Assume $a\sim a_1$. The derivative poset $(\mathfrak{M}',\Theta')$ of the poset $(\mathfrak{M},\Theta)$ with respect to the triple (a,b_1,b_2) is obtained as follows:

1) between the subsets B and $\{b_1,b_2\}$ a new point p is inserted such that $B\subset p^{\triangledown}$ and $p<b_i$ $(i=1,2)$;
2) a relation $a<p$ is added (and induced relations);
3) the point a_1 is replaced by a "subset-rhomb" $\{a_1,\xi_1,\xi_2,p_1\}$,

where $a_1 < \xi_i < p_1$ for $i=1,2$;

4) Θ_2' is obtained from Θ_2 by an addition the class $\{p, p_1\}$.

Denote $A = a^{\Delta}_{\mathfrak{M}} \setminus a$; $A' = \mathfrak{M}' \setminus \{b_1, b_2\}^{\nabla}$; $B' = \mathfrak{M}' \setminus a^{\Delta}$. Consider the categories

$$R = \langle (\mathfrak{M}, \Theta)\text{-sp} \mid \alpha(a); \gamma(b_1, b_2) \rangle = \langle (\mathfrak{M}, \Theta)\text{-sp} \mid a^+ \subset \{b_1, b_2\}^+; b_1 b_2 = B^+ \rangle;$$

$$R' = \langle (\mathfrak{M}', \Theta')\text{-sp} \mid \alpha(a); \beta(p); \gamma(b_1, b_2) \rangle = \langle (\mathfrak{M}', \Theta')\text{-sp} \mid a^+ \subset (B')^+; \; b_1 b_2 = p^+; \; p_1^- = \{\xi_1, \xi_2\}^+ \rangle$$

and define the functor $'$: $R \to R'$ which maps an object $(U, U_K)_{K \in \Theta}$ of the category R into an object $(U', U'_K)_{K \in \Theta'}$ of the category R' by the rule:

$$\begin{aligned}
&U' = U;\\
&U'_{(a,a_1)} = U_{(a,a_1)} \cap (U_B^+, U);\\
\text{(IV.1)}\quad &U'_{b_i} = U_{b_i} + U_a^+, \quad i=1,2;\\
&U'_{\xi_i} = (U_{(a,a_1)} \cap (U_{b_i}, U))\pi_{a_1}, \quad i=1,2;\\
&U'_{(p,p_1)} = \{(z_1, y)\} + (U_a^+, 0) = \{(z_2, y)\} + (U_a^+, 0), \text{ where}\\
&\qquad z_i \in U_{b_i};\ z_1 - z_2 = x \text{ and } (x, y) \in U_{(a,a_1)}.
\end{aligned}$$

As usually $\varphi' = \varphi$ for $\varphi \in \mathcal{M}or\, R$.

Obviously $(k(a, b_i))' = k(A', p)$. The objects of the category R, not containing direct summands $k(a, b_i)$, $i=1,2$, are called reduced. By $U^{\downarrow}$ denote a direct summand of the object U' satisfying the equality $U' = U^{\downarrow} \oplus k^m(A', p)$ where $m = \dim(U_a^+ + U_B^+)/U_B^+$.

A procedure of building $U^{\downarrow}$ for a given U is standard: if $U = E \oplus W$ for any subspaces E and W with coditions $E \subset U_a^+$; $W \supset U_B^+$ and $E + U_B^+ = U_a^+ + U_B^+$, then $U^{\downarrow} = (W, W_K)_{K \in \Theta'}$ where $W_K = U'_K \cap W^{|K|}$.

For a construction of $W^{\uparrow}$ for arbitrary $W \in \mathrm{Ob}\, R'$ suppose $W_{(p,p_1)} = \underline{W}_{(p,p_1)} \oplus \{(f_1, g_1), \dots, (f_n, g_n)\}$, $W_{\xi_1} = \underline{W}_{\xi_1} \oplus \{u_1, \dots, u_r\}$ and $W_{\xi_2} = \underline{W}_{\xi_2} \oplus \{v_1, \dots, v_s\}$, where in the brackets bases of some complements are written, and consider new vector spaces $E_1 = \{u'_1, \dots, u'_r\}$, $E_2 = \{v'_1, \dots, v'_s\}$ and $E_p = \{e_1, \dots, e_n\}$. Let $E = E_1 \oplus E_2 \oplus E_p$. Then $W^{\uparrow} = (U, U_K)_{K \in \Theta}$, where

$U = W \oplus E;$

$U_{(a,a_1)} = W_{(a,a_1)} \oplus E^{A \cap \{a_1\}} \oplus \{(e_1,g_1),\dots,(e_n,g_n)\} \oplus$
$+\{(u_1',u_1),\dots,(u_r',u_r)\} \oplus \{(v_1',v_1),\dots,(v_s',v_s)\};$

(IV.2) $U_{b_1} = W_{b_1} \oplus E_1 \oplus \{e_1+f_1,\dots,e_n+f_n\};$

$U_{b_2} = W_{b_2} \oplus E_2 \oplus \{f_1,\dots,f_n\};$

$U_K = W_K + E^{K \cap A}$ for others $K \in \Theta$.

One can easily establish that $W^{\uparrow}$ is reduced and $(W^{\uparrow})' = W + k^m(A',p)$, where $m = \dim E = r+s+n$. Moreover, for a reduced $U \in \mathrm{Ob}\, R$, we have $U^{\downarrow\uparrow} \cong U$, and therefore Theorem 1 is true for this differentiation too. In the dual situation one can describe the dual procedure IV*.

11. DIFFERENTIATION V. A quadruple of incomparable points (a_1,a_2,b_1,b_2) of a poset $(\mathfrak{M},\Theta)$ is called V-suitable if $\mathfrak{M} = \{a_1, a_2\}^{\Delta} + \{b_1,b_2\}^{\nabla}$, where $a_1^{\Delta} \setminus a_1 = a_2^{\Delta} \setminus a_2 = A$ and $b_1^{\nabla} \setminus b_1 = b_2^{\nabla} \setminus b_2 = B$. Consider three kinds of this differentiation: V_1, V_2, V_3.

CASE V_1 (which in fact can be reduced to Differentiation I with n=1). The idea is the following: add conditionally a new maximal small point b, for which $b_i < b$ $(i=1,2)$, then apply Differentiation 1 with respect to the pair (a_2,b), then throw the point b off.

In this case the poset $(\mathfrak{M}',\Theta')$ is obtained from $(\mathfrak{M},\Theta)$ by an addition a new small point ξ which is inserted between $\{a_1,a_2\}$ and A, i.e., $a_i < \xi$ $(i=1,2)$ and $\xi^{\Delta} \setminus \xi = A$ (clearly $\Theta_2' = \Theta_2$).

The derivative object (U',U_K') is built for an object (U,U_K) in such a way: $U' = U$; $U_{\xi}' = U_{a_1} + U_{a_2}$; $U_{a_1}' = U_{a_1} \cap (U_{b_1} + U_{b_2})$ and $U_K' = U_K$ for others $K \in \Theta'$. Denote $R = (\mathfrak{M},\Theta)\text{-sp}$; $R' = \langle (\mathfrak{M}',\Theta')\text{-sp} \mid a_1 \subset b_1 + b_2 \rangle$. Assuming n=1, $a = a_2$, $c_1 = a_1$ (in $\mathfrak{M}$), $c_1^{+} = \xi$, $c_1^{-} = a_1$ (in $\mathfrak{M}'$) and substituting everywhere $U_{b_1} + U_{b_2}$ for U_b (we consider $U_b' = U_{b_1}' + U_{b_2}'$) we define the functor $': R \to R'$, the reduced objects and the operations $\downarrow$ and $\uparrow$ in exactly the same way as it was done above for

Differentiation 1. We obtain the same result (Theorem 1) for Differentiation V_1.

CASE V_2. Here $\mathfrak{M}'$ is obtained from $\mathfrak{M}$ as follows:

1) every point a_i is replaced by a two-element chain $a_i^- < a_i^+$ consisting of small points ($i=1,2$);

2) relations $a_i^- < a_j^+$ and $a_i^- < b_j$ are added for all $i,j \in \{1,2\}$.

The two-element classes aren't changed. Applying, if necessary, Differentiations $\mathcal{J}$ and V_1 we can reduce a consideration of objects of the category $(\mathfrak{M},\Theta)$-sp to the case when the relations $a_1+a_2 \subset b_1+b_2$ and $b_1 b_2 = B^+$ are satisfied. Denote

$R = \langle (\mathfrak{M},\Theta)\text{-sp} \mid a_1+a_2 \subset b_1+b_2\ ;\ b_1 b_2 = B^+ \rangle$,

$R' = \langle (\mathfrak{M}',\Theta')\text{-sp} \mid a_1^+ + a_2^+ \subset b_1+b_2\ ;\ b_1 b_2 = B^+;\ a_1^- + a_2^- \subset B^+ \rangle$ and define an action of the functor $'$: $R \to R'$ on objects $(U,U_K)_{K\in\Theta}$ by such a rule:

$$
\text{(V.1)}\qquad
\begin{aligned}
&U' = U;\\
&U'_{a_i^-} = U_{a_i} \cap U_B^+ ;\\
&U'_{a_i^+} = U_{a_i} + U_{a_j} \cap U_{b_j} \quad (i \neq j);\\
&U'_K = U_K \text{ for the rest of } K.
\end{aligned}
$$

Also put $\varphi' = \varphi$ for a morphism φ.

We have $(k(a_i,b_i))' = (k(a_1,a_2,b_i))' = k(a_1^+,a_2^+,b_i)$. The objects of the category R without direct summands $k(a_i,b_i)$ and $k(a_1,a_2,b_i)$ will be called reduced ($i=1,2$).

It is natural to choose the object $U^{\downarrow}$ as a direct summand of U' for which $U' = U^{\downarrow} + k^{m_1}(a_1^+,a_2^+,b_1) \oplus k^{m_2}(a_1^+,a_2^+,b_2)$ where $m_i = \dim U_{a_i} \cap U_{b_i} / U_{a_i} \cap U_B^+$. For a direct building of $U^{\downarrow}$ we present the space $U_{a_i} \cap U_{b_i}$ in the form $U_{a_i} \cap U_{b_i} = (U_{a_i} \cap U_B^+) \oplus E_i$ for some complement E_i, and the space U_{b_i} in the form $U_{b_i} = U_B^+ \oplus E_i \oplus X_i$ for some X_i (note that $(E_1 \oplus X_1) \cap (E_2 \oplus X_2) = 0$). Let $E = E_1 \oplus E_2$ and $X = X_1 \oplus X_2 \oplus U_B^+$. Let W be a space for which $U = E \oplus W$ and $W \supset X$. Then $U^{\downarrow} = (W, W_K)_{K\in\Theta'}$ where $W_K - U'_K \cap W^{|K|}$ for all $K \in \Theta'$.

Conversely, assume $W \in \operatorname{Ob} R$ and $W_{a_i^+} = W_{a_i^-} + H_i + F_i$, where $F_i \subset W_{b_i}$ and $H_i \cap W_{b_i} = 0$. Denote by $f_1^j, \ldots, f_{m_j}^j$ a basis of F_i (we specially take $j \neq i$), choose new vector spaces $E = \{e_1^i, \ldots, e_{m_i}^i\}$, $i=1,2$, and suppose $E=E_1 \oplus E_2$. Then $W^{\uparrow}=(U,U_K)_{K\in\Theta}$ where

$$
\begin{aligned}
&U=W\oplus E;\\
&U_{b_i}=W_{b_i}\oplus E_i;\\
(V.2)\qquad &U_{a_i}=W_{a_i^-}\oplus H_i\oplus E_i\oplus\{e_1^j+f_1^j,\ldots,e_{m_j}^j+f_{m_j}^j\},\ j\neq i;\\
&U_K=W_K\oplus E^{K\cap A}\ \text{for others } K.
\end{aligned}
$$

One can easily verify that the defined primitive object $W^{\uparrow}$ is reduced, $(W^{\uparrow})' \cong W\oplus(\bigoplus_{i=1}^{2} k^{m_i}(a_1^+,a_2^+,b_i))$ and $U^{\downarrow\uparrow} \cong U$ for a reduced $U \in \mathrm{Ob}\, R$.

Hence, Theorem 1 is true in the situation V_2.

CASE V_3. Here the poset $(\mathfrak{M}',\Theta')$ is obtained from $(\mathfrak{M},\Theta)$ as follows:

1) between the subsets $\{a_1,a_2\}$ and A a new point a_o is inserted (i.e. $a_i < a_o$ for $i=1,2$ and $a_o^{\Delta} \supset A$), and between B and $\{b_1,b_2\}$ a new point b_o;
2) relations $a_o < b_i$ and $a_i < b_o$ $(i=1,2)$ are added;
3) a relation $a_o \sim b_o$ is added.

An application of Differentiations $\mathcal{J}$, V_1, V_2 and V_2^* allows to deal only with representations of $(\mathfrak{M},\Theta)$ satisfying the following relations $b_1b_2=B^+$; $a_1+a_2 \subset b_1+b_2$; $a_ib_j \subset B^+$ and $A^- \subset a_i+b_j$, in particular, $a_k \subset a_i+b_j$. We use some of them only. Denote

$$
R=\langle(\mathfrak{M},\Theta)\text{-sp}\mid b_1b_2=B^+;\ a_1\subset b_1+b_2;\ a_ib_1\subset B^+;\ a_i\subset a_j+b_1;\ a_2\subset a_1+b_2;\ i=1,2;\ j\neq i\rangle,
$$

$R'=\langle(\mathfrak{M}',\Theta')\text{-sp}\mid b_1b_2=b_o^+;\ a_1+a_2\subset B^+\rangle$ and define the functor $'\colon R \to R'$ on objects in such a way:

$$
\begin{aligned}
&U'=U;\\
&U'_{a_i}=U_{a_i}\cap U_B^+;\\
&U'_{b_i}=U_{b_i}+U_{a_1};\\
&U'_{(a_o,b_o)}=\{(x,y)\mid x\in U_{a_2};\ y\in U_{b_2};\ x-y\in U_{b_1}\}+U_{a_1}^2;\\
&U'_K=U_K\ \text{for others } K
\end{aligned}
$$

(naturally $\varphi'=\varphi$ if $\varphi\in \mathcal{M}or\, R$).

We have $(k(a_1,a_2,b_2))'=k(a_o,b_o)$, and the reduced objects of the category R are those which don't contain direct summands

$k(a_1,a_2,b_2)$. For a construction of U choose such subspaces E and W for which $U=E\oplus W$; $E\subset U^+_{a_1}$; $W\supset U_{b_1}$ and $E\oplus U_{b_1}=U_{a_1}+U_{b_1}$. Then $U^{\downarrow}=(W,W_K)_{K\in\Theta'}$ where $W_K=U'_K\cap W^{|K|}$ for all $K\in\Theta'$ (note $W_{a_0}\subset W^+_{\mathcal{B}_0}$).

For the inverse construction of $W^{\uparrow}$ (where $W\in \mathrm{Ob}R'$) write the space $W_{(a_0,b_0)}$ in the form $W_{(a_0,b_0)}=\underline{W}(a_0,b_0)\oplus F$, where F is some complement with a basis $(\alpha_1,\beta_1),\dots,(\alpha_m,\beta_m)$. Also present the space $W_{\mathcal{B}_2}$ in the form $W_{\mathcal{B}_2}=\{\beta_1,\dots,\beta_m\}\oplus H$ for some complement H. Then consider a new space E with a basis $e_1,\dots,e_m$ and take $W^{\uparrow}=(U,U_K)_{K\in\Theta}$, where

$$U=W\oplus E;$$
$$U_{a_1}=W_{a_1}\oplus E;$$
$$\text{(V.4)}\qquad U_{a_2}=W_{a_2}\oplus\{e_1+\alpha_1,\dots,e_m+\alpha_m\};$$
$$U_{b_1}=W_{b_1};$$
$$U_{b_2}=H\oplus\{e_1+\beta_1,\dots,e_m+\beta_m\};$$
$$U_K=W_K\oplus E^{K\cap A}\ \text{for others } K\in\Theta.$$

The obtained object $W^{\uparrow}$ is reduced and such that $(W^{\uparrow})'\cong W\oplus k^m(a_0,b_0)$. For a reduced $U\in\mathrm{Ob}R$ we have $U^{\downarrow\uparrow}\cong U$ as well. That's why Theorem 1 takes place for the case V_3 too.

12. SOME CONSEQUENCES. First we shall discuss the change of dimentions by differentiations. Let U be a representation of a poset $(\mathfrak{M},\Theta)$ and $\underline{\dim}U=(h,h_K)_{K\in\Theta}$. If $h_K>0$ for each $K\in\Theta$ then U is called exact. Clearly, if $U^+_x\subset U^-_y$ for some $x,y\in\mathfrak{M}$, then U is in fact a representation of the poset $\overline{\mathfrak{M}}$ obtained from $\mathfrak{M}$ by an addition a relation $x<y$.That's why it is naturally to consider representations U with a condition

$$(*)\qquad U^+_x\subset U^-_y\quad\Rightarrow\quad x\leqslant y$$

It follows easily from the description of the algorithm that, for any exact U with the condition (*), the inequality $\dim U^{\downarrow}<\dim U$ holds for each of Differentiations I,...,IV,V_3 or $\overline{\alpha}$. An application of Differentiations α,β,γ may not decrease dimU but decreases the sum of all coordinates of $\underline{\dim}$U (or the sum of coordinates related to big points). The Differentiations V_1,V_2 and V_2^* can be cosidered (in the case of unchangable dimU) as preparing operations for Differentiation V_3 (and together with

V_3 form a general Differentiation V). So, we have

CONSEQUENCE 1. If U is an exact object of the category R satisfying the condition (*), then $\dim U^{\downarrow} < \dim U$ for any of Differentiations I,...,V or $\bar{\alpha}$.

In conclusion some words about a behavior of series by differentiations. When considering series, representations (S, S_K) of $(\mathfrak{m}, \Theta)$ over the plynomial ring k[t] are introduced. They are defined analogously to the case of a field (instead of k-spaces free k[t]-modules of a finite rank are considered).

We say that an ordinary representation $(U, U_K)_{K\in\Theta}$ is generated by a k[t]-representation $(S, S_K)_{K\in\Theta}$ if there exist such a k[t]-module A (having a finite k-dimention) for which $U = S\otimes A$ and $U_K = \mathrm{Im}(S_K \otimes A \xrightarrow{\delta_K \otimes 1} S^{|K|} \otimes A)$, where $\delta_K : S_K \to S^{|K|}$ is a canonical inclusion (all tensor products are taken over k[t]).

The set of all ordinary representations of a given dimension -vector d, generated by one k[t]-representation, is called a series of representations (of a fixed dimension).

The operations of differentiation and integration ($\downarrow$ and $\uparrow$) of k[t]-representations are defined in a standard way with help of transition to rational envelopes over the quotient field k(t) (see [6]). Using completely analogous to [6] considerations, we establish that for any of the mentioned above differentiations, any $n \in \mathbb{N}$ and any k[t]-representations S,T of the posets $(\mathfrak{m}, \Theta)$ and $(\mathfrak{m}', \Theta')$ respectively (the rational envelopes of which satisfy the relations in R and R') it holds the following:

$$(S\otimes A)^{\downarrow} \cong S^{\downarrow}\otimes A \text{ and } (T\otimes A)^{\uparrow} = T^{\uparrow}\otimes A \text{ for almost all } A \in \mathcal{A}_n,$$

where $\mathcal{A}_n$ is a full set pairwise non-isomorphic indecomposable k[t]-modules of a dimention n over k. Moreover, $S^{\downarrow\uparrow}$ and S ($T^{\uparrow\downarrow}$ and T) generate almost the same representations over k (where the rational envelope of S is supposed to be reduced). So, we have

CONSEQUENCE 2. When differentiating or integrating (in any of the cases I,...,V, α, $\bar{\alpha}$, β, γ and dual to them) a series of indecomposables go into a series, i.e. almost all representations of a given dimension, generated by one k[t]-representation, go again into representations generated by one k[t]-representation. Moreover, each series of the initial poset is obtained by integration from some series of the derivative poset exept of the standard series $\Phi(\lambda, n)$ of the pencil (a,b) in the case of Differentiation III.

13. APPENDIX. We present here the matrix differentiation (integration) for the main cases: I, II, III, IV, V_2, V_3. Everywhere ⧅ denotes a unit square matrix. The notation [↕] ([↔]) means that a matrix is non-singular in rows (in columns). Please, remember that representations satisfy the relations of the categories R and R'.

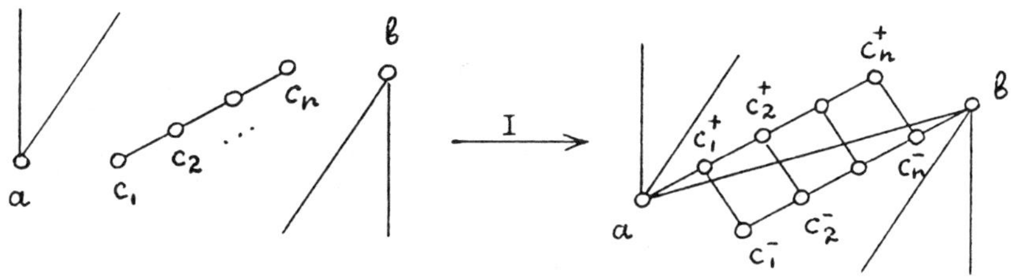

without summands k(a)

to reduce $a, c_1, \dots, c_n$ above U_b

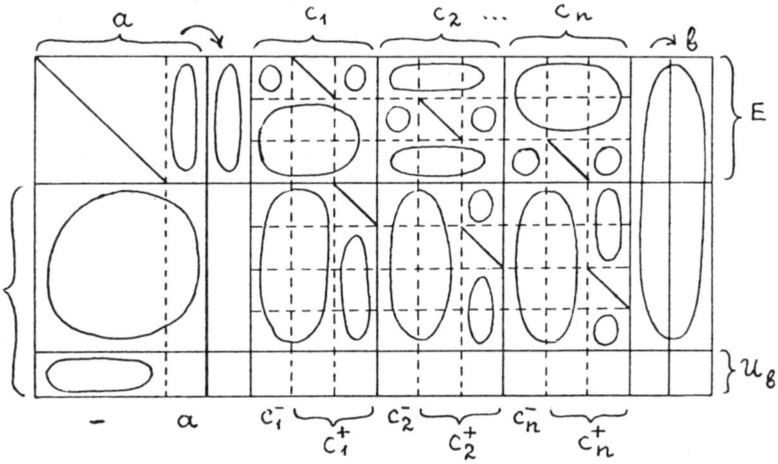

possibly $a_1 \in b^{\triangledown}$, $b_1 \in a^{\triangle}$

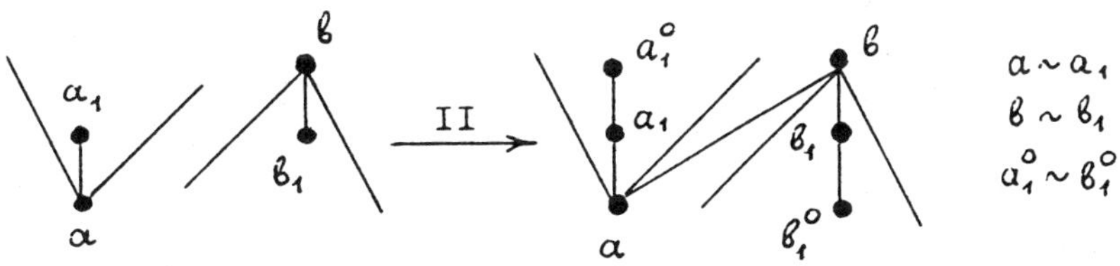

to reduce a, b above U_B^+

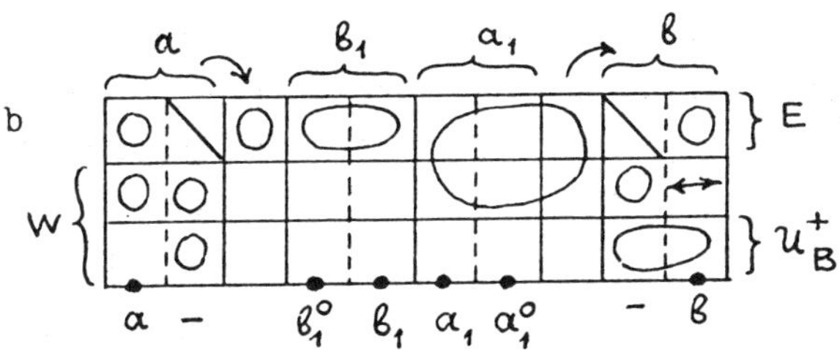

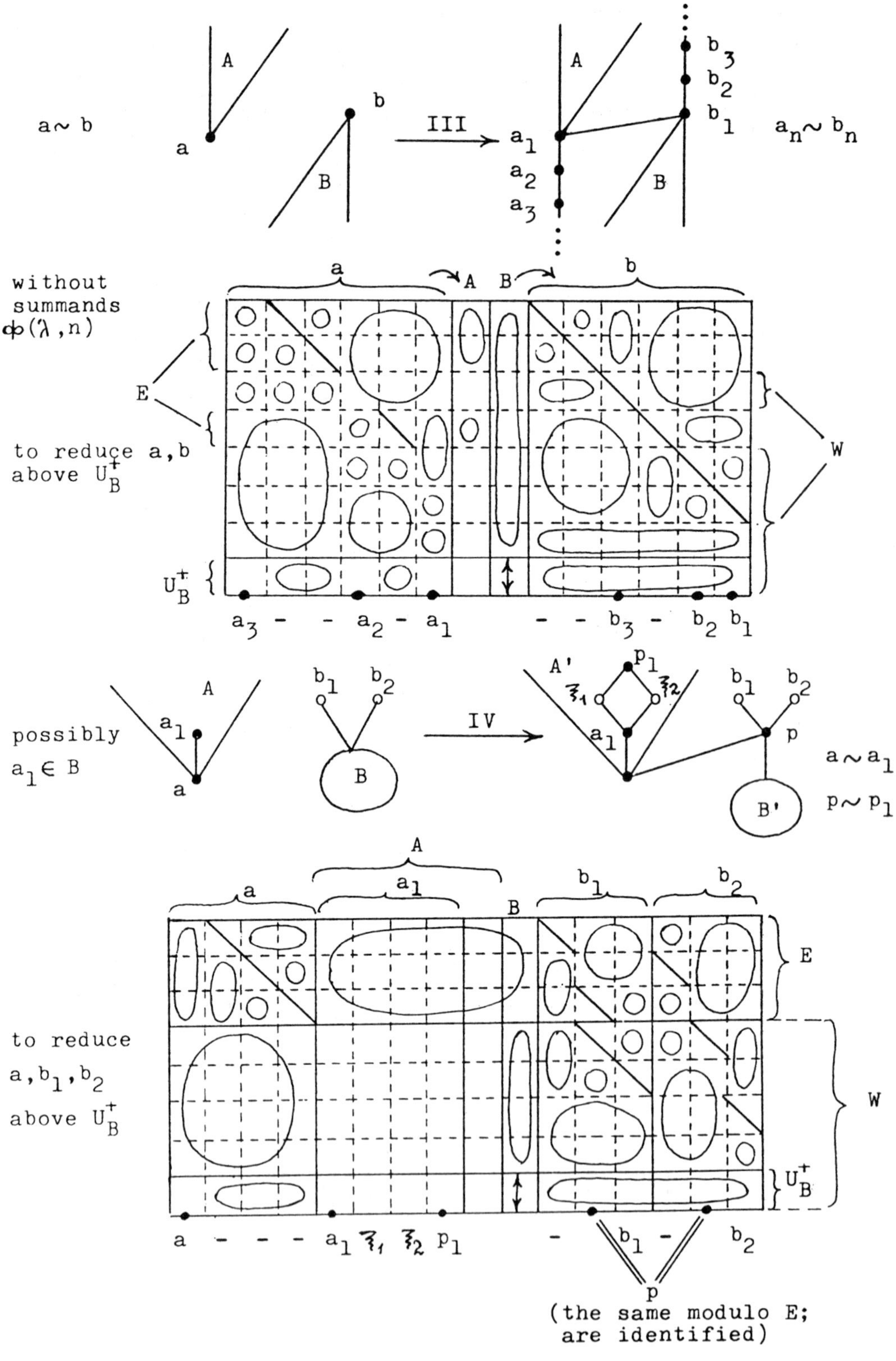

a∼ b
A
a
b
B
III
a_1
a_2
a_3
b_3
b_2
b_1
$a_n \sim b_n$
without summands $\Phi(\lambda, n)$
E
to reduce a, b above U_B^+
U_B^+
W
a_3 - - a_2 - a_1
- - b_3 - b_2 b_1
possibly $a_1 \in B$
IV
A'
p_1
ξ_1
ξ_2
p
B'
$a \sim a_1$
$p \sim p_1$
to reduce a, b_1, b_2 above U_B^+
a - - - a_1 ξ_1 ξ_2 p_1
- b_1 - b_2
p
(the same modulo E; are identified)

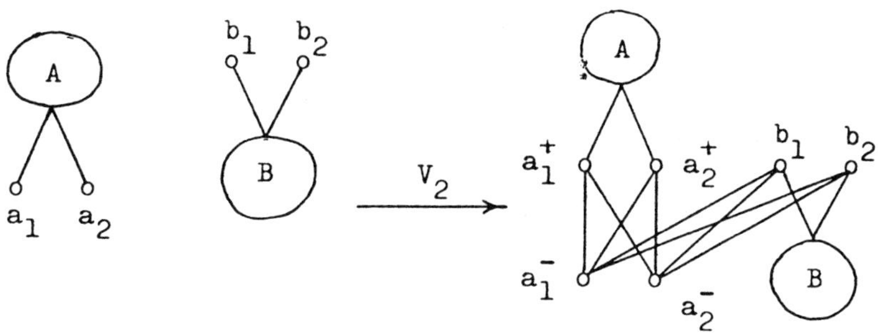

without summunds $k(a_i, b_i)$

to reduce b_1, b_2 and partly a_1, a_2 above U_B^+

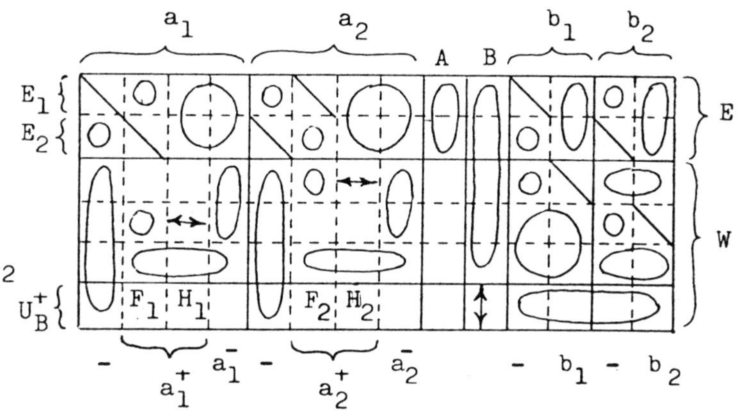

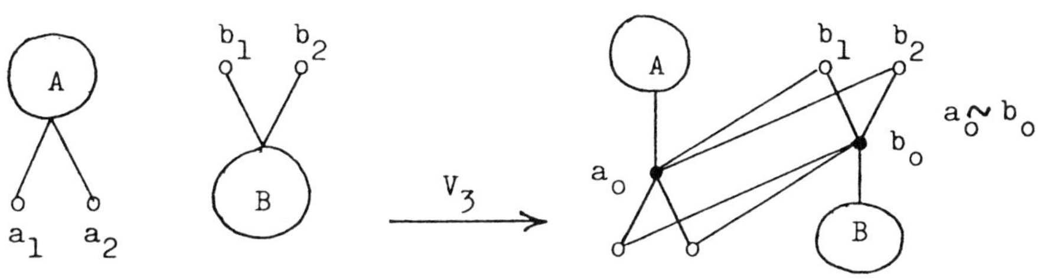

to reduce a_1, b_1 and partly a_2, b_2 above U_B^+

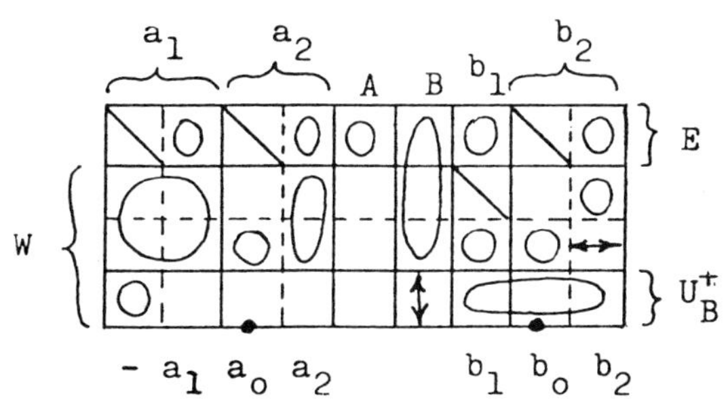

BIBLIOGRAPHY

1. Gabriel P., "Representations indecomposables des ensembles ordonnes", Sem. Dubreil (Algebra), 13(1972/73), 1301-1304.

2. Nazarova L.A., Roiter A.V., "Representations of posets", Zap. Naucn. Sem. LOMI, 28(1972), 5-31.

3. Nazarova L.A., Roiter A.V., "On a problem of I.M.Gelfand", Funk. Anal. i Pril., 7(1973), 4, 54-69.

4. Nazarova L.A., Roiter A.V., "Categorical matrix problems and the Brauer-Thrall conjecture", Preprint 73.9, Inst. Mat. AN UkSSR, Kiev, 1973, 1-100.

5. Zavadskij A.G., "Differentiation with respect to a pair of points", Matrix problems, Kiev, 1977, 115-121.

6. Zavadskij A.G., Nazarova L.A., "On finiteness and boundness of the parameter's number", Preprint 81.27, Inst. Mat. AN UkSSR, Kiev, 1981, 21-29.

Department of Mathematics
KISI
pr. Wozduchoflotskij 31,
Kiev-37, 252037, USSR